Product Performance Evaluation using CAD/CAE

Dedication

To my wife Sheng-Mei (盛美)
for her endless support, patience,
and encouragement.
She is always there for me.

Product Performance Evaluation using CAD/CAE

The Computer Aided Engineering Design Series

Kuang-Hua Chang

AMSTERDAM • BOSTON • HEIDELBERG • LONDON
NEW YORK • OXFORD • PARIS • SAN DIEGO
SAN FRANCISCO • SINGAPORE • SYDNEY • TOKYO

Academic Press is an imprint of Elsevier

Academic Press is an imprint of Elsevier
The Boulevard, Langford Lane, Kidlington, Oxford, OX5 1GB, UK
225 Wyman Street, Waltham, MA 02451, USA

First published 2013

Notices

Knowledge and best practice in this field are constantly changing. As new research and experience broaden our understanding, changes in research methods, professional practices, or medical treatment may become necessary.

Practitioners and researchers must always rely on their own experience and knowledge in evaluating and using any information, methods, compounds, or experiments described herein. In using such information or methods they should be mindful of their own safety and the safety of others, including parties for whom they have a professional responsibility.

To the fullest extent of the law, neither the Publisher nor the authors, contributors, or editors, assume any liability for any injury and/or damage to persons or property as a matter of products liability, negligence or otherwise, or from any use or operation of any methods, products, instructions, or ideas contained in the material herein.

British Library Cataloguing in Publication Data
A catalogue record for this book is available from the British Library

Library of Congress Cataloguing in Publication Data
A catalog record for this book is available from the Library of Congress

ISBN: 978-0-12-398460-9

For information on all Academic Press publications visit
our website at **store.elsevier.com**

Trademarks/Registered Trademarks
Brand names mentioned in this book are protected by their
respective trademarks and are acknowledged

Printed and bound in the United States of America

13 14 15 16 10 9 8 7 6 5 4 3 2 1

Working together to grow
libraries in developing countries

www.elsevier.com | www.bookaid.org | www.sabre.org

ELSEVIER   BOOK AID International   Sabre Foundation

Contents

Preface

The conventional product development process employs a design-build-test philosophy. The sequentially executed product development process often results in a prolonged lead time and an elevated product cost. The e-Design paradigm presented in the *Computer Aided Engineering Design* series employs IT-enabled technology, including computer-aided design, engineering, and manufacturing (CAD/CAE/CAM) tools, as well as advanced prototyping technology to support product design from concept to detailed designs, and ultimately manufacturing. This e-Design approach employs virtual prototyping (VP) technology to support a cross-functional team in analyzing product performance, reliability, and manufacturing costs early in the product development stage and in conducting quantitative trade-offs for design decision making. Physical prototypes of the product design are then produced using rapid prototyping (RP) technique mainly for design verification. The e-Design approach holds potential for shortening the overall product development cycle, improving product quality, and reducing product cost.

The *Computer Aided Engineering Design* series intends to provide readers with a comprehensive coverage of essential elements for understanding and practicing the e-Design paradigm in support of product design, including design method and process, and computer-based tools and technology. The book series consists of four books: *Product Design Modeling using CAD/CAE, Product Performance Evaluation using CAD/CAE, Product Manufacturing and Cost Estimating using CAD/CAE,* and *Design Theory and Methods using CAD/CAE.* The *Product Design Modeling using CAD/CAE* book discusses virtual mockup of the product that is first created in the CAD environment. The critical design parameterization that converts the product solid model into parametric representation, enabling the search for better designs, is an indispensable element of practicing the e-Design paradigm, especially in the detailed design stage. The second book, *Product Performance Evaluation using CAD/CAE,* focuses on applying numerous computer-aided engineering (CAE) technologies and software tools to support evaluation of product performance, including structural analysis, fatigue and fracture, rigid body kinematics and dynamics, and failure probability prediction and reliability analysis. The third book, *Product Manufacturing and Cost Estimating using CAD/CAE,* introduces computer-aided manufacturing (CAM) technology to support manufacturing simulations and process planning, RP technology and computer numerical control (CNC) machining for fast product prototyping, as well as manufacturing cost estimate that can be incorporated into product cost calculations. The product performance, reliability, and cost calculated can then be brought together to the cross-functional team for design trade-offs based on quantitative engineering data obtained from simulations. Design trade-off is one of the key topics included in the fourth book, *Design Theory and Methods using CAD/CAE.* In addition to conventional design optimization methods, the fourth book discusses decision theory, utility theory, and decision-based design. Simple examples are included to help readers understand the fundamentals of concepts and methods introduced in this book series.

In addition to the discussion on design principles, methods, and processes, this book series offers detailed review on the commercial off-the-shelf software tools for the support of modeling, simulations, manufacturing, and product data management and data exchanges. Tutorial style lessons on using commercial software tools are provided together with project-based exercises. Two suites of engineering software are covered: they are Pro/ENGINEER-based, including Pro/MECHANICA Structure, Pro/ENGINEER Mechanism Design, and Pro/MFG; and SolidWorks-based, including SolidWorks Simulation, SolidWorks Motion, and CAMWorks. These tutorial lessons are designed to help readers gain hands-on experiences to practice the e-Design paradigm.

The book you are reading, *Product Performance Evaluation using CAD/CAE,* is the first book of the *Computer Aided Engineering Design* series, but is the first of the series to publish. The objective of this book is to provide readers with fundamental understanding in product performance evaluation, and to enable them to apply the principles, methods, and software tools to support practical design applications. In Chapter 1, a brief introduction to the e-Design paradigm and tool environment will be given. Following this introduction, important topics in product performance evaluation, including structural performance of critical components, kinematics and dynamics of mechanical systems, fatigue and fracture, as well as product reliability analysis at both component and system levels will be discussed.

Chapter 2 focuses on structural analysis, including both analytical methods and finite element analysis (FEA), in which the essential elements in using FEA for modeling and analysis of structural performance are discussed. In addition, two companion projects are included: Project S3 Structural FEA and Fatigue Analysis Using SolidWorks Simulation and Project P3 Structural FEA and Fatigue Analysis Using Pro/MECHANICA Structure. These two projects offer tutorial lessons that should help readers to learn and be able to use the software tools for solving problems that are beyond hand calculations using analytical methods. Example files needed for going through the tutorial lessons are available for download from the book's website: http://booksite.elsevier.com/9780123984609. The goal of this chapter is to help readers become confident and competent in using FEA for creating adequate models and obtaining reasonably accurate results to support product design.

Chapter 3 provides an overview on motion analysis. Again, both analytical and computer-aided methods, that is, the so-called computer-aided kinematic and dynamic analyses, are included. General concept and process in carrying out motion simulation for kinematic and dynamic analysis are included in this chapter. In order to support readers to use the computer-aided analysis capability for general design applications, we have provided two companion projects: Project S2 Motion Analysis Using SolidWorks Motion and Project P2 Motion Analysis Using Pro/ENGINEER Mechanism Design. Tutorial lessons of these two projects should help readers to carry out motion simulations. Again, the goal of this chapter is to help the reader become confident and competent in using motion software tools for engineering design.

Chapter 4 offers a brief discussion on structural fatigue and fracture, which is one of the most technically challenging issues facing aerospace and mechanical engineers. In addition to basic theory, this chapter provides a brief review on the computational methods that support structural fatigue and fracture analysis in various stages. Similar to the previous chapters, tutorial lessons that provide details in using SolidWorks Simulation and Pro/MECHANICA Structure for crack initiation calculations are offered. You may find these lessons in Projects S3 and P3. The goal of this chapter is to enable readers to create adequate models and obtain reasonable results that support design involving fatigue and fracture.

In engineering design, there are uncertainties we must consider. Uncertainties exist in loading, material properties, geometric size, and material strength. Mechanical engineers must understand the importance of the probabilistic aspect in product design and must be able to apply adequate reliability analysis methods to solve engineering problems. Chapter 5 provides a brief overview on reliability analysis, which calculates failure probability of a prescribed performance measure considering uncertainties. This chapter also touches on design from a probabilistic perspective and compares the effectiveness of the probabilistic approach with conventional methods, such as safety factor and worst-case scenario. The goal of this chapter is to provide basic probabilistic theory and reliability analysis methods that enable readers to deal with basic engineering problems involving uncertainties.

As you may notice, any individual chapter included in this book can easily be expanded to a full textbook. Please keep in mind, this book is not intended to provide you an in-depth and thorough discussion on the respective subjects, but offer readers the concept and process of applying the computer-aided engineering technology and software tools to solve various aspects of engineering problems.

This *Product Performance Evaluation using CAD/CAE* book should serve well for a half-semester (8 weeks) instruction in engineering colleges of general universities. Typically, a three-hour lecture and one-hour laboratory exercise per week are desired. This book (and the book series) aims at providing engineering senior and first-year graduate students a comprehensive reference to learn advanced technology in support of engineering design using IT-enabled technology. Typical engineering courses that the book serves include Engineering Design, Integrated Product and Process Development, Concurrent Engineering, Design and Manufacturing, Modern Product Design, Computer-Aided Engineering, as well as Senior Capstone Design. In addition to classroom instruction, this book should support practicing engineers who wish to learn more about the e-Design paradigm at their own pace.

Resources available with this book:

For Instructors using this book for a course, an instructor manual and set of powerpoint slides are available by registering at www.textbooks.elsevier.com.

For readers of this book, updates and other resources related to the book will be posted from time to time at http://booksite.elsevier.com/9780123984609.

About the Author

Dr Kuang-Hua Chang is a Williams Companies Foundation Presidential Professor at the University of Oklahoma (OU), Norman, OK, USA. He received his PhD in mechanical engineering from the University of Iowa in 1990. Since then, he joined the Center for Computer-Aided Design (CAD) at Iowa as a research scientist and Computer-Aided Engineering (CAE) Technical Area Manager. In 1996, he joined Northern Illinois University as an assistant professor. In 1997, he joined the OU. He teaches mechanical design and manufacturing, in addition to conducting research in computer-aided modeling and simulation for design and manufacturing of mechanical systems. He has worked with aerospace and automotive industries and served as technical consultant to US industry and foreign companies. His work has been published in several books and more than 130 articles in international journals and conference proceedings.

He has received numerous awards for his teaching and research, including the Presidential Professorship in 2005 for *meeting the highest standards of excellence in scholarship and teaching,* OU Regents Award for Superior Accomplishment in Research and Creative Activity in 2004, OU BP AMOCO Foundation Good Teaching Award in 2002, and OU Regents Award for Superior Teaching in 2010. He is a five-time recipient of the CoE Alumni Teaching Award, given to top teachers in the College of Engineering at OU. His research paper was given a Best Paper Award at the iNEER Conference for Engineering Education and Research in 2005 (iCEER-2005). In 2006, he was awarded a Ralph R. Teetor Educational Award by the Society of Automotive Engineers (SAE) *in recognition of significant contributions to teaching, research and student development.* He was honored by the Oklahoma City Mayor's Committee on Disability Concerns with the 2009 Don Davis Award, which is *the highest honor granted in public recognition of extraordinarily meritorious service which has substantially advanced opportunities for people with disabilities by removing social, attitudinal and environmental barriers in the greater Oklahoma City area.*

About the Cover

The picture shown on the cover is the solid model of a Formula SAE (Society of Automotive Engineers) style racecar designed and built by engineering students at the University of Oklahoma (OU) during 2004–2005. To the author's knowledge, this was the first such detailed CAD model built by engineering students for a racecar. This model was built in *Pro/ENGINEER* with about 1400 parts and assemblies. Even though this was a team effort, most parts and assemblies were created and managed by then Senior mechanical engineering student, Mr Dave Oubre (we called him Super Dave). His dedication in creating such a detailed and accurate racecar solid model is admirable. His effort is highly appreciated.

Each year engineering students throughout the world design and build Formula-style racecars and participate in the annual Formula SAE competitions (http://students.sae.org/ competitions/formulaseries). The result is a great experience for young engineers in an engineering project as well as the opportunity to work in a team of multiple engineering disciplines. The OU team has been competitive in the Formula SAE competitions. The team won numerous awards throughout the years and finished 12th and 8th overall at the Formula SAE and Formula SAE West competitions, respectively, in 2006. Their 2005 racecar design shown on the cover won the prestigious 2005 PTC Award in the Education, Colleges, and Universities category this worldwide competition is sponsored by Parametric Technology Corporation. The OU team improved from a mediocre to a top-ten competitor between 2004 and 2006; e-Design has been one of the major contributing factors that turned the team around in such a short time.

Acknowledgments

I would like to first thank Mr Joseph P. Hayton for recognizing the need of such an engineering design book series that offers knowledge in modern engineering design principles, methods, and tools to mechanical engineering students. His enthusiasm in moving the book idea forward and eventually publishing the book series is highly appreciated. Mr Hayton's colleagues at Elsevier, Ms Lisa Jones, Ms Chelsea Johnson, and Ms Fiona Geraghty, have made significant contributions in transforming the original manuscripts into a well-organized and professionally-polished books that is suitable and presentable to our readers.

We are also thankful to Mr Yunxiang Wang and Mr Matthew Majors, mechanical engineering students at the University of Oklahoma (OU), for their help in preparing tutorial examples included in this book. Their contributions to this book are greatly appreciated.

I am also grateful to my current and former graduate students, Dr Mangesh Edke, Dr Qunli Sun, Dr Sung-Hwan Joo, Dr Xiaoming Yu, Dr Hsiu-Ying Hwang, Mr Trey Wheeler, Mr Yunxiang Wang, and Mr Javier Silver, for their excellent efforts in conducting research on numerous aspects of engineering design. Ideas and results that came out of their research have been largely incorporated into this book. Their dedication to the research in developing computer-aided approaches for support of product performance evaluation is acknowledged and is highly appreciated.

Introduction to e-Design

Chapter Outline

Conventional product development employs a design-build-test philosophy. The sequentially executed development process often results in prolonged lead times and elevated product costs. The proposed e-Design paradigm employs IT-enabled technology for product design, including virtual prototyping (VP) to support a cross-functional team in analyzing product performance, reliability, and manufacturing costs early in product development, and in making quantitative trade-offs for design decision making. Physical prototypes of the product design are then produced using the rapid prototyping (RP) technique and computer numerical control (CNC) to support design verification and functional prototyping, respectively.

e-Design holds potential for shortening the overall product development cycle, improving product quality, and reducing product costs. It offers three concepts and methods for product development:

- Bringing product performance, quality, and manufacturing costs together early in design for consideration.
- Supporting design decision making based on quantitative product performance data.
- Incorporating physical prototyping techniques to support design verification and functional prototyping.

1.1 Introduction

A conventional product development process that is usually conducted sequentially suffers the problem of the *design paradox* (Ullman 1992). This refers to the dichotomy or mismatch between the design engineer's knowledge about the product and the number of decisions to be made (flexibility) throughout the product development cycle (see Figure 1.1). Major design decisions are usually made in the early design stage when the product is not very well understood. Consequently, engineering changes are frequently requested in later product

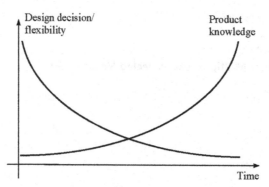

Figure 1.1: The design paradox.

development stages, when product design evolves and is better understood, to correct decisions made earlier.

Conventional product development is a design-build-test process. Product performance and reliability assessments depend heavily on physical tests, which involve fabricating functional prototypes of the product and usually lengthy and expensive physical tests. Fabricating prototypes usually involves manufacturing process planning and fixtures and tooling for a very small amount of production. The process can be expensive and lengthy, especially when a design change is requested to correct problems found in physical tests.

In conventional product development, design and manufacturing tend to be disjoint. Often, manufacturability of a product is not considered in design. Manufacturing issues usually appear when the design is finalized and tests are completed. Design defects related to manufacturing in process planning or production are usually found too late to be corrected. Consequently, more manufacturing procedures are necessary for production, resulting in elevated product cost.

With this highly structured and sequential process, the product development cycle tends to be extended, cost is elevated, and product quality is often compromised to avoid further delay. Costs and the number of engineering change requests (ECRs) throughout the product development cycle are often proportional according to the pattern shown in Figure 1.2. It is reported that only 8% of the total product budget is spent for design; however, in the early stage, design determines 80% of the lifetime cost of the product (Anderson 1990). Realistically, today's industries will not survive worldwide competition unless they introduce new products of better quality, at lower cost, and with shorter lead times. Many approaches and concepts have been proposed over the years, all with a common goal—to shorten the product development cycle, improve product quality, and reduce product cost.

A number of proposed approaches are along the lines of virtual prototyping (Lee 1999), which is a simulation-based method that helps engineers understand product behavior and make

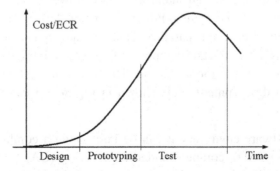

Figure 1.2: Cost/ECR versus time in a conventional design cycle.

design decisions in a virtual environment. The virtual environment is a computational framework in which the geometric and physical properties of products are accurately simulated and represented. A number of successful virtual prototypes have been reported, such as Boeing's 777 jetliner, General Motors' locomotive engine, Chrysler's automotive interior design, and the Stockholm Metro's Car 2000 (Lee 1999). In addition to virtual prototyping, the concurrent engineering (CE) concept and methodology have been studied and developed with emphasis on subjects such as product life cycle design, design for X-abilities (DFX), integrated product and process development (IPPD), and Six Sigma (Prasad 1996).

Although significant research has been conducted in improving the product development process, and successful stories have been reported, industry at large is not taking advantage of new product development paradigms. The main reason is that small and mid-size companies cannot afford to develop an in-house computer tool environment like those of Boeing and the Big-Three automakers. On the other hand, commercial software tools are not tailored to meet the specific needs of individual companies; they often lack proper engineering capabilities to support specific product development needs, and most of them are not properly integrated. Therefore, companies are using commercial tools to support segments of their product development without employing the new design paradigms to their full advantage.

The e-Design paradigm does not supersede any of the approaches discussed. Rather, it is simply a realization of concurrent engineering through virtual and physical prototyping with a systematic and quantitative method for design decision making. Moreover, e-Design specializes in performance and reliability assessment and improvement of complex, large-scale, compute-intensive mechanical systems. The paradigm also uses design for manufacturability (DFM), design for manufacturing and assembly (DFMA), and manufacturing cost estimates through virtual manufacturing process planning and simulation for design considerations.

The objective of this chapter is to present an overview of the e-Design paradigm and the sample tool environment that supports a cross-functional team in simulating and designing mechanical products concurrently in the early design stage. In turn, better-quality products can be designed and manufactured at lower cost. With intensive knowledge of the product gained from simulations, better design decisions can be made, breaking the aforementioned design paradox. With the advancement of computer simulations, more hardware tests can be replaced by computer simulations, thus reducing cost and shortening product development time. The desirable cost and ECR distributions throughout the product development cycle shown in Figure 1.3 can be achieved through the e-Design paradigm.

A typical e-Design software environment can be built using a combination of existing computer-aided design (CAD), computer-aided engineering (CAE), and computer-aided manufacturing (CAM) as the base, and integrating discipline-specific software tools that

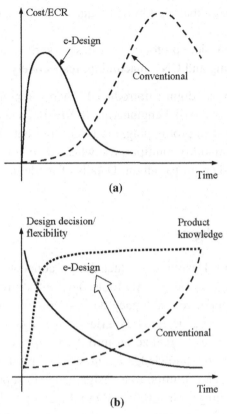

Figure 1.3: (a) Cost/ECR versus e-Design cycle time; (b) product knowledge versus e-Design cycle time.

are commercially available for specific simulation tasks. The main technique in building the e-Design environment is tool integration. Tool integration techniques, including product data models, wrappers, engineering views, and design process management, have been developed (Tsai et al. 1995) and are described in *Design Theory and Methods using CAD/CAE*, a book in The Computer Aided Engineering Design Series. This integrated e-Design tool environment allows small and mid-size companies to conduct efficient product development using the e-Design paradigm. The tool environment is flexible so that additional engineering tools can be incorporated with a lesser effort.

In addition, the basis for tool integration, such as product data management (PDM), is well established in commercial CAD tools and so no wheel needs to be reinvented. The e-Design paradigm employs three main concepts and methods for product development:

- Bringing product performance, quality, and manufacturing cost for design considerations in the early design stage through virtual prototyping.

- Supporting design decision making through a quantitative approach for both concept and detail designs.
- Incorporating product physical prototypes for design verification and functional tests via rapid prototyping and CNC machining, respectively.

In this chapter the e-Design paradigm is introduced. Then components that make up the paradigm, including knowledge-based engineering (KBE) (Gonzalez and Dankel 1993), virtual prototyping, and physical prototyping, are briefly presented. Designs of a simple airplane engine and a high-mobility multipurpose wheeled vehicle (HMMWV) are briefly discussed to illustrate the e-Design paradigm. Details of modeling and simulation are provided in later chapters.

1.2 The e-Design Paradigm

As shown in Figure 1.4, in e-Design, a product design concept is first realized in solid model form by design engineers using CAD tools. The initial product is often established based on the designer's experience and legacy data of previous product lines. It is highly desirable to capture and organize designer experience and legacy data to support decision making in a discrete form so as to realize an initial concept. The KBE (Gonzalez and Dankel 1993) that computerizes knowledge about specific product domains to support design engineers in arriving at a solution to a design problem supports the concept design. In addition, a KBE system integrated with a CAD tool may directly generate a solid model of the concept design that directly serves downstream design and manufacturing simulations.

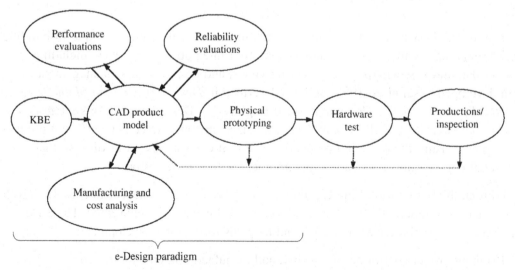

Figure 1.4: The e-Design paradigm.

With the product solid model represented in CAD, simulations for product performance, reliability, and manufacturing can be conducted. The product development tasks and the cross-functional team are organized according to engineering disciplines and expertise. Based on a centralized computer-aided design product model, simulation models can be derived with proper simplifications and assumptions. However, a one-way mapping that governs changes from CAD models to simulation models must be established for rapid simulation model updates (Chang et al. 1998). The mapping maintains consistency between CAD and simulation models throughout the product development cycle.

Product performance, reliability, and manufacturing can then be simulated concurrently. Performance, quality, and costs obtained from multidisciplinary simulations are brought together for review by the cross-functional team. Design variables—including geometric dimensions and material properties of the product CAD models that significantly influence performance, quality, and cost—can be identified by the cross-functional team in the CAD product model. These key performance, quality, and cost measures, as well as design variables, constitute a product design model. With such a model, a systematic design approach, including a parametric study for concept design and a trade-off study for detail design, can be conducted to improve the product with a minimum number of design iterations.

The product designed in the virtual environment can then be fabricated using rapid prototyping machines for physical prototypes directly from product CAD solid models, without tooling and process planning. The physical prototypes support the cross-functional team for design verification and assembly checking. Change requests that are made at this point can be accommodated in the virtual environment without high cost and delay.

The physics-based simulation technology potentially minimizes the need for product hardware tests. Because substantial modeling and simulations are performed, unexpected design defects encountered during the hardware tests are reduced, thus minimizing the feedback loop for design modifications. Moreover, the production process is smooth since the manufacturing process has been planned and simulated. Potential manufacturing-related problems will have been largely addressed in earlier stages.

A number of commercial CAD systems provide a suite of integrated CAD/CAE/CAM capabilities (e.g., Pro/ENGINEER and SolidWorks®). Other CAD systems, including CATIA® and NX, support one or more aspects of the engineering analysis. In addition, third-party software companies have made significant efforts in connecting their capabilities to CAD systems. As a representative example, CAE and CAM software companies worked with SolidWorks and integrated their software into SolidWorks environments such as CAMWorks®. Each individual tool is seamlessly integrated into SolidWorks.

In this book, Pro/ENGINEER and SolidWorks, with a built-in suite of CAE/CAM modules, are employed as the base for the e-Design environment. In addition to their superior solid modeling capability based on parametric technology (Zeid 1991), Pro/MECHANICA® and SolidWorks Simulation support simulations of nominal engineering, including structural and thermal problems. Mechanism Design of Pro/ENGINEER and SolidWorks Motion support motion simulation of mechanical systems. Moreover, CAM capabilities implemented in CAD, such as Pro/MFG (Parametric Technology Corp., www.ptc.com), and CAMWorks, provide an excellent basis for manufacturing process planning and simulations. Additional CAD/CAE/CAM tools introduced to support modeling and simulation of broader engineering problems encountered in general mechanical systems can be developed and added to the tool environment as needed.

1.3 Virtual Prototyping

Virtual prototyping is the backbone of the e-Design paradigm. As presented in this chapter, VP consists of constructing a parametric product model in CAD, conducting product performance simulations and reliability evaluations using CAE software, and carrying out manufacturing simulations and cost estimates using CAM software. Product modeling and simulations using integrated CAD/CAE/CAM software are the basic and common activities involved in virtual prototyping. However, a systematic design method, including parametric study and design trade-offs, is indispensable for design decision making.

1.3.1 Parameterized CAD Product Model

A parametric product model in CAD is essential to the e-Design paradigm. The product model evolves to a higher-fidelity level from concept to detail design stages (Chang et al. 1998). In the concept design stage, a considerable portion of the product may contain non-CAD data. For example, when the gross motion of the mechanical system is sought the non-CAD data may include engine, tires, or transmission if a ground vehicle is being designed. Engineering characteristics of the non-CAD parts and assemblies are usually described by engineering parameters, physics laws, or mathematical equations. This non-CAD representation is often added to the product model in the concept design stage for a complete product model. As the design evolves, non-CAD parts and assemblies are refined into solid-model forms for subsystem and component designs as well as for manufacturing process planning.

A primary challenge in conducting product performance simulations is generating simulation models and maintaining consistency between CAD and simulation models through mapping. Challenges involved in model generation and in structural and dynamic simulations are discussed next, in which an airplane engine model in the detail design stage, as shown in Figure 1.5, is used for illustration.

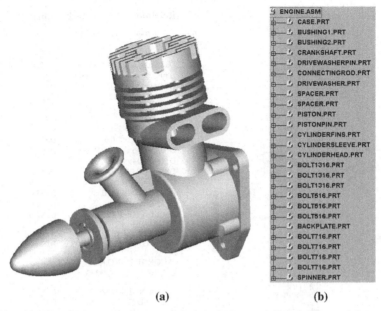

ENGINE.ASM
CASE.PRT
BUSHING1.PRT
BUSHING2.PRT
CRANKSHAFT.PRT
DRIVEWASHERPIN.PRT
CONNECTINGROD.PRT
DRIVEWASHER.PRT
SPACER.PRT
SPACER.PRT
PISTON.PRT
PISTONPIN.PRT
CYLINDERFINS.PRT
CYLINDERSLEEVE.PRT
CYLINDERHEAD.PRT
BOLT1316.PRT
BOLT1316.PRT
BOLT1316.PRT
BOLT516.PRT
BOLT516.PRT
BOLT516.PRT
BACKPLATE.PRT
BOLT716.PRT
BOLT716.PRT
BOLT716.PRT
BOLT716.PRT
SPINNER.PRT

(a)   (b)

Figure 1.5: Airplane engine model: (a) CAD model and (b) model tree.

Parameterized Product Model

A parameterized product model defined in CAD allows design engineers to conveniently explore design alternatives for support of product design. The CAD product model is parameterized by defining dimensions that govern the geometry of parts through geometric features and by establishing relations between dimensions within and across parts. Through dimensions and relations, changes can be made simply by modifying a few dimensional values. Changes are propagated automatically throughout the mechanical product following the dimensions and relations. A single-piston airplane engine with a change in its bore diameter is shown in Figure 1.6, so as illustrating change propagation through parametric dimensions and relationships. More in-depth discussion of the modeling and parameterization of the engine example can be found in *Product Design Modeling using CAD/CAE*, a book in The Computer Aided Engineering Design Series.

Analysis Models

For product structural analysis, finite element analysis (FEA) is often employed. In addition to structural geometry, loads, boundary conditions, and material properties can be conveniently defined in the CAD model. Most CAD tools are equipped with fully automatic mesh generation capability. This capability is convenient but often leads to large FEA models with some geometric discrepancy at the part boundary. Plus, triangular and tetrahedral elements are often the only elements supported. An engine connecting rod example meshed using Pro/MESH (part of Pro/MECHANICA) with default mesh parameters is shown in

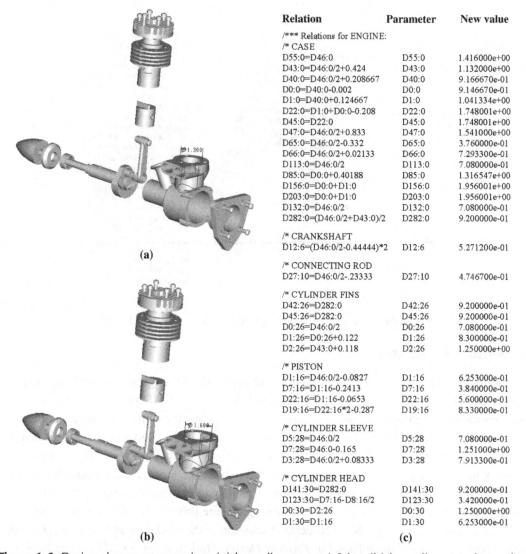

Relation	Parameter	New value
/*** Relations for ENGINE:		
/* CASE		
D55:0=D46:0	D55:0	1.416000e+00
D43:0=D46:0/2+0.424	D43:0	1.132000e+00
D40:0=D46:0/2+0.208667	D40:0	9.166670e-01
D0:0=D40:0-0.002	D0:0	9.146670e-01
D1:0=D40:0+0.124667	D1:0	1.041334e+00
D22:0=D1:0+D0:0-0.208	D22:0	1.748001e+00
D45:0=D22:0	D45:0	1.748001e+00
D47:0=D46:0/2+0.833	D47:0	1.541000e+00
D65:0=D46:0/2-0.332	D65:0	3.760000e-01
D66:0=D46:0/2+0.02133	D66:0	7.293300e-01
D113:0=D46:0/2	D113:0	7.080000e-01
D85:0=D0:0+0.40188	D85:0	1.316547e+00
D156:0=D0:0+D1:0	D156:0	1.956001e+00
D203:0=D0:0+D1:0	D203:0	1.956001e+00
D132:0=D46:0/2	D132:0	7.080000e-01
D282:0=(D46:0/2+D43:0)/2	D282:0	9.200000e-01
/* CRANKSHAFT		
D12:6=(D46:0/2-0.44444)*2	D12:6	5.271200e-01
/* CONNECTING ROD		
D27:10=D46:0/2-.23333	D27:10	4.746700e-01
/* CYLINDER FINS		
D42:26=D282:0	D42:26	9.200000e-01
D45:26=D282:0	D45:26	9.200000e-01
D0:26=D46:0/2	D0:26	7.080000e-01
D1:26=D0:26+0.122	D1:26	8.300000e-01
D2:26=D43:0+0.118	D2:26	1.250000e+00
/* PISTON		
D1:16=D46:0/2-0.0827	D1:16	6.253000e-01
D7:16=D1:16-0.2413	D7:16	3.840000e-01
D22:16=D1:16-0.0653	D22:16	5.600000e-01
D19:16=D22:16*2-0.287	D19:16	8.330000e-01
/* CYLINDER SLEEVE		
D5:28=D46:0/2	D5:28	7.080000e-01
D7:28=D46:0-0.165	D7:28	1.251000e+00
D3:28=D46:0/2+0.08333	D3:28	7.913300e-01
/* CYLINDER HEAD		
D141:30=D282:0	D141:30	9.200000e-01
D123:30=D7:16-D8:16/2	D123:30	3.420000e-01
D0:30=D2:26	D0:30	1.250000e+00
D1:30=D1:16	D1:30	6.253000e-01

(a)

(b)

(c)

Figure 1.6: Design change propagation: (a) bore diameter = 1.3 in.; (b) bore diameter changed to 1.6 in.; (c) relations of geometric dimensions.

Figure 1.7. The FEA model consists of 1,270 nodes and 4,800 tetrahedron elements, yet it still reveals discrepancy to the true CAD geometry. Moreover, mesh distortion due to large deformation of the structure, such as hyperelastic problems, often causes FEA to abort prematurely. Semiautomatic mesh generation is more realistic; therefore, tools such as MSC/Patran® (MacNeal-Schwendler Corp., www.mscsoftware.com) and HyperMesh® (Altair® Engineering, Inc., www.altair.com) are essential to support the e-Design environment for mesh generation.

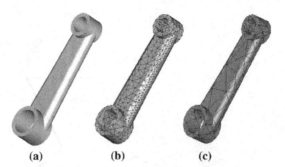

Figure 1.7: Finite element meshes of a connecting rod: (a) CAD solid model, (b) h-version finite element mesh, and (c) p-version finite element mesh.

In general, p-version FEA (Szabó and Babuška 1991) is more suitable for structural analysis in terms of minimizing the gap in geometry between CAD and finite element models, and in lessening the tendency toward mesh distortion. It also offers capability in convergence analysis that is superior to regular h-version FEA. As shown in Figure 1.7c, the same connecting rod is meshed with 568 tetrahedron p-elements, using Pro/MECHANICA with a default setting. A one-way mapping between changes in CAD geometric dimensions and finite element mesh for both h- and p-version FEAs can be established through a design velocity field (Haug et al. 1986), which allows direct and automatic generation of the finite element mesh of new designs.

Another issue worth considering is the simplification of 3D solid models to surface (shell) or curve (beam) models for analysis. Capabilities that semiautomatically convert 3D thin-shell solids to surface models are available in, for example, Pro/MECHANICA and SolidWorks Simulation.

Motion Simulation Models

Generating motion simulation models involves regrouping parts and subassemblies of the mechanical system in CAD as bodies and often introducing non-CAD components to support a multibody dynamic simulation (Haug 1989). Engineers must define the joints or force connections between bodies, including joint type and reference coordinates. Mass properties of each body are computed by CAD with the material properties specified. Integration between Mechanism Design and Pro/ENGINEER, as well as between SolidWorks Motion (Chang 2008) and SolidWorks, is seamless. Design changes made in geometric dimensions propagate to the motion model directly. In addition, simulation tools, such as Dynamic Analysis and Design Systems (DADS) (LMS, www.lmsintl.com/DADS) and communication and data systems integration, are also integrated with CAD with proper parametric mapping from CAD to simulation models that support parametric study. As an example, the motion inside an airplane engine is modeled as a slider-crank mechanism in Mechanism Design, as shown in Figure 1.8.

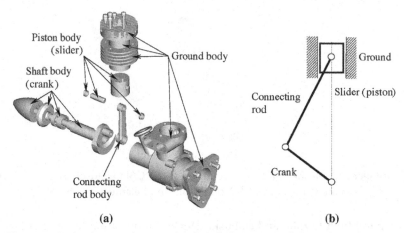

Figure 1.8: Engine motion model: (a) definition and (b) schematic view.

A common mistake made in creating motion simulation models is selecting improper joints to connect bodies. Introducing improper joints creates an invalid or inaccurate model that does not simulate the true behavior of the mechanical system. Intelligent modeling capability that automatically specifies joints in accordance with assembly relations defined between parts and subassemblies in solid models is available in, for example, SolidWorks Motion.

1.3.2 Product Performance Analysis

As mentioned earlier, product performance evaluation using physics-based simulation in the computer environment is usually called, in a narrow sense, virtual prototyping, or VP. With the advancement of simulation technology, more engineering questions can be answered realistically through simulations, thus minimizing the needs for physical tests. However, some key questions cannot be answered for sophisticated engineering problems—for example, the crashworthiness of ground vehicles. Although VP will probably never replace hardware tests completely, the savings it achieves for less sophisticated problems is significant and beneficial.

Motion Analysis

System motion simulations include workspace analysis (kinematics), rigid- and flexible-body dynamics, and inverse dynamic analysis. Mechanism Design and SolidWorks Motion, based on theoretical work (Kane and Levinson 1985), mainly support kinematics and rigid-body simulations for mechanical systems. They do not properly support mechanical system simulation such as a vehicle moving on a user-defined terrain. General-purpose dynamic simulation tools, such as DADS (www.lsmintl.com) or Adams® (www.mscsoftware.com), are more desirable for simulation of general mechanical systems.

Structural Analysis

Pro/MECHANICA supports linear static, vibration, buckling, fatigue, and other such analyses, using p-version FEA. General-purpose finite element codes, such as MSC/Nastran® (MacNeal-Schwendler Corp., www.mscsoftware.com) and ANSYS® (ANSYS Analysis Systems, Inc., www.ansys.com) are ideal for the e-Design environment to support FEA for a broad range of structural problems—for example, nonlinear, plasticity, and transient dynamics. Meshless methods developed in recent years (for example, Chen et al. 1997) hold promise for avoiding finite element mesh distortion in large-deformation problems. Multiphase problems (e.g., acoustic and aero-structural) are well supported by specialized tools such as LMS® SYSNOISE (Numerical Integration Technologies 1998). LS-DYNA® (Hallquist 2006) is currently one of the best codes for nonlinear, plastic, dynamics, friction-contact, and crashworthiness problems. These special codes provide excellent engineering analysis capabilities that complement those provided in CAD systems.

Fatigue and Fracture Analysis

Fatigue and fracture problems are commonly encountered in mechanical components because of repeated mechanical or thermal loads. MSC Fatigue® (MacNeal-Schwendler Corp., www.mscsoftware.com), with an underlying computational engine developed by nCode® (www.ncode.com) is one of the leading fatigue and fracture analysis tools. It offers both high- and low-cycle fatigue analyses. A critical plane approach is available in MSC Fatigue for prediction of fatigue life due to general multiaxial loads.

Note that the recently developed extended finite element method (XFEM) supports fracture propagation without remeshing (Moës et al. 2002). XFEM was recently integrated in ABAQUS®. Also note that additional capabilities, such as thermal analysis, computational fluid dynamics (CFD) and combustion, can be added to meet specific needs in analyzing mechanical products. Integration of additional engineering disciplines are briefly discussed in Section 1.3.4.

Product Reliability Evaluations

Product reliability evaluations in the e-Design environment focus on the probability of specific failure events (or failure mode). The failure event corresponds to a product performance measure, such as the fatigue life of a mechanical component. For the reliability analysis of a single failure event, the failure event or failure function is defined as (Madsen et al. 1986)

$$g(X) = \psi^u - \psi(X) \tag{1.1}$$

where

ψ is a product performance measure
ψ^u is the upper bound (or design requirement) of the product performance
X is a vector of random variables

When product performance does not meet the requirement—that is, when $\psi^u \leq \psi(X)$, the event fails. Therefore, the probability of failure P_f of the particular event $g(X) \leq 0$ is

$$P_f = P[g(X) \leq 0] \tag{1.2}$$

where $P[\bullet]$ is the probability of event $\bullet$.

Given the joint probability density function $f_X(x)$ of the random variables X, the probability of failure for a single event of a mechanical component can be expressed as

$$P_f = P[g(X) \leq 0] = \underset{g(X) \leq 0}{\int \int ... \int} f_X(x)dx \tag{1.3}$$

The probability of failure in Eq. 1.3 is commonly evaluated using the Monte Carlo method or the first- or second-order reliability method (FORM or SORM) (Wu and Wirsching 1984, Yu et al. 1998).

Once the probabilities of several failure events in subsystems or components are computed, system reliability can be obtained by, for example, fault-tree analysis (Ertas and Jones 1993). No general-purpose software tool for reliability analysis of general mechanical systems is commercially available yet. Numerical evaluation of stochastic structures under stress (NESSUS®) (www. nessus.swri.org), which is currently in development can be a good candidate for incorporation into the e-Design environment. With the probability of failure, critical quality design criteria, such as mean time between failure (MTBF), can be computed (Ertas and Jones 1993).

Two main challenges exist in reliability analysis: One, realistic distribution data are difficult to acquire and often are not available in the early stage; two, failure probability computations are often expensive. The first challenge may be alleviated by employing legacy data from previous product lines. Approximation techniques (e.g., Yu et al. 1998) can be employed to make the computation affordable even for an individual failure event within a mechanical component.

1.3.3 *Product Virtual Manufacturing*

Virtual manufacturing addresses issues of design for manufacturability (DFM) (Prasad 1996) and design for manufacturing and assembly (DFMA) (Boothroyd et al. 1994) early in product development. In the e-Design paradigm, DFM and DFMA are performed by conducting virtual manufacturing and assembly using, for example, Pro/MFG. DFM and DFMA of the product are verified through animations of the virtual manufacturing and assembly process.

Pro/MFG is a Pro/ENGINEER module supporting the virtual machining process, including milling, drilling, and turning. By incorporating part design and also defining workpieces, workcells, fixtures, cutting tools, and cutting parameters, Pro/MFG automatically generates a tool path (see Figure 1.9a), which simulates the machining process (Figure 1.9b), calculates machining time, and produces cutter location (CL) data. The CL data can be post-processed for CNC codes. In addition, casting, sheet metal, molding, and welding can be simulated using Pro/CASTING, Pro/SHEETMETAL, Pro/MOLD, and Pro/WELDING, respectively.

With such virtual manufacturing process planning and animation, manufacturability of the product design can, to some extent, be verified. The DFMA tool (Boothroyd et al. 1994)

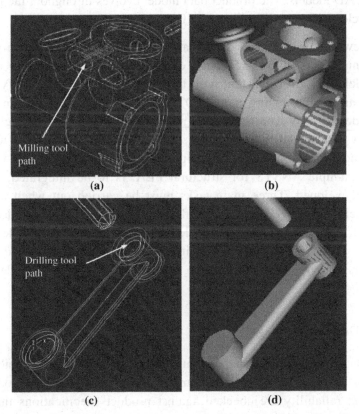

Figure 1.9: Virtual machining process: (a) engine case—milling tool path; (b) milling simulation; (c) connecting rod—drilling tool path; (d) drilling simulation.

developed by Boothroyd Dewhurst, Inc., assists the cross-functional team in quantifying product assembly time and labor costs. It also challenges the team to simplify product structure, thereby reducing product as well as assembly costs.

One of the limitations in using virtual manufacturing tools (e.g., Pro/MFG) is that chip formation (Fang and Jawahir 1996), a primary consideration in computer numerical control (CNC), is not incorporated into the simulation. In addition, machining parameters, such as power consumption, machining temperature, and tool life, which contribute to manufacturing costs are not yet simulated.

1.3.4 Tool Integration

Techniques developed to support tool integration (Chang et al. 1998) include parameterized product data models, engineering views, tool wrappers, and design process management. Parameterized product data models represent engineering data that are needed for conducting virtual prototyping of the mechanical system. The main sources of the product data model are CAD and non-CAD models. The product data model evolves throughout the product development cycle as illustrated in Figure 1.10.

Engineering views allow engineers from various disciplines to view the product from their own technical perspectives. Through engineering views, engineers create simulation models that are consistent with the product model by simplifying the CAD representation, as needed adding non-CAD product representation and mapping. Tool wrappers provide two-way data translation and transmission between engineering tools and the product data model. Design process management provides the team leader with a tool to monitor and manage the design process. When a new tool of an existing discipline, for example ANSYS for structural FEA, is to be integrated, a wrapper for it must be developed. Three main tasks must be carried out when a new engineering discipline, say computational fluid dynamics (CFD), is added to the environment. First, the product data model must be extended to include engineering data needed to support CFD. Second, engineering views must be added to allow design engineers to generate CFD models. Finally, wrappers must be developed for specific CFD tools.

1.3.5 Design Decision Making

Product performance, reliability, and manufacturing cost that are evaluated using simulations can be brought to the cross-functional team for review. Product performance and reliability are checked against product specifications that have been defined and have evolved from the beginning of the product development process.

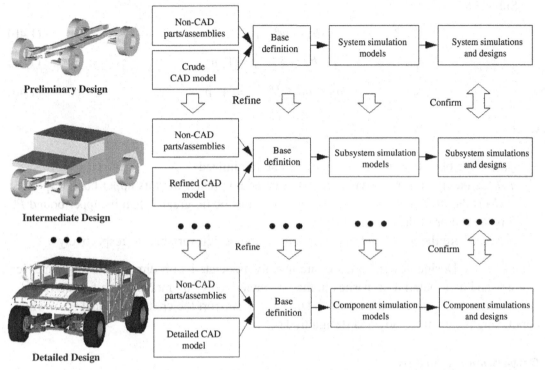

Figure 1.10: Hierarchical product models evolved through the e-Design process.

Manufacturing cost derived from the virtual manufacturing simulations can be added to product cost. The cross-functional team must address areas of concern identified in product performance, reliability, and manufacturability, and it must identify a set of design variables that influence these areas. Design modifications can then be conducted. In the past, quality functional deployment (QFD) (Ertas and Jones 1993) was largely employed in design modification to assign qualitative weighting factors to product performance and design changes. e-Design employes a systematic and quantitative approach to design modifications (for example, Yu et al. 1997).

Design Problem Formulation

Before a design can be improved, design problems must be defined. A design problem is often presented in a mathematical form, typically as

$$\text{Minimize } \varphi(\boldsymbol{b}) \tag{1.4a}$$

Subject to

$$\psi_i(\boldsymbol{b}) \leq \psi_i^u \quad i = 1, m \tag{1.4b}$$

$$P_{f_j}(\boldsymbol{b}) \leq P_{f_j}^u \quad j = 1, n \tag{1.4c}$$

$$b_k^l \leq b_k \leq b_k^u \quad k = 1, p \tag{1.4d}$$

where

$\varphi(\boldsymbol{b})$ is the objective (or cost) function to be minimized

$\psi_i(\boldsymbol{b})$ is the ith constraint function that must be no greater than its upper bound ψ_i^u

$P_{f_j}(\boldsymbol{b})$ is the jth failure probability index that must be no greater than its upper bound $P_{f_j}^u$

$\boldsymbol{b}$ is the vector of design variables

b_k^l and b_k^u are the lower and upper bounds of the design variable b_k, respectively

Note that in e-Design design variables are usually associated with dimensions of geometric features and part material properties in the parameterized CAD models. The feature-based design parameters serve as the common language to support the cross-functional team while conducting parametric study and design trade-offs.

Design Sensitivity Analysis

Before quantitative design decisions can be made, there must be a design sensitivity analysis (DSA) that computes derivatives of performance measures, including product performance, failure probability, and manufacturing cost, with respect to design variables. Dependence of performance measures on design variables is usually implicit. How to express product performance in terms of design variables in a mathematical form is not straightforward. Analytical DSA methods combined with numerical computations have been developed mainly for structural responses (Haug et al. 1986) and fatigue and fracture (Chang et al. 1997). DSA for failure probability with respect to both deterministic and random variables has also been developed (Yu et al. 1997). In addition, DSA and optimization using meshless methods have been developed for large-deformation problems (Grindeanu et al. 1999). More details about the analytical DSA for structural responses also referred to Haug et al. (1985).

For problems such as motion and manufacturing cost, where premature or no analytical DSA capability is available, the finite difference method is the only choice. The finite difference method is expressed in the following equation:

$$\frac{\partial \psi}{\partial b_j} \approx \frac{\psi(\boldsymbol{b} + \Delta b_j) - \psi(\boldsymbol{b})}{\Delta b_j} \tag{1.5}$$

where Δb_j is a perturbation in the jth design variable. With sensitivity information, parametric study and design trade-offs can be conducted for design improvements at the concept and detail stages, respectively.

Parametric Study

A parametric study that perturbs design variables in the product design model to explore various design alternatives can effectively support product concept designs. The parametric study is simple and easy to perform as long as the mapping between CAD and simulation models has been established. The mapping supports fast simulation model generation for performance analyses. It also supports DSA using the finite difference method. The parametric study is possible for concept design because the number of design variables to perturb is usually small. A spreadsheet with a proper formula defined among cells is well suited to support the parametric study. The use of Microsoft Excel is illustrated in Figure 1.11.

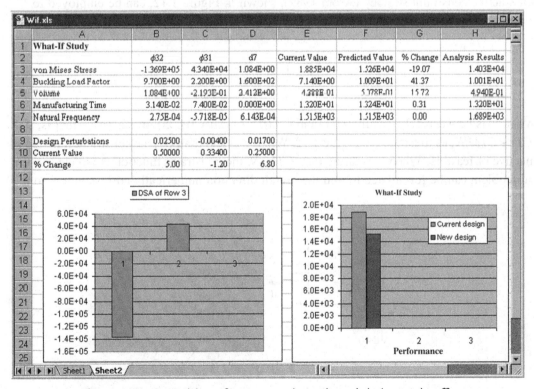

Figure 1.11: Spreadsheet for parametric study and design trade-offs.

Design Trade-Off Analysis

With design trade-off analysis, the design engineer can find the most appropriate design search direction for the design problem formulated in Eq. 1.4, using four possible algorithms:

- Reduce cost.
- Correct constraint neglecting cost.
- Correct constraint with a constant cost.
- Correct constraint with a cost increment.

As a general rule, the first algorithm, reduce cost, can be chosen when the design is feasible; in other words, all constraint functions are within the desired limits. When the design is infeasible, generally one may start with the third algorithm, correct constraint with a constant cost. If the design remains infeasible, the fourth algorithm, correct constraint with a cost increment—say 10%—may be appropriate. If a feasible design is still not found, the second algorithm, correct constraint neglecting cost, can be selected. A quadratic programming (QP) subproblem can be formulated to numerically find the search direction that corresponds to the algorithm selected.

An ε-active constraint strategy (Arora 1989), shown in Figure 1.12, can be employed to support design trade-offs. The constraint functions in Eq. 1.4 are normalized by

$$y_i = \frac{\psi_i}{\psi_i^u} - 1 \le 0, \quad i = 1, m \tag{1.6}$$

When y_i is between *CT* (usually 0.03) and *CTMIN* (usually 0.005), it is active—that is, $\varepsilon = |CT| + CTMIN$, as illustrated in Figure 1.12. When y_i is less than *CT*, the constraint function is inactive or feasible. When y_i is larger than *CTMIN*, the constraint function is violated. A QP subproblem can be formulated to find the search direction numerically corresponding to the

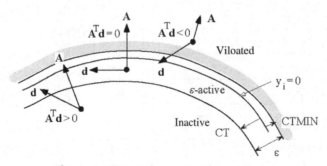

Figure 1.12: ε-active constraint strategy.

option selected. For example, the QP subproblem for the first algorithm (cost reduction) can be formulated as

Minimize $\qquad\qquad\qquad\qquad$ $\mathbf{c}^{\mathrm{T}}\mathbf{d} + 0.5\, \mathbf{d}^T\mathbf{d}$

Subject to $\qquad\qquad\qquad\qquad$ $\mathbf{A}^{\mathrm{T}}\mathbf{d} \leq \mathbf{y}$ $\qquad\qquad\qquad$ (1.7)

$$\mathbf{b}^L - \mathbf{b}^{(k)} \leq \mathbf{d} \leq \mathbf{b}^U - \mathbf{b}^{(k)}$$

where

$$\mathbf{c} = [c_1, c_2, ..., c_{n1+n2}]^T, \quad c_i = \partial\varphi/\partial b_i$$

$\mathbf{d}$ is the search direction to be determined.

$$A_{ij} = \partial P_{y_i}/\partial b_j; \quad \mathbf{b} = [b_1, b_2, ...b_n]^T$$

k is the current design iteration.

The objective of the design trade-off algorithm is to find the optimal search direction $\mathbf{d}$ under a given circumstance. Details are discussed in *Design Theory and Methods using CAD/CAE*, a book in The Computer Aided Engineering Design Series.

What-If Study

After the search direction $\mathbf{d}$ is found, a number of step sizes α can be used to perturb the design along the direction $\mathbf{d}$. Objective and constraint function values, represented as ψ_i, at a perturbed design $\mathbf{b} + \alpha\mathbf{d}$ can be approximated using the first-order sensitivity information of the functions by Taylor series expansion about the current design $\mathbf{b}$ without going through simulations; that is,

$$\psi_i(\mathbf{b} + \alpha\mathbf{d}) \approx \psi_i(\mathbf{b}) + \frac{\partial\psi_i}{\partial\mathbf{b}}\alpha\mathbf{d} \qquad\qquad (1.8)$$

Note that since there is no analysis involved, the what-if study can be carried out very efficiently. This allows the design engineer to explore design alternatives more effectively.

Once a satisfactory design is identified, after trying out different step sizes α in an approximation sense, the design model can be updated to the new design and then simulations of the new design can be conducted. Equation 1.8 also supports parametric study, in which the design perturbation $\delta\mathbf{b}$ is determined by engineers based on sensitivity information. To ensure a reasonably accurate function prediction using Eq. 1.8, the step sizes must be small so that the perturbation $\partial\psi_i/(\partial b)(\alpha d)$ is, as a rule of thumb, less than 10% of the function value $\psi_i(b)$.

1.4 Physical Prototyping

In general, two techniques are suitable for fabricating physical prototypes of the product in the design process: rapid prototyping (RP) and computer numerical control (CNC) machining. RP systems, based on solid freeform fabrication (SFF) technology (Jacobs 1994), fabricate physical prototypes of the structure for design verification. The CNC machining fabricates functional parts as well as the mold or die for mass production of the product.

1.4.1 Rapid Prototyping

The Solid Freeform Fabrication (SFF) technology, also called Rapid Prototyping (RP), is an additive process that employs a layer-building technique based on horizontal cross-sectional data from a 3D CAD model. Beginning with the bottommost cross-section of the CAD model, the rapid prototyping machine creates a thin layer of material by slicing the model into so-called 2½ D layers. The system then creates an additional layer on top of the first based on the next higher cross-section. The process repeats until the part is completely built. It is illustrated using an engine case in the example shown in Figure 1.13. Rapid prototyping systems are capable of creating parts with small internal cavities and complex geometry.

Most important, SFF follows the same layering process for any given 3D CAD models, so it requires neither tooling nor manufacturing process planning for prototyping, as required by conventional manufacturing methods. Based on CAD solid models, the SFF technique fabricates physical prototypes of the product in a short turnaround time for design verification. It also supports tooling for product manufacturing, such as mold or die fabrications, through, for example, investment casting (Kalpakjian 1992).

Note that there are various types of SFF systems commercially available, such as the SLA®-7000 and Sinterstation® by 3D Systems (Figures 1.14a and 1.14b). In this chapter, the Dimension 1200 sst® machine (www.stratasys.com), as shown in Figure 1.14c, is presented. More details about it as well as other RP systems will be discussed in *Product Manufacturing and Cost Estimating using CAD/CAE*, a book in The Computer Aided Engineering Design Series.

The CAD solid model of the product is first converted into a stereolithographic (STL) format (Chua and Leong 1998), which is a faceted boundary representation uniformly accepted by the industry. Both the coarse and refined STL models of an engine case are shown in Figure 1.15. Even though the STL model is an approximation of the true CAD geometry, increasing the number of triangles can minimize the geometric error effectively. This can be achieved by

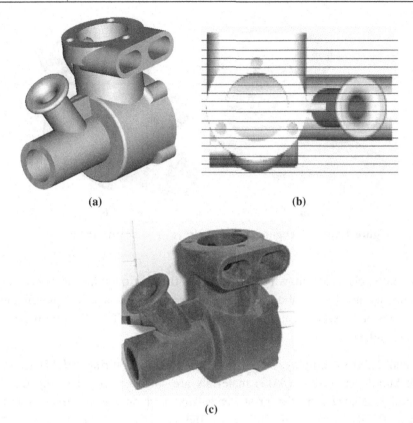

(a)   (b)

(c)

Figure 1.13: SFF: layered manufacturing: (a) 3D CAD model, (b) 2-1/2D slicing, and (c) physical model.

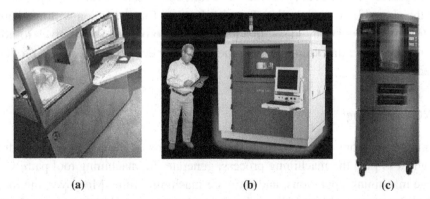

(a)   (b)   (c)

Figure 1.14: Commercial RP systems: (a) 3D Systems' SLA 7000, (b) SinterStation 2500 (*Source: 3D Systems Corporation, USA*), and (c) Stratasys Inc.'s Dimension 1200 sst (*Source: Stratasys Ltd*).

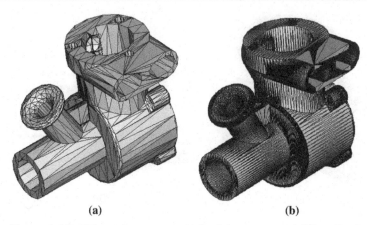

| (a) | (b) |

Figure 1.15: STL engine case models: (a) coarse and (b) refined.

specifying a smaller chord length, which is defined as the maximum distance between the true geometric boundary and the neighboring edge of the triangle. The faceted representation is then sliced into a series of 2D sections along a prespecified direction. The slicing software is SFF-system dependent.

The Dimension 1200 sst employs fused deposition manufacturing (FDM) technology. Acrylonitrile butadiene styrene (ABS) materials are softened (by elevating temperature), squeezed through a nozzle on the print heads, and laid on the substrate as build and support materials, respectively, following the 2D contours sliced from the 3D solid model (Figure 1.16). Note that various crosshatch options are available in CatalystEX® software (www.dimensionprinting.com), which comes with the rapid prototyping system.

The physical prototypes are mainly for the cross-functional team to verify the product design and check the assembly. However, they can also be used for discussion with marketing personnel to develop marketing ideas. In addition, the prototypes can be given to potential customers for feedback, thus bringing customers into the design loop early in product development.

1.4.2 CNC Machining

The machining operations of virtual manufacturing, such as milling, turning, and drilling, allow designers to plan the machining process, generate the machining tool path, visualize and simulate machining operations, and estimate machining time. Moreover, the tool path generated can be converted into CNC codes (M-codes and G-codes) (Chang et al. 1998, McMahon and Browne 1998) to fabricate functional parts as well as a die or mold for production.

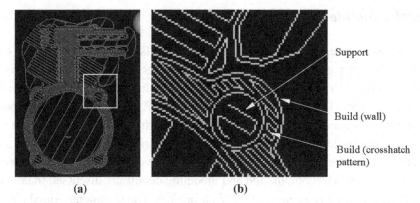

Figure 1.16: Crosshatch pattern of a typical cut-out layer: (a) overall and (b) enlarged.

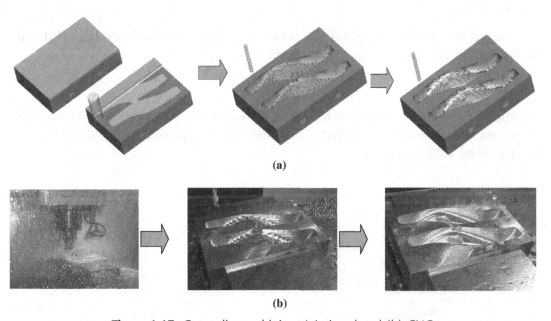

Figure 1.17: Cover die machining: (a) virtual and (b) CNC.

For example, the cover die of a mechanical part is machined from an 8 in. × 5.25 in. × 2 in. steel block, as shown in Figure 1.17a. The cutter location data files generated from virtual machining are post-processed into machine control data (MCD)—that is, G- and M-codes, for CNC machining, using post-processor UNCX01.P11 in Pro/MFG. In addition to volume milling and contour surface milling, drilling operations are conducted to create the waterlines. A 3-axis CNC mill, HAAS VF-series (HAAS Automation, Inc. 1996), is employed for fabricating the die for casting the mechanical part (Figure 1.17b).

1.5 Example: Simple Airplane Engine

A single-piston, two-stroke, spark-ignition airplane engine (shown in Figure 1.5) is employed to illustrate the e-Design paradigm and tool environment. The cross-functional team is asked to develop a new model of the engine with a 30% increment in both maximum torque and horsepower at 1,215 rpm. The design of the new engine will be carried out at two interrelated levels: system and component. At the system level, the performance measure is the power output; at the component level, the structural integrity and manufacturing cost of each component are analyzed for improvement. Note that only a very brief discussion is provided in this introductory chapter. The computation and modeling details are discussed in later chapters and *Product Design Modeling using CAD/CAE*, a book in The Computer Aided Engineering Design Series.

System-Level Design

Power is proportional to the rotational speed of the crankshaft (N), the swept volume (V_s), and the brake mean effective pressure (P_b) (Taylor 1985):

$$W_b = P_b \, V_s \, N \tag{1.9}$$

The effective pressure P_b applied on top of the piston depends on, among other factors, the swept volume and the rotational speed of the crankshaft. The pressure is limited by the integrity of the engine structure.

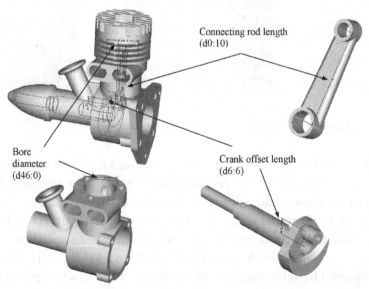

Figure 1.18: Engine assembly with design variables at the system level.

Design variables at the system level include bore diameter (d46:0) and stroke, defined as the distance between the top face of the piston at the bottom and top dead-center positions. In the CAD model, the stroke is defined as the sum of the crank offset length (d6:6) and the connecting rod length (d0:10), as shown in Figure 1.18. To achieve the requirement for system performance, these three design variables are modified as listed in Table 1.1. The design variable values were calculated following theory and practice for internal combustion engines (Taylor 1985). Details of the computation can be found in Silva (2000).

Table 1.1: Changes in design variables at the system level

Design Variable	Current Value (in.)	New Value (in.)	Change (in.)	% Change
Bore diameter (d46:0)	1.416	1.6	0.164	11.6
Crank length (d6:6)	0.5833	0.72	0.1567	26.9
Connecting rod length (d0:10)	2.25	2.49	0.24	10.7

The solid models of the entire engine are automatically updated and properly assembled using the parametric relations established earlier (refer to Figure 1.6b). The change causes P_b to increase from 140 to 180 lbs, so the peak load increases from 400 to 600 lbs. The load magnitude and path applied to the major load-carrying components, such as the connecting rod and crankshaft, are therefore altered. Results from motion analysis show that the system performs well kinematically. Reaction forces applied to the major load-carrying components are computed—for example, for the connecting rod shown in Figure 1.19. The change also affects manufacturing time for some components.

Component-Level Design

Structural performance is evaluated and redesigned to meet the requirements. In addition, virtual manufacturing is conducted for components with significant design changes. Build materials (volume) and manufacturing times constitute a significant portion of the product cost. In this section, the design of the connecting rod is presented to demonstrate the design decision-making method discussed.

Because of the increased load transmitted through the piston and the increased stroke length, the connecting rod can experience buckling failure during combustion. In addition, because changes in stroke length, stiffness, and mass vary, the natural frequency of the rod may be different. Moreover, load is repeatedly applied to the connecting rod, potentially leading to fatigue failure. Structural FEA are conducted to evaluate performance. In addition, virtual manufacturing is carried out to determine the machining cost of the rod.

Because of the increment of the connecting rod length (d0:10) and the magnitude of the external load applied (see Figure 1.20), the rod's maximum von Mises stress increases

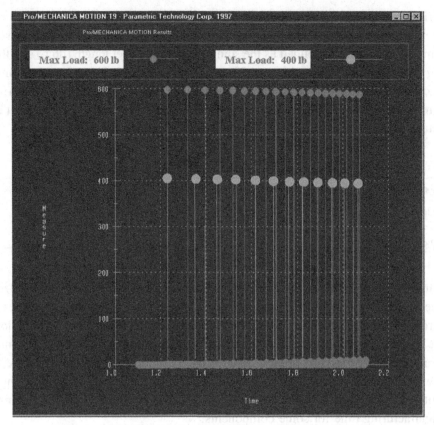

Figure 1.19: Dynamic load applied to the connecting rod.

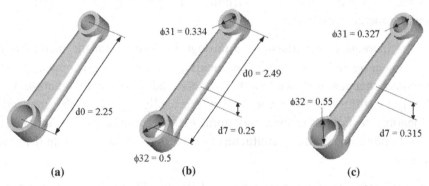

Figure 1.20: Engine connecting rod: (a) original design; (b) changes at the system level; (c) changes at the component level.

from 13,600 to 18,850 psi and the buckling load factor decreases from 33 to 7. The first natural frequency is 1,515 Hz. The machining time estimated for the connecting rod is 13.2 minutes using hole-drilling and face-milling operations (shown earlier in Figure 1.9d).

Design Trade-Off

The design trade-off method discussed in Section 1.3.5 is applied to the components, with significant changes resulting from the system-level design. Only the design trade-off conducted for the connecting rod is discussed.

Performance measures for the connecting rod, including buckling load factor, fatigue life, natural frequency, volume, and machining costs (time), are brought together for design trade-off. Three design variables, $\phi32$, $\phi31$, and d7, are identified, as shown in Figure 1.20b. The objective is to minimize volume and manufacturing time subject to maximum allowable von Mises stress, operating frequency, and minimum allowable buckling load factor. The engine is designed to work at 21 kHz, and the minimum allowable buckling load factor for the connecting rod is assumed to be 10.

Sensitivity coefficients for performance and cost measures with respect to design variables are calculated (refer to Figure 1.11) using the finite difference method. Design trade-offs are conducted followed by a what-if study. When a satisfactory design is found, the solid model of the rod is updated for performance evaluation and virtual manufacturing. This process is repeated twice when all the requirements are met. The design change is summarized in Tables

Table 1.2: Changes in design variables at the component level

Design Variable	Current Value (in.)	New Value (in.)	% Change
Diameter of the large hole ($\phi32$)	0.50	0.55	10
Diameter of the small hole ($\phi31$)	0.334	0.32728	−2.01
Thickness (d7)	0.25	0.31484	25.9

Table 1.3: Changes in performance measures at the component level

Performance Measure	Current Value	New Value	% Change
VM stress	18.9 ksi	10.5 ksi	−44.4
Buckling load factor	7.1	14.2	100
Volume	0.438813 in.^{3}	0.5488 in.^{3}	25.1
Machining time	13.2 min	13.2 min	0
Natural frequency	1515 Hz	1840 Hz	21.5

Figure 1.21: Physical prototypes of engine parts.

1.2 and 1.3, which show that the machining time is maintained and a small volume increment is needed to achieve the required performance.

Rapid Prototyping

When the design is finalized through virtual prototyping, rapid prototyping is used to fabricate a physical prototype of the engine, as shown in Figure 1.21. The prototype can be used for design verification as well as tolerance and assembly checking.

1.6 Example: High-Mobility Multipurpose Wheeled Vehicle

The overall objective of the high-mobility multipurpose wheeled vehicle (HMMWV) design is to ensure that the vehicle's suspension is durable and reliable after accommodating an additional armor loading of 2,900 lb. A design scenario using a hierarchical product model (see Figure 1.10) that evolves during the design process is presented in this section.

In the preliminary design stage, vehicle motion is simulated and design changes are performed to improve the vehicle's gross motion. At this stage, the dynamic behavior of the HMMWV's suspension is simulated and designed. The specific objectives of the preliminary design are to avoid the problem of metal-to-metal contact in the shock absorber due to added armor load, and to improve the driver's comfort by reducing vertical acceleration at the HMMWV driver's seat.

By modifying the spring constant to improve the HMMWV suspension design at the preliminary design stage, the load path generated in HMMWV dynamics simulation is affected in the suspension unit. In the detail design stage, the objective is to assess and redesign the durability, reliability, and structural performance of selected suspension components affected by the added armor load that result in changes in load path and load magnitude.

Note that only a very brief discussion is provided in this introductory chapter. The computation and modeling details are discussed in later chapters.

Hierarchical Product Model

In this particular case, a hierarchical product model is employed to support the HMMWV's design. In all models, nonsuspension parts, such as instrument panel, seats, and lights, are not modeled. Important vehicle components, such as engine and transmission, are modeled using engineering parameters without depending on CAD representation. A low-fidelity CAD model consisting of 18 parts (Figure 1.22) is created using Pro/ENGINEER to support the preliminary design. This model has accurate joint definition and fairly accurate mass property, but less accurate geometry. The goal of the low-fidelity model is to support vehicle dynamic simulation. It is created using substantially less effort compared to that required for the detailed model.

The detailed product model, consisting of more than 200 parts and assemblies (Figure 1.23), is created to support the detail design of suspension components. The detailed model is

Figure 1.22: HMMWV CAD model for preliminary design.

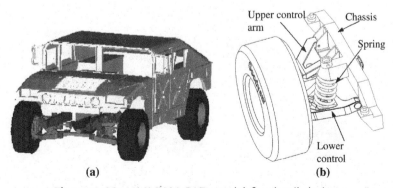

(a)　　　　(b)

Figure 1.23: HMMWV CAD model for detail design.

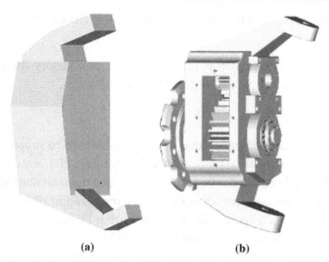

(a)   (b)

Figure 1.24: HMMWV gear hub assembly models: (a) preliminary and (b) detailed.

derived from the preliminary model by (1) breaking an entity into more parts and assemblies (e.g., the gear hub assembly, shown in Figure 1.24) to simulate and design detailed parts, and (2) refining the geometry of mechanical components to support structural FEA (e.g., the lower control arm, shown in Figure 1.25).

Preliminary Design

The HMMWV is driven repeatedly on a virtual proving ground, as shown in Figure 1.26, with a constant speed of 20 MPH for a period of 23 seconds. A dynamic simulation model, shown in Figure 1.27, is first derived from the low-fidelity CAD solid model of the HMMWV (refer to Figure 1.22). A more in-depth discussion of the HMMWV vehicle dynamic model is provided in Chapter 3.

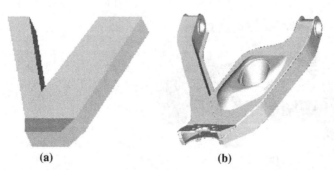

(a)   (b)

Figure 1.25: HMMWV lower control arm models: (a) preliminary and (b) detailed.

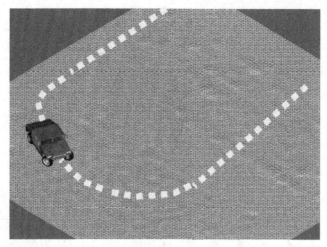

Figure 1.26: HMMWV dynamic simulation.

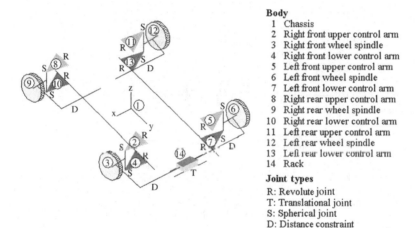

Body

1 Chassis
2 Right front upper control arm
3 Right front wheel spindle
4 Right front lower control arm
5 Left front upper control arm
6 Left front wheel spindle
7 Left front lower control arm
8 Right rear upper control arm
9 Right rear wheel spindle
10 Right rear lower control arm
11 Left rear upper control arm
12 Left rear wheel spindle
13 Left rear lower control arm
14 Rack

Joint types

R: Revolute joint
T: Translational joint
S: Spherical joint
D: Distance constraint

Figure 1.27: HMMWV dynamic model.

Using DADS, severe metal-to-metal contact is identified within the shock absorber, caused by the added armor load and rough driving conditions, as shown in Figure 1.28. The spring constant is adjusted to avoid any contact problems; it is increased in proportion to the mass increment of the added armor to maintain the vehicle's natural frequency. This design change not only eliminates the contact problem (see Figure 1.28) but also reduces the amplitude of vertical acceleration at the driver's seat, which improves driving comfort (see Figure 1.29). However, the change alters the load path in the components of the suspension subsystem—for example, the shock absorber force acting on the control arm increases about 75%, as shown in Figure 1.30.

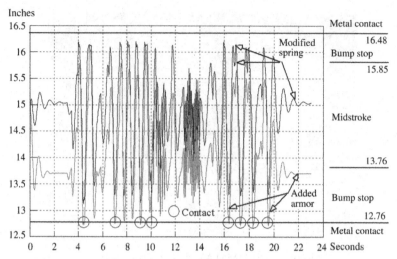

Figure 1.28: Shock absorber operation distance (in inches).

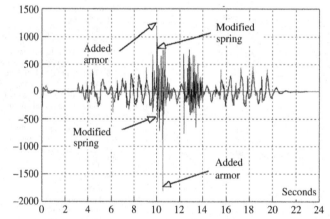

Figure 1.29: HMMWV driver seat vertical accelerations (in./sec^2).

Detail Design

Simulations are carried out for fatigue, vibration, and buckling of the lower control arm (Figure 1.30); reliability of gears in the gear hub assembly (refer to Figure 1.24b); the spring of the shock absorber (see Figure 1.23); and the bearings of the control arm (see Figure 1.30).

Using ANSYS, the first natural frequency of the lower control arm is obtained as 64 Hz, which is far away from vehicle vibration frequency, eliminating concern about resonance. The buckling load factor is analyzed using the peak load at time 10.05 seconds in the 23-second simulation period. The result shows that the control arm will not buckle even under

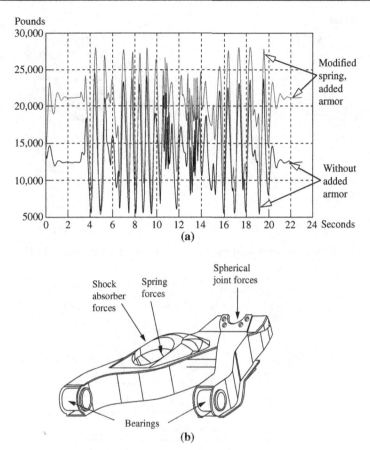

Figure 1.30: History of shock absorber forces (lbs): (a) force history with and without added armor load, (b) locations of force application

the most severe load. Therefore, the current design is acceptable as far as buckling and resonance of the lower control arm are concerned.

Results obtained from fatigue analyses show that fatigue life (crack initiation) of the lower control arm degrades significantly—for example, from 6.61E+09 to 1.79E+07 blocks (one block is 20 seconds) at critical areas (see Figure 1.31b)—because of the additional armor load and change of load path. Therefore, the design must be altered to improve control arm durability. Reliability of the bearing, gear, and spring at a 99% fatigue failure rate is 2.18E+07, 3.36E+06, and 1.27E+02 blocks, respectively. Note that the fatigue life of the spring at the required reliability is not desirable.

Design Trade-Off

Eleven design parameters, including geometric dimensions (d1 and d2 in Figure 1.32a), material property (cyclic strength coefficient K' of the lower control arm), and thickness of

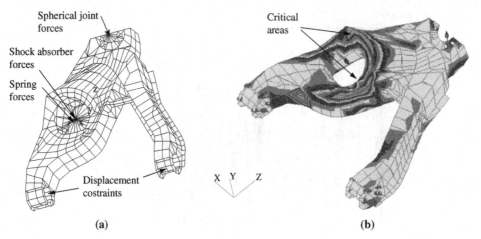

Spherical joint forces

Shock absorber forces

Spring forces

Displacement costraints

Critical areas

X Y Z

(a)

(b)

Figure 1.31: HMMWV lower control arm models: (a) finite element and (b) fatigue life prediction.

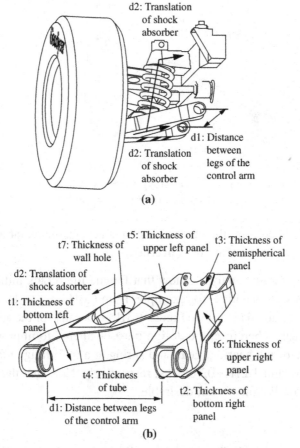

d2: Translation of shock absorber

d2: Translation of shock absorber

d1: Distance between legs of the control arm

(a)

t7: Thickness of wall hole

t5: Thickness of upper left panel

t3: Thickness of semispherical panel

d2: Translation of shock adsorber

t1: Thickness of bottom left panel

t6: Thickness of upper right panel

t4: Thickness of tube

t2: Thickness of bottom right panel

d1: Distance between legs of the control arm

(b)

Figure 1.32: Design parameters defined for the control arm: (a) suspension geometric dimensions and (b) thickness dimensions.

the control arm sheet metal (t1 to t7 in Figure 1.32b) are defined to support design modification.

A global design trade-off that involves changes in more than one component is conducted first. Geometric design parameters d1 and d2 are modified to reduce loads applied to the control arm, bearing, spring, and gears in the gear hub so that the durability and reliability

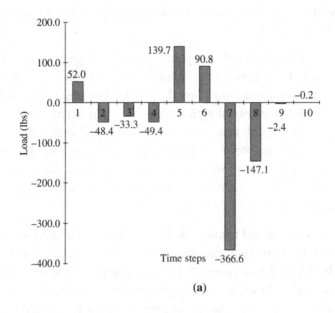

(a)

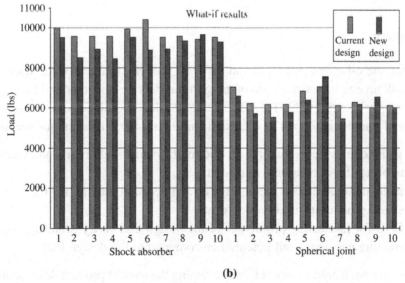

(b)

Figure 1.33: Sensitivity of load on the spherical joint of control arm w.r.t d2 at 10 time steps (a) design sensitivity display and (b) what-if study.

of these components can be improved. Changes in d1 and d2 affect not only the lower control arm but also the upper control arm and the chassis frame. Sensitivity coefficients of loads at discretized time steps (a total of 10 selected time steps) with respect to parameters d1 and d2 are calculated using a finite difference method. Sensitivity coefficients can be displayed in bar charts (see Figure 1.33a) to guide design modifications. A what-if study is carried out with a design perturbation of 0.6 and 0.3 in. for d1 and d2, respectively, to obtain a reduction in loads. An example of the what-if results is shown in Figure 1.33b.

A local design trade-off that involves design parameters of a single component is carried out for the lower control arm. Thickness design parameters t1 to t7 and the material design parameter K' are modified to increase the control arm's fatigue life. Fatigue life at ten nodes of its finite element model in the critical area is measured. Sensitivity coefficients of control arm fatigue life at these nodes with respect to the thickness and material parameters are calculated. A design trade-off method using a QP algorithm is employed because of the large number of design parameters and performance measures involved. An improved design obtained shows that with a 0.6% weight increment, fatigue life at the critical area increases about ten times: from 1.79 E+07 to 1.68 E+08 blocks.

A dynamic simulation is performed again with the detailed model and modified design to ensure that the metal contact problem, encountered in the preliminary design stage, is eliminated as a result of model refinement and design changes in the detail design stage. The global design trade-off reduces the load applied to the shock absorber spring. This reduction significantly increases the spring fatigue life to the desired level.

1.7 Summary

In this chapter, the e-Design paradigm and software tool environment were discussed. The e-Design paradigm employs virtual prototyping for product design and rapid prototyping and computer numerical control (CNC) for fabricating physical prototypes of a design for design verification and functional tests. The e-Design paradigm offers three unique features:

- The VP technique, which simulates product performance, reliability, and manufacturing costs; and brings these measures to design.
- A systematic and quantitative method for design decision making for the parameterized product in solid model forms.
- RP and CNC for fabricating prototypes of the design that verify product design and bring marketing personnel and potential customers into the design loop.

The e-Design approach holds potential for shortening the overall product development cycle, improving product quality, and reducing product costs. With intensive knowledge of the product gained from simulations, better design decisions can be made, thereby overcoming what is known as the *design paradox*. With the advancement of computer simulations, more hardware tests can be replaced by them, reducing cost and shortening product development time. Manufacturing-related issues can be largely addressed through virtual manufacturing in early design stages. Moreover, manufacturing process planning conducted in virtual manufacturing streamlines the production process.

Questions and Exercises

1.1. In this assignment, you are asked to search and review articles (such as in *Mechanical Engineering* magazine) that document successful stories in industry that involve employing the e-Design paradigm and/or employing CAD/CAE/CAM technology for product design.
   - Briefly summarize the company's history and its main products.
   - Briefly summarize the approach and process that the company adopted for product development in the past.
   - Why must the company make changes? List a few factors.
   - Which approach and process does the company currently employ?
   - What is the impact of the changes to the company?
   - In which journal, magazine, or website was the article published?
1.2. In this chapter we briefly discussed rapid prototyping technology and the Dimension 1200 sst machine. The sst uses fused deposition manufacturing technology for support of layer manufacturing. Search and review articles to understand the FDM technology and machines that employ such technology other than the Dimension series.

References

Anderson, D.M., 1990. Design for Manufacturability: Optimizing Cost, Quality, and Time to Market. CIM Press.

Arora, J.S., 1989. Introduction to Optimal Design. McGraw-Hill.

Boothroyd, G., Dewhurst, P., Knight, W., 1994. Product Design for Manufacturing and Assembly. Marcel Dekker.

Chang, K.H., Choi, K.K., Wang, J., Tsai, C.S., Hardee, E., 1998. A multi-level product model for simulation-based design of mechanical systems. Concurrent Engineering Research and Application (CERA) Journal 6 (2), 131–144.

Chang, K.H., Yu, X., Choi, K.K., 1997. Shape design sensitivity analysis and optimization for structural durability. International Journal of Numerical Methods in Engineering 40, 1719–1743.

Chang, K.H., 2011. Dynamic Simulation and Mechanism Design with SolidWorks Motion 2011. Schroff Development Corporation.

Chang, T.-C., Wysk, R.A., Wang H-, P., 1998. Computer-Aided Manufacturing, 2nd ed. Prentice Hall.

Chen, J.S., Pan, C., Wu, T.C., 1997. Large deformation analysis of rubber based on a reproducing kernel particle method. Computational Mechanics 19, 153–168.

Chua, C.K., Leong, K.F., 1998. Rapid Prototyping: Principles and Applications in Manufacturing. John Wiley.

Ertas, A., Jones, J.C., 1993. The Engineering Design Process. John Wiley.

Fang, X.D., Jawahir, I.S., 1996. A hybrid algorithm for predicting chip form/chip breakability in machining. International Journal of Machine Tools and Manufacture 36 (10), 1093–1107.

Gonzalez, A.J., Dankel, D.D., 1993. The Engineering of Knowledge-Based Systems, Theory and Practice. Prentice Hall.

Grindeanu, I., Choi, K.K., Chen, J.S., Chang, K.H., 1999. Design sensitivity analysis and optimization of hyperelastic structures using a meshless method. AIAA Journal 37 (8), 990−997.

HAAS Automation Inc., 1996. VF Series Operations Manual.

Hallquist, J.O., 2006. LS-DYNA3D Theory Manual. Livermore Software Technology Corp.

Haug, E.J., Choi, K.K., Komkov, V., 1986. Design Sensitivity Analysis of Structural Systems. Academic Press.

Haug, E.J., Luh, C.M., Adkins, F.A., Wang, J.Y., 1996. Numerical algorithms for mapping boundaries of manipulator workspaces. Journal of Mechanical Design 118, 228−234.

Haug, E.J., 1989. Computer-Aided Kinematics and Dynamics of Mechanical Systems, vol. I: Basic Methods. Allyn and Bacon.

Jacobs, P.F., 1994. StereoLithography and Other RP&M Technologies. ASME Press.

Kalpakjian, S., 1992. Manufacturing Engineering and Technology, second ed. Addison-Wesley.

Kane, T.R., Levinson, D.A., 1985. Dynamics: Theory and Applications. McGraw-Hill.

Lee, W., 1999. Principles of CAD/CAM/CAE Systems. Addison-Wesley Longman.

Madsen, H.O., Krenk, S., Lind, N.C., 1986. Methods of Structural Safety. Prentice Hall.

McMahon, C., Browne, J., 1998. CADCAM, second ed. Addison-Wesley.

Moës, N., Gravouil, A., Belytschko, T., 2002. Nonplanar 3D crack growth by the extended finite element and level sets. Part I: Mechanical model. International Journal for Numerical Methods in Engineering 53 (11), 2549−2568.

Numerical Integration Technologies, 1998. SYSNOISE 5.0.

Prasad, B., 1996. Concurrent Engineering Fundamentals, vols. I and II: Integrated Product and Process Organization. Prentice Hall.

Silva, J., 2000. Concurrent Design and Manufacturing for Mechanical Systems. MS thesis. University of Oklahoma.

Szabó, B., Babuška, I., 1991. Finite Element Analysis. John Wiley.

Taylor, C., 1985. The Thermal-Combustion Engine in Theory and Practice, vol. I: Thermodynamics, Fluid Flow, Performance, second ed. MIT Press.

Tsai, C.S., Chang, K.H., Wang, J., 1995. Integration infrastructure for a simulation-based design environment, Proceedings, Computers in Engineering Conference and the Engineering Data Symposium. ASME Design Theory and Methodology Conference.

Ullman, D.G., 1992. The Mechanical Design Process. McGraw-Hill.

Wu, Y.T., Wirsching, P.H., 1984. Advanced reliability method for fatigue analysis. Journal of Engineering Mechanics 110, 536−563.

Yu, X., Chang, K.H., Choi, K.K., 1998. Probabilistic structural durability prediction. AIAA Journal 36 (4), 628−637.

Yu, X., Choi, K.K., Chang, K.H., 1997. A mixed design approach for probabilistic structural durability. Journal of Structural Optimization 14, 81−90.

Zeid, I., 1991. CAD/CAM Theory and Practice. McGraw-Hill.

Sources

Adams: www.mscsoftware.com

ANSYS: www.ansys.com

CAMWorks: www.camworks.com

CatalystEX: www.dimensionprinting.com

LMS DADS: http://lsmintl.com

Dimension sst: www.stratasys.com

HAAS VF-Series: www.haascnc.com

HyperMesh: www.altairhyperworks.com

LS-DYNA: www.lstc.com

MSC/Nastran, MSC/Patran: www.mscsoftware.com

nCode: www.ncode.com

NESSUS: www.nessus.swri.org

Pro/ENGINEER, Pro/MECHANICA, Pro/MFG, Pro/SHEETMETAL, Pro/WELDING, etc.: www.ptc.com

SLA-7000, Sinterstation: www.3dsystems.com

SolidWorks Motion, SolidWorks Simulation: www.solidworks.com

SYSNOISE 5.0: www.lmsintl.com/SYSNOISE

Structural Analysis

Chapter Outline

Structural analysis is an essential part of product design. Mechanical components must be strong and durable so that the entire mechanical system is able to sustain operating loads for its intended operating conditions. The structural integrity of the entire system must be ensured.

In structural analysis, different types of problems are solved depending on the operating conditions and intended use of the product. These include static, buckling, vibration, transient dynamics, frequency responses, and others. Except for simple elastostatic problems, very few of these can be solved in closed form using analytical equations. Most, if not all, encountered in engineering design must rely on numerical methods for solutions, such as finite element method. Although finite element methods (FEMs) are powerful, understanding theories and analytical methods is crucial for two reasons. One, the underlying mechanics and physics of the analytical methods help us understand how structures behave under specific conditions. Second, it is critical to verify numerical results obtained from methods such as FEM. Very often, it is possible to use analytical solutions to check FEA results for more complex problems. The bottom line is that a solid background in structural mechanics is essential to use FEM software correctly and effectively. As an analogy, one must know basic addition, subtraction, multiplication, and division before punching keys on a calculator.

In addition to analytical methods, in this chapter we will devote major discussion to finite element methods. Note that our focus will not be on presenting comprehensive FEM theory. Instead, we will focus on the discussion of essential elements in practical use of FEM software for modeling and analysis. The goal of this chapter is to help the reader become confident and competent in using FEM for creating adequate models and obtaining reasonably accurate results to support product design. Those who are interested in FEM theory may refer to excellent books, such as Pilky and Wunderlich (2002), Szabó and Babuška (1991), and Bathe (2007). Major FEM software packages, both general-purpose and specialized codes, will be briefly discussed to provide a general understanding of software availability and options as well as sufficient information for making reasonable choices. In addition, advanced methods based on the finite element concept, such as the meshless method, will be introduced briefly.

A human middle ear model is included as a case study, in which a mechanics study was carried out using FEA that explains how the ossicles work to eventually facilitate hearing aid development. In addition, two practice examples, a cantilever beam and a thin-walled tube modeled with Pro/MECHANICA® Structure and SolidWorks© Simulation, are offered. These examples can also be found on this book's companion website (http://booksite.elsevier. com/9780123984609). Detailed instructions on bringing up these models and steps for carrying out FEA are given in Projects P3 and S3.

Overall the objectives of this chapter are (1) to provide basic FEM theory using simple examples to explain how the method works, (2) to help the reader become familiar with FEM modeling and analysis to effectively use these tools for design, (3) to familiarize the reader with existing commercial software, and (4) to help the reader use Pro/MECHANICAL Structure and/or SolidWorks Simulations for basic applications after going through the tutorial lessons.

2.1 Introduction

Structural analysis comprises the set of mechanics theories that obey physical laws required to study and predict the behavior of structures. The subjects of structural analysis are engineering artifacts whose integrity is judged largely on their ability to withstand loads. Structural analysis incorporates the fields of mechanics and dynamics as well as the many failure theories. From a theoretical perspective, the primary goal of structural analysis is the computation of deformations, internal forces, and stresses. In practice, structural analysis reveals the structural performance of the engineering design and ensures the soundness of structural integrity in design without dependence on direct testing.

In the mechanical and aerospace industries, engineers often confront the challenge of designing mechanical systems and components that can sustain operating loads, meet functional requirements, and last longer. It is imperative that these structures contain a minimum of material to reduce cost and increase efficiency of the mechanical system, such as in terms of fuel consumption. The geometry of these load-bearing structural components is usually complicated because of strength and efficiency requirements. Three of such structural components are shown in Figure 2.1.

A number of approaches have been developed to support design of structural components and systems, such as shape optimization (Chang and Edke 2010), topology optimization (Bendsoe and Sigmund 2003), and reliability-based design (Choi et al. 2007). Many have been employed to create designs for challenging applications, such as those shown in Figure 2.1. The success of these design methods is largely attributed to accurate underlying analysis methods that are built on fundamental mechanics theory and physics laws.

(a)   **(b)**

Figure 2.1: Structural components of highly complex geometry: (a) automotive suspension component and engine block and (b) airplane landing gear strut.

To perform an accurate analysis a structural engineer must determine such information as structural loads, geometry, support conditions, and materials properties. The results of his or her analysis typically include support reactions, stresses, and displacements. This information is then compared to criteria that indicate the conditions of failure. Advanced structural analysis may examine dynamic response, stability, and nonlinear behavior.

Analytical methods make use of analytical formulations that apply mostly to simple linear elastic models, lead to closed-form solutions, and can often be solved by hand. Such methods include strength of materials, energy methods, and linear elasticity. Numerical approaches, such as FEM, are more applicable to structures of arbitrary size and complexity; they employ numerical algorithms for solving differential equations built upon theories of mechanics.

Regardless of approach, the formulation is based on the same three fundamental relations: equilibrium, constitutive, and compatibility. The solutions are approximate when any of these relations is only approximately satisfied or is only an approximation of reality. There always exist uncertainties in modeling and analysis of structural components. It is imperative to account for these uncertainties in order to design more reliable mechanical systems.

In this chapter, we start by briefly reviewing analytical methods and then discuss finite element methods. Essential ingredients in using FEM for modeling and analysis of structural problems, such as mesh generation and boundary conditions, are discussed in detail. We also discuss commercial FEA software, both general-purpose and specialized. Example problems modeled and solved using commercial codes are presented. In this chapter, we also include advanced FEA methods that might be of interest in solving more complex and specialized problems.

2.2 Analytical Methods

Analytical methods employ analytical formulations that apply mostly to simple linear elastic problems and lead to closed-form solutions. The solutions are based on linear isotropic infinitesimal elasticity. In other words, they contain assumptions (among others) that the materials in question are elastic, that stress is related linearly to strain, that all deformations are small, and that the material behaves identically regardless of the direction of the applied load (that is, it is *isotropic*). As with any simplifying assumptions in engineering, the more the model strays from reality, the less useful the result. In general, analytical solutions cannot be directly used for support of design. However, they may provide references that support verification of numerical solutions of complex structures.

Three analytical methods commonly employed for structural analysis are briefly discussed. They are often considered in courses such as strength of materials, energy method, and elasticity. We also consider the design criteria of various failure modes, as well as safety factors to address uncertainties and variations.

2.2.1 Strength of Materials

The strength of materials method is available for simple structural members subject to specific loadings, such as axially loaded bars, prismatic beams in a state of pure bending, bars under direct shear forces, and shafts subject to torsion (Figure 2.2). The solutions can be superimposed using the superposition principle to analyze a member undergoing combined loading following the linear elasticity assumption. Solutions for special cases exist for common structures such as thin-walled pressure vessels.

For the analysis of components and systems other than standard structural members, this method can be used in conjunction with force equilibrium for analytical solutions. An example is the inclined cantilever beam shown in Example 2.1 with its tip constrained to move along the vertical direction. The stress and displacement can be calculated by first decomposing the reaction force at the tip into the axial and transverse directions, and then calculating stress and displacement using standard equations following the superposition principle.

Example 2.1

Calculate the displacement at the tip of the inclined cantilever beam shown in the figure below. Note that the tip of the beam is constrained to move vertically. E, A, I, and ℓ are the modulus, cross-sectional area, moment of inertia, and length of the beam, respectively. The cross-section is solid circular with a radius r.

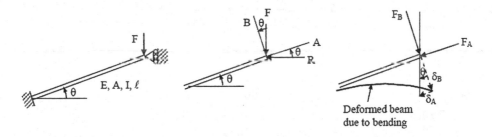

Deformed beam due to bending

Solutions

This is a statically indeterminate problem. The reaction force R at the tip cannot be solved by force/moment equilibrium equations alone. Instead, it can be obtained by adding the condition imposed by the displacement constraint at the tip. We first draw a free-body diagram for bending force F_B and axial force F_A, and use standard equations for displacements at the tip due to bending and axial forces. We then impose the boundary condition at the tip to solve for the reaction force followed by the displacements.

From force equilibrium, the bending force F_B and axial force F_A can be obtained as

$$F_B = F \cos \theta - R \sin \theta \text{ and } F_A = F \sin \theta + R \cos \theta$$

or simply

$$F_B = Fc - Rs \text{ and } F_A = Fs + Rc$$

From equations provided in Strength of Materials (or Solid Mechanics) textbooks, the displacements due to bending and axial forces are, respectively,

$$\delta_B = \frac{F_B \ell^3}{3EI} = \frac{(Fc - Rs)\ell^3}{3EI} \quad \text{and} \quad \delta_A = \frac{F_A \ell}{EA} = \frac{(Fs + Rc)\ell}{EA}$$

Imposing the displacement constraint at the tip gives the relation between δ_B and δ_A as the following:

$$\frac{\delta_A}{\delta_B} = \tan\theta = \frac{\sin\theta}{\cos\theta} = \frac{s}{c}$$

Thus

$$c\delta_A = s\delta_B \quad \text{or} \quad \frac{c(Fs + Rc)\ell}{EA} = \frac{s(Fc - Rs)\ell^3}{3EI}$$

From the preceding equation, R can be obtained as

$$R = \frac{Fsc(A\ell^2 - 3I)}{Q}$$

where $Q = 3Ic^2 + A\ell^2 s^2$, and

$$\delta_B = \frac{(Fc - Rs)\ell^3}{3EI} = \frac{Fc\ell^3}{EQ} \quad \text{and} \quad \delta_A = \frac{Fs\ell^3}{EQ}$$

Therefore, the vertical displacement at the tip is

$$\delta = \frac{\delta_B}{c} = \frac{F\ell^3}{EQ} = \frac{F\ell^3}{E(3Ic^2 + A\ell^2 s^2)}$$

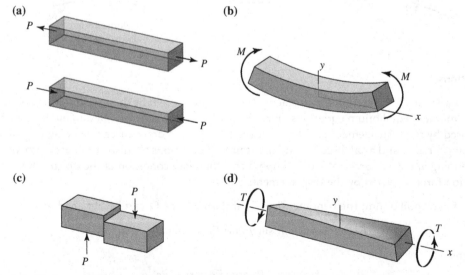

Figure 2.2: Basic structural members under loads: (a) axially loaded bars, (b) prismatic beam in a state of pure bending, (c) bar under direct shear, and (d) shaft subject to torsion.

Note that stress in the beam can be calculated by superposing bending and axial stresses due to the respective bending force F_B and axial force F_A. The stress calculation is left as an exercise (Exercise 2.1).

2.2.2 Energy Method

Energy principles in structural mechanics express the relationships between stresses, strains or deformations, displacements, material properties, and external effects in the form of energy or work done by internal and external forces. Since energy is a scalar quantity, these relationships provide convenient and alternative means for formulating the governing equations of deformable bodies in solid mechanics. The energy method is very general and is applicable to many structural problems, especially for dynamic responses and nonlinear analysis. One of its simplest forms is Castigliano's theorem, which is powerful for solving more complex elastostatic problems.

Strain energy must first be calculated before applying Castigliano's theorem. The strain energy for standard structural members is given in Figure 2.3. For a structure component loaded with a force P or torque/moment M, the theorem states:

> The partial derivative of the strain energy, considered as a function of the applied forces (and moments) acting on a linearly elastic structure, with respect to one of these forces (or moments), is equal to the displacement (or rotation angle) in the direction of the force (or moment) of its point of application.

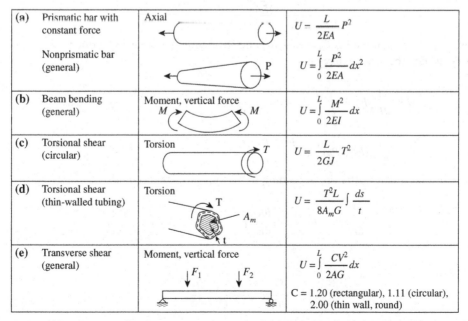

Figure 2.3: Strain energy of basic structural members: (a) truss of axial load; (b) beam of pure bending; (c) shaft of pure torsion; (d) torsion for thin-walled shaft; and (e) transverse shear.

Mathematically, the theorem states:

$$\delta = \frac{\partial U}{\partial P} \quad \text{or} \quad \theta = \frac{\partial U}{\partial M} \tag{2.1}$$

where

δ is the displacement (or θ the rotation angle) in the direction of force P (or moment M) of its point of application

U is the strain energy.

Note that for a built-up structure (consisting of multiple components) U is the sum of the strain energy of its constituent components.

Example 2.2 illustrates how the energy theory can be applied to solve a structural problem.

Example 2.2

Calculate the displacement at the tip of the inclined cantilever beam discussed in Example 2.1 using Castigliano's theorem.

Solutions

The tip displacement can be obtained using $\delta = \dfrac{\partial U}{\partial F}$, where U is the strain energy of the beam due to the applied force F and the reaction force R. Note that the beam bends and shrinks as a result of these two forces, respectively; the strain energy U can be found using (see Figure 2.3)

$$U = \frac{\ell}{2EA}P^2 + \int\limits_0^\ell \frac{M^2}{2EI}\,dx$$

In this case, the axial force P is $F_A = F_S + R_C$, and the bending moment M is due to the transverse force F_B. Hence the strain energy is

$$U = \frac{\ell}{2EA}F_A^2 + \int\limits_0^\ell \frac{(F_Bx)^2}{2EI}\,dx = \frac{\ell}{2EA}(F_S + R_C)^2 + \int\limits_0^\ell \frac{[(F_C - R_S)x]^2}{2EI}\,dx$$

$$= \frac{\ell}{2EA}(F_S + R_C)^2 + \frac{(F_C - R_S)^2\ell^3}{6EI}$$

and the displacement at the tip, according to Castigliano's theorem is

$$\delta = \frac{\partial U}{\partial F} = \frac{\ell}{EA}(F_S + R_C)\left(s + \frac{\partial R}{\partial F}c\right) + \frac{\ell^3}{3EI}(F_C - R_S)\left(c - \frac{\partial R}{\partial F}s\right) = \frac{F\ell^3}{EQ}$$

where

$$\frac{\partial R}{\partial F} = \frac{sc(A\ell^2 - 3I)}{Q}$$

As can be seen in Example 2.2, the formulation of the energy method for structural analysis is more general and concise. Even though the algebra involved may be about the same as in the strength of materials approach, the energy method is more powerful and is able to solve complex problems—for example, a quarter-arc beam under a point force as shown in Example 2.3.

Example 2.3
Calculate the horizontal displacement δ_B at the tip of the curved cantilever beam shown in the figure below. We assume that the diameter of the cross-section d is small compared with the radius R of the arc beam, so we may ignore the transverse shear effect.

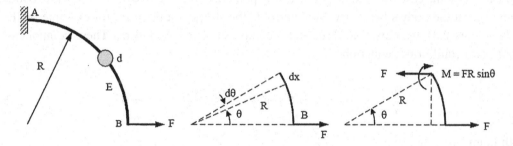

Solutions
The strain energy for the curved beam can be easily converted from the Cartesian coordinate system to the polar coordinate system as

$$U - \int_0^\ell \frac{M^2}{2EI}\,dx - \int_0^\theta \frac{M^2}{2EI}\,Rd\theta$$

where $dx = Rd\theta$. Furthermore, from the free-body diagram, the internal moment of the beam is $M = FR\sin\theta$. Therefore, the horizontal displacement at the tip B is

$$h_B = \frac{\partial U}{\partial F} = \frac{\partial}{\partial F}\int_0^{\pi/2} \frac{(FR\sin\theta)^2}{2EI}\,Rd\theta = \frac{FR^3}{EI}\int_0^{\pi/2}(\sin\theta)^2\,d\theta$$

$$= \frac{FR^3}{EI}\int_0^{\pi/2}\frac{1}{2}(1 - \cos 2\theta)\,d\theta = \frac{FR^3}{2EI}\left(\theta - \frac{1}{2}\sin 2\theta\right)_0^{\pi/2}$$

$$= \frac{\pi FR^3}{4EI}$$

Note that it would be difficult to solve this problem using the basic strength of materials approach.

2.2.3 Linear Elasticity

Elasticity is the mathematical study of how solid objects deform and become internally stressed as a result of prescribed loading conditions. Elasticity relies on the continuum hypothesis and is applicable at macroscopic length scales. Linear elasticity is a simplification of the more general nonlinear theory of elasticity and is a branch of continuum mechanics. The fundamental "linearizing" assumptions of linear elasticity are: infinitesimal strains or "small" deformations (or strains) and the linear relationships between the components of stress and strain.

The goal of solving a problem in elasticity is usually to find the stress distribution in an elastic body and, in some cases, to find the strain at any point due to given body forces and prescribed conditions at the body's boundary. To determine the stress at a point in, for example, a 2D body (Figure 2.4), we must find three stress components, σ_x, σ_y, and τ_{xy}. These components satisfy two equilibrium equations:

$$\frac{\partial \sigma_x}{\partial x} + \frac{\partial \tau_{xy}}{\partial y} + f_x = 0 \quad \text{and} \quad \frac{\partial \sigma_y}{\partial y} + \frac{\partial \tau_{xy}}{\partial x} + f_y = 0 \tag{2.2a}$$

with boundary conditions

$$\tau_{xy} n_y + T_x = 0 \quad \text{and} \quad \tau_{xy} n_x + T_y = 0 \text{ on } \Gamma_1 \tag{2.2b}$$

$$u_x = u_y = 0 \text{ on } \Gamma_0 \tag{2.2c}$$

where f_x and f_y are the body forces (or gravitational forces), n is the surface normal vector at the traction boundary (Γ_1), and $T = [T_x, T_y]^T$ is the traction force. Note that at boundary Γ_0, the displacements are $u_x = u_y = 0$.

Since two equations (Eq. 2.2a) are not sufficient to solve for the three unknowns (σ_x, σ_y, and τ_{xy}), we introduce the three strain components ε_x, ε_y, and γ_{xy}. At the same time we have the

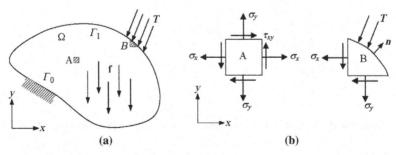

Figure 2.4: Loaded 2D structure: (a) load and boundary conditions and (b) stress elements A and B.

three relations defining the strain components in terms of the two displacement components u and v in the x- and y-directions, respectively:

$$\varepsilon_x = \frac{\partial u}{\partial x}, \quad \varepsilon_y = \frac{\partial v}{\partial y}, \quad \gamma_{xy} = \frac{\partial u}{\partial y} + \frac{\partial v}{\partial x} \tag{2.3}$$

Equation 2.3 presents three equations for only two unknowns (u and v); we cannot expect that in general that these equations will have a solution if the strain components are arbitrarily prescribed. The strains must satisfy the so-called compatibility condition, implying that the underlying displacement functions governed by the strains ensure a continuously deformed body. For a 2D planar structure, the compatibility equation is

$$\frac{\partial^2 \varepsilon_x}{\partial y^2} + \frac{\partial^2 \varepsilon_y}{\partial x^2} = \frac{\partial^2 \gamma_{xy}}{\partial x \partial y} \tag{2.4}$$

In addition, we have three stress-strain relations, for example, for a plane stress problem:

$$\varepsilon_x = \frac{1}{E}(\sigma_x - \nu \sigma_y), \quad \varepsilon_y = \frac{1}{E}(\sigma_y - \nu \sigma_x), \quad \gamma_{xy} = \frac{\tau_{xy}}{G} \tag{2.5}$$

Thus we have altogether eight unknowns (σ_x, σ_y, τ_{xy}, u, v, ε_x, ε_y, and γ_{xy}) and eight equations, combining Eqs. 2.2 through 2.5. This system of equations is generally sufficient for the solution of an elasticity problem.

Formulation of the elasticity problem is very general regardless of the geometry of the structure. However, solving this system of equations is not straightforward. One of the possibilities is to introduce a stress function $\psi(x,y)$ and focus only on solving the three stress components. The stress function must satisfy Eq. 2.2, such that

$$\sigma_x = \frac{\partial^2 \psi}{\partial y^2}, \quad \sigma_y = \frac{\partial^2 \psi}{\partial x^2}, \quad \tau_{xy} = -\frac{\partial^2 \psi}{\partial x \partial y} \tag{2.6}$$

neglecting the gravitational force f.

Also, the stress function must represent the strain components that satisfy the compatibility equation. If we insert Eq. 2.6 into Eq. 2.5 and then into Eq. 2.4, we have

$$\frac{\partial^4 \psi}{\partial x^4} + 2\frac{\partial^4 \psi}{\partial x^2 \partial y^2} + \frac{\partial^4 \psi}{\partial y^4} = \nabla^4 \psi = 0 \tag{2.7}$$

Our problem thus reduces to the determination of the stress function ψ with appropriate boundary conditions. Once the stress function is found, the stress components can be determined by Eq. 2.6. Example 2.4 illustrates how the elasticity method can be applied to solve structural problems.

Example 2.4

A cantilever beam of narrow rectangular cross-section under an end load P is shown in the figure below. With its width b small compared with its depth h, the loaded beam may be regarded as an example of plane stress. The boundary conditions are that the upper and lower edges are free from load and that the resulting shear force at $x = 0$ is equal to P. Find the three stress components, σ_x, σ_y, and τ_{xy}.

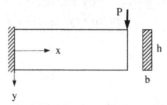

Solutions

The key to solving this problem using the elasticity formulation is selecting an adequate stress function ψ. Since the bending stress σ_x is linear in terms of both x and y, it is reasonable to assume a trial function ψ as

$$\sigma_x = \frac{\partial^2 \psi}{\partial y^2} = c_1 xy$$

where c_1 is a constant to be determined using boundary conditions. Integrating twice over y yields

$$\psi = \frac{c_1}{6} xy^3 + yf_1(x) + f_2(x)$$

where f_1 and f_2 are unknown functions of x. Substitution of this expression into Eq. 2.7 yields

$$y\frac{d^4 f_1}{dx^4} + \frac{d^4 f_2}{dx^4} = 0$$

Since f_1 and f_2 are functions of x alone, the second term here is independent of y. This is possible only if

$$\frac{d^4 f_1}{dx^4} = 0 \quad \text{and} \quad \frac{d^4 f_2}{dx^4} = 0$$

or if $f_1 = c_2 x^3 + c_3 x^2 + c_4 x + c_5$ and $f_2 = c_6 x^3 + c_7 x^2 + c_8 x + c_9$, where c_2 through c_9 are constants of integration. Therefore, the stress function ψ is

$$\psi = \frac{c_1}{6} xy^3 + y\left(c_2 x^3 + c_3 x^2 + c_4 x + c_5\right) + \left(c_6 x^3 + c_7 x^2 + c_8 x + c_9\right)$$

From Eq. 2.6,

$$\sigma_y = \frac{\partial^2 \psi}{\partial x^2} = 6(c_2 y + c_6)x + 2(c_3 y + c_7)$$

$$\tau_{xy} = -\frac{\partial^2 \psi}{\partial x \partial y} = -\frac{c_1}{2}y^2 - 3c_2 x^2 - 2c_3 x - c_4$$

The boundary conditions require that $\sigma_y = 0$ at $y = \pm h/2$, yielding $c_2 = c_3 = c_6 = c_7 = 0$. Thus,

$$\tau_{xy} = -\frac{c_1}{2}y^2 - c_4$$

Again, imposing boundary conditions for $\tau_{xy} = 0$ at $y = \pm h/2$ yields $c_4 = -c_1 h^2/8$. Also, at the loaded end of the beam the sum of the shear force must be equal to P:

$$\int\limits_{-h/2}^{h/2} \tau_{xy}b\,dy = \int\limits_{-h/2}^{h/2} \frac{c_1}{8}b\left(4y^2 - h^2\right)dy = P$$

from which $c_1 = -\frac{12P}{bh^3}$. Note that $I = bh^3/12$ is the moment of inertia of the rectangular cross-section.

The final expressions for the stress components are therefore

$$\sigma_x = -\frac{Pxy}{I} \quad \sigma_y = 0 \quad \text{and} \quad \tau_{xy} = -\frac{P}{2I}\left(\frac{h^2}{4} - y^2\right)$$

This coincides with the solutions given by the strength of materials approach.

Once the stress components are solved, the strain components can be obtained from Eq. 2.5 and then the displacement functions from Eq. 2.3. Note that, in solving displacements, integration constants must be determined by displacement conditions. This is left as an exercise (Exercise 2.3).

Question: What if an incorrect stress function were selected? What would happen if

$$\sigma_x = \frac{\partial^2 \psi}{\partial y^2} = c_1 x^2 y$$

or

$$\sigma_x = \frac{\partial^2 \psi}{\partial y^2} = c_1 x^2$$

were chosen? Again, this is left as exercise (Exercise 2.4).

2.2.4 Failure Criteria

It is well known that failure of a tensile member occurs when the stress caused by the actual load reaches the stress limit—that is, the strength of the member's material (usually yield strength S_y for ductile materials and ultimate tensile strength S_u for brittle materials). Correlation of the actual stress with the material strength, which is the maximum stress that the structural member is able to bear without a failure, is straightforward in this case because they are both uniaxial. But how can we correlate the biaxial or triaxial stress state in a component—whose material strength is measured in uniaxial tests—to assess failure tendency? This is where failure theories come into the picture. One of the most common in design is the maximum shear stress theory, which states:

> Failure will occur in a complex part if any of the maximum shear stresses τ_{max} exceeds the material shear yield strength S_{sy} that causes failure in the simple, uniaxial test.

That is, the maximum shear stress must be smaller than the shear yield strength to ensure a safe design.

Mathematically, the theorem says:

$$\tau_{max} \leq S_{sy} \tag{2.8}$$

What is S_{sy}? For the uniaxial stress case shown in Figure 2.5a, when the stress of a stress element exceeds the stress limit—for example, the yield strength for ductile materials S_y—the structure fails. If we rotate the stress element $45°$ clockwise (i.e., equivalently rotate the stress point $90°$ clockwise from X to X' on Mohr's circle) as shown in Figure 2.5b, the shear stress τ_{max} becomes

$$\tau_{max} = \frac{\sigma_1}{2} \leq \frac{S_y}{2}$$

Therefore, referring to Eq. 2.8, we have $S_{sy} = S_y/2$. Also, from strength of materials, the maximum stress for a biaxial stress element is

$$\tau_{max} = \frac{1}{2}|\sigma_1 - \sigma_2| \tag{2.9}$$

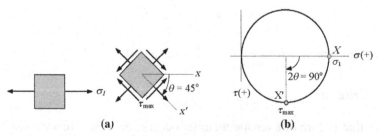

Figure 2.5: (a) Stress element under uniaxial load and (b) rotated stress element with Mohr's circle.

where σ_1 and σ_2 are the two principal stresses.

Graphically, the maximum shear stress theory for a biaxial stress state can be depicted in a Tresca's hexagon (Figure 2.6), which states a safe zone as

$$|\sigma_1| \quad \text{and} \quad |\sigma_2| \le S_y \tag{2.10a}$$

for stress points (σ_1, σ_2) falling in the first and third quadrants, and

$$\tau_{\max} = \frac{1}{2}|\sigma_1 - \sigma_2| \le \frac{S_y}{2} \tag{2.10b}$$

for those in the second and fourth quadrants. That the principal stresses fall inside the hexagon indicates a safe design. The smaller hexagon indicates a safe zone when a larger safety factor n is considered.

Note that the maximum shear stress theory is applicable to ductile materials. Others are applicable to brittle materials. Figure 2.7 provides useful and popular theories, as well as how criteria should be selected. As a rule of thumb, ductile and brittle materials are separated by strain at fracture. For ductile materials, the strain is greater than 5% at fracture. Usually brittle materials fracture with a strain less than 5%. More details about the failure criteria shown in Figure 2.7 can be found in either a strength of materials textbook, such as Beer et al. (2002), or a design of mechanical components textbook—for example Hamrock et al. (2002).

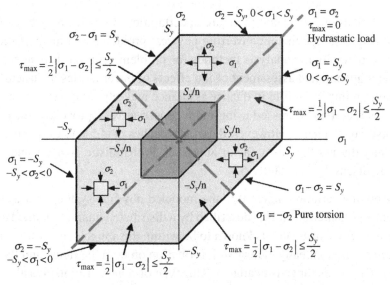

Figure 2.6: Tresca's hexagon.

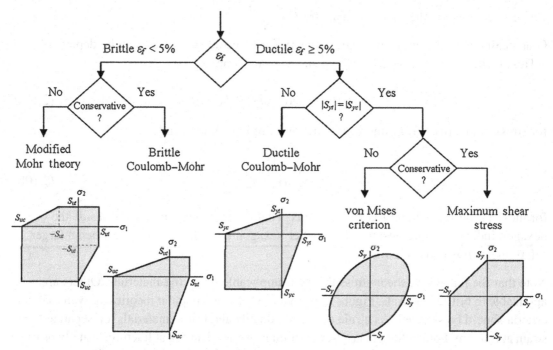

Figure 2.7: Flowchart for choosing proper failure criteria for design.

2.2.5 Uncertainties, Variations, and Safety Factors

Whether analytical or numerical, structural analysis assumes no variations in physical parameters and no uncertainties in physical conditions. In reality, loads, material properties, and geometric dimensions vary, and there are uncertainties in determining boundary conditions, making assumptions in converting physical problems to a mathematically solvable form, and employing analysis methods for problem solution. It is now widely recognized that a quantitative assessment of the effects of uncertainties in structural analysis and design plays an important role in quality assurance and reliability estimation. Variations and uncertainties are often addressed using more conservative approaches, such as safety factor and worst-case scenario. However, these approaches could lead to overdesign. In this section we briefly discuss the safety factor approach. Probabilistic analysis and reliability estimate will be discussed in Chapter 5.

Variability covers the variation inherent to the modeled physical system or the environment under consideration. Generally, this is described by a distributed quantity defined over a range of possible values. The exact value is known to be within this range, but it varies from unit to unit or from time to time. Ideally, objective information on both the range and the likelihood of the quantity within this range is available. This type of nondeterminism is also referred as aleatory, irreducible, stochastic, or due to chance.

Uncertainty is a potential deficiency in any phase or activity of the modeling process that derives from lack of knowledge. The word *potential* emphasizes that the deficiency may or may not occur. This definition basically states that uncertainty is caused by incomplete information resulting from vagueness, nonspecificity, or dissonance. Vagueness characterizes information that is imprecisely defined, unclear, or indistinct. It is typically the result of human opinion regarding unknown quantities. In the literature, this type of nondeterminism is also referred as epistemic, reducible, or subjective.

On a relative scale, variability is easier to quantify. Recently developed reliability analysis methods have been largely employed to incorporate stochastic or variability of physical parameters in failure probability estimates for product design. However, quantifying and addressing uncertainty in engineering design, especially that caused by incomplete information, is not as straightforward.

The simplest and most widely used approach in addressing variability and uncertainty in engineering design is referred to as factor of safety. In employing this approach designers often overdesign the product in hope of overcoming potential problems caused by variability and uncertainty.

The sources of variability and uncertainty are classified into geometric simplification, material modeling, level of sophistication of analysis, and human error. Determining an adequate safety factor for product design often requires numerous considerations, such as the degree of uncertainty about loading, the degree of uncertainty about material strength, the uncertainties in relating applied loads to material strength via stress analysis, the consequences of failure (e.g., human safety and economics), and the cost of providing a large safety factor (Norton 2011). Determining safety factors for engineering design is not completely science. Industry has developed certain guidelines for safety factors based on the past experience and failure rate. The following are general guidelines for engineering students (Juvinall and Marshek 2000).

- $n = 1 \sim 2$ for reliable materials used under stable conditions subject to reliable loads (hence variability and uncertainly in stress).
- $n = 2 \sim 3$ for less tried materials used under average conditions, subject to fairly reliable loads.
- $n = 3 \sim 4$ for at most one uncertainty in materials, conditions, and loads.
- $n = 1 \sim 6$ for fatigue (applied to endurance limit).
- $n = 3 \sim 6$ for impact.

2.3 Finite Element Methods

The finite element method, or analysis, (FEM or FEA) is a numerical technique for finding approximate solutions of partial differential equations (PDEs) or sometimes integral equations that govern a physical problem.

The term *finite element* was first coined by Clough in 1960. In the early 1960s, engineers used the method for approximate solutions of problems in stress analysis, fluid flow, heat transfer, and other areas. The first book on FEM, by Zienkiewicz and Chung, was published in 1967. In the late 1960s and early 1970s, FEM was applied to a wide variety of engineering problems.

The finite element method originated from the need for solving complex elasticity and structural analysis problems in civil and aeronautical engineering. Its development began in the middle to late 1950s for airframe and structural analysis. By the late 1950s, the key concepts of stiffness matrix and element assembly existed essentially in the form used today. NASA issued a request for proposals for the development of the finite element software Nastran in 1965. The method was provided with a rigorous mathematical foundation in 1973 (Strang and Fix 1973) and has since been generalized into a branch of applied mathematics for numerical modeling of physical systems in a wide variety of engineering disciplines, such as electromagnetism and fluid dynamics. Recently the method was further extended in various mathematical forms, such as p-version FEM (Szabó and Babuška 1991), the meshless method (Li and Liu 2004, Belytschko and Chen 2007), and extended FEM (Moës et al. 1999).

Although FEM has been generalized and applied to solve various engineering problems, in this chapter we focus on structural problems. We start with a simple example to illustrate the basic concept of FEA, and then discuss general formulations for both its h- and p-versions. The important concept of solution convergence, in the h- and p-methods, as well as adaptations, is also treated.

2.3.1 A Simple Example

The formulation of the finite element method starts with the principle of virtual work. Virtual work is the work done by real forces in moving through virtual displacements. A virtual displacement is any displacement consistent with the constraints of the structure—that is, it satisfies the boundary conditions at the supports, or the *essential boundary conditions*. The principle of virtual work states that if a solid is in equilibrium then the virtual work done in any virtual displacement that satisfies the essential boundary conditions is zero.

For the cantilever beam example shown in Figure 2.8, the governing differential equation is known from strength of materials:

$$\frac{\partial^2}{\partial x^2}\left(EI\frac{\partial^2 w}{\partial x^2}\right) = q + P\delta(\ell - x) \tag{2.11a}$$

where $\delta(\ell-x)$ is the Dirac delta function, defined as $\displaystyle\int_{-\infty}^{\infty} \delta(\ell - x)f(x)dx = f(\ell)$. The boundary conditions are

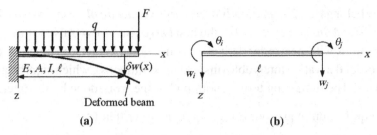

Figure 2.8: Simple cantilever beam: (a) load and virtual displacement and (b) beam element with displacement and rotation degrees of freedom.

$$w(0) = 0 \quad \text{and} \quad \frac{\partial w}{\partial x}(0) = 0 \tag{2.11b}$$

$$EI\frac{\partial^2 w}{\partial x^2}(\ell) = 0 \quad \text{and} \quad \frac{\partial}{\partial x}\left(EI\frac{\partial^2 w}{\partial x^2}\right)_{x=\ell} = 0 \tag{2.11c}$$

Note that Eq. 2.11b shows the essential boundary conditions that the virtual displacement must satisfy. Certainly, the true solutions must satisfy both conditions stated in Eqs. 2.11b and 2.11c.

Multiply the virtual displacement $\delta w(x)$ on both sides of Eq. 2.11a, and integrate over the beam length. On the right side, we have

$$\int_0^\ell (q + P\delta(\ell - x))\delta w dx = \int_0^\ell q\delta w dx + \int_0^l P\delta(\ell - x)\delta w(x)dx = \int_0^l q\delta w dx + P\delta w(\ell) \tag{2.12}$$

This is the virtual work done by the externally applied forces in moving through the virtual displacement $\delta w(x)$.

Twice integrating the term on the left side by parts yields

$$\int_0^l \left(\frac{\partial^2}{\partial x^2}\left(EI\frac{\partial^2 w}{\partial x^2}\right)\right)\delta w dx = \frac{\partial}{\partial x}\left(EI\frac{\partial^2 w}{\partial x^2}\right)\delta w\Big|_0^\ell - \int_0^\ell \left(\frac{\partial}{\partial x}\left(EI\frac{\partial^2 w}{\partial x^2}\right)\right)\frac{\partial \delta w}{\partial x}dx$$

$$= \frac{\partial}{\partial x}\left(EI\frac{\partial^2 w}{\partial x^2}\right)\delta w\Big|_0^\ell - EI\frac{\partial^2 w}{\partial x^2}\frac{\partial \delta w}{\partial x}\Big|_0^\ell + \int_0^\ell EI\frac{\partial^2 w}{\partial x^2}\frac{\partial^2 \delta w}{\partial x^2}dx \tag{2.13}$$

The first two terms at $x = \ell$ vanish because of the boundary conditions of Eq. 2.11c that require that the true solutions comply; these are the so-called natural boundary conditions. If the virtual displacement δw belongs to a *kinematically admissible virtual displacement space Z*, defined as $Z = \{u \in C^1, u = 0 \text{ and } \partial u/\partial x = 0 \text{ at } x = 0\}$, the first two terms at $x = 0$ also vanish. Note that C^1 is a space of functions with first-order derivative continuous. In fact, the function space can be relaxed to first-order derivative integrable (instead of continuous), which significantly expands the function space. The remaining term is nothing but the work done by the internal forces.

Therefore, the equilibrium equation, Eq. 2.11a, can be written as

$$\int_0^\ell EI \frac{\partial^2 w}{\partial x^2} \frac{\partial^2 \delta w}{\partial x^2} dx = \int_0^\ell q \delta w dx + P \delta w(\ell), \quad \text{for all } \delta w \in Z, \tag{2.14}$$

These generalized formulations are often referred to as *weak*, which is essentially what the principle of virtual work states.

To solve Eq. 2.14, we convert the integral equation into a set of linear equations in matrix form by introducing interpolation functions (or shape functions) to represent the displacement fields in the finite elements.

For a beam element under bending, shown in Figure 2.8b, there are two degrees of freedom at each end point, or *node*: the displacement w and the rotation angle θ. From strength of materials, displacement of a cantilever beam with a point load at the tip is a cubic function in the length parameter. Therefore, in FEA a cubic shape function is employed for a beam element. For a cantilever beam with a point load at the tip, one beam element is sufficient to provide an exact solution. However, for a beam with a uniformly distributed load, the exact displacement solution is a fourth-order function in beam length. One element of the cubic shape function does not give an exact solution. In fact, for the example shown in Figure 2.8, where a distributed load is present, exact solutions are obtained at nodes but not in-between.

Using a cubic shape function, the displacement function $w(x)$ in the beam element can be interpolated as

$$w(x) = N^T W = [N_1 \ N_2 \ N_3 \ N_4] \begin{bmatrix} w_i \\ \theta_i \\ w_j \\ \theta_j \end{bmatrix} \tag{2.15}$$

where

$$N_1 = 1 - 3\frac{x^2}{\ell^2} + 2\frac{x^3}{\ell^3}, \quad N_2 = x - 2\frac{x^2}{\ell} + \frac{x^3}{\ell^2},$$

$$N_3 = 3\frac{x^2}{\ell^2} - 2\frac{x^3}{\ell^3}, \quad N_4 = -\frac{x^2}{\ell} + \frac{x^3}{\ell^2} \tag{2.16}$$

and

$$\frac{\partial^2 w}{\partial x^2} = \frac{\partial^2 N^T}{\partial x^2} W = \left[\frac{6}{\ell^2} + 12\frac{x}{\ell^3} \quad \frac{-4}{\ell} + 6\frac{x}{\ell^2} \quad \frac{6}{\ell^2} - 12\frac{x}{\ell^3} \quad \frac{-2}{\ell} + 6\frac{x}{\ell^2} \right] \begin{bmatrix} w_i \\ \theta_i \\ w_j \\ \theta_j \end{bmatrix}$$

Similarly, $\dfrac{\partial^2 \delta w}{\partial x^2} = \dfrac{\partial^2 N^T}{\partial x^2} \delta W$. Inserting these into Eq 2.14 and carrying out the integrations, we have

$$\int_0^\ell EI \frac{\partial^2 w}{\partial x^2} \frac{\partial^2 \delta w}{\partial x^2} dx = \delta W^T K W$$

where K is the so-called *stiffness matrix*; that is,

$$K = \frac{EI}{\ell^3} \begin{bmatrix} 12 & 6\ell & -12 & 6\ell \\ 6\ell & 4\ell^2 & -6\ell & 2\ell^2 \\ -12 & -6\ell & 12 & -6\ell \\ 6\ell & 2\ell^2 & -6\ell & 4\ell^2 \end{bmatrix}$$

and

$$\int_0^\ell q\delta w dx + P\delta w(\ell) = \delta W^T \int_0^\ell qN dx + \delta W^T \begin{bmatrix} 0 \\ 0 \\ P \\ 0 \end{bmatrix} = \delta W^T \begin{bmatrix} \frac{1}{2}q\ell \\ \frac{1}{12}q\ell^2 \\ \frac{1}{2}q\ell + P \\ -\frac{1}{12}q\ell^2 \end{bmatrix} = \delta W^T F \qquad (2.18)$$

where F is the *force vector*. Equation 2.14 becomes $\delta W^T K W = \delta W^T F$. Since δW is a virtual displacement, we have

$$KZ = F$$

or

$$\frac{EI}{\ell^3}\begin{bmatrix} 12 & 6\ell & -12 & 6\ell \\ 6\ell & 4\ell^2 & -6\ell & 2\ell^2 \\ -12 & -6\ell & 12 & -6\ell \\ 6\ell & 2\ell^2 & -6\ell & 4\ell^2 \end{bmatrix}\begin{bmatrix} w_i \\ \theta_i \\ w_j \\ \theta_j \end{bmatrix} = \begin{bmatrix} \frac{1}{2}q\ell \\ \frac{1}{12}q\ell^2 \\ \frac{1}{2}q\ell + P \\ -\frac{1}{12}q\ell^2 \end{bmatrix} \tag{2.19}$$

If we impose the boundary conditions—in this case, $w_i = \theta_i = 0$—the reduced matrix equations become

$$\frac{EI}{\ell^3}\begin{bmatrix} 12 & -6\ell \\ -6\ell & 4\ell^2 \end{bmatrix}\begin{bmatrix} w_j \\ \theta_j \end{bmatrix} = \begin{bmatrix} \frac{1}{2}q\ell + P \\ -\frac{1}{12}q\ell^2 \end{bmatrix}$$

Invert the matrix, and solve the equations:

$$\begin{bmatrix} w_j \\ \theta_j \end{bmatrix} = \frac{\ell^3}{EI}\begin{bmatrix} 12 & -6\ell \\ -6\ell & 4\ell^2 \end{bmatrix}^{-1}\begin{bmatrix} \frac{1}{2}q\ell + P \\ -\frac{1}{12}q\ell^2 \end{bmatrix} = \frac{\ell}{12EI}\begin{bmatrix} 4\ell^2 & 6\ell \\ 6\ell & 12 \end{bmatrix}\begin{bmatrix} \frac{1}{2}q\ell + P \\ -\frac{1}{12}q\ell^2 \end{bmatrix} = \begin{bmatrix} \frac{q\ell^4}{8EI} + \frac{P\ell^3}{3EI} \\ \frac{q\ell^3}{6EI} + \frac{P\ell^2}{2EI} \end{bmatrix} \tag{2.20}$$

This is identical to the exact solutions at the tip while combining the tip displacements of the separate loading cases: point load P and distributed load q.

The finite element solution for displacement $w(x)$ can then be obtained from Eq. 2.15:

$$w(x) = N^T W = [N_3\ N_4]\begin{bmatrix} w_j \\ \theta_j \end{bmatrix} = \left[3\frac{x^2}{\ell^2} - 2\frac{x^3}{\ell^3},\ -\frac{x^2}{\ell} + \frac{x^3}{\ell^2}\right]\begin{bmatrix} \frac{q\ell^4}{8EI} + \frac{P\ell^3}{3EI} \\ \frac{q\ell^3}{6EI} + \frac{P\ell^2}{2EI} \end{bmatrix}$$

$$= \frac{q\ell}{24EI}(5x^2\ell - 2x^3) + \frac{P}{6EI}(3x^2\ell - x^3)$$

where the first and second terms represent the beam displacements due to the distributed load q and the point force P, respectively. From strength of materials, the second term is exact, but the exact solution for the distributed load q is

$$w^q(x) = \frac{q}{24EI}\left(6x^2\ell^2 - 4x^3\ell + x^4\right)$$

Apparently, the finite element method does not give an exact solution for $w^q(x)$ because the element shape function is a cubic function and the exact solution is a fourth-order polynomial function. However, when more elements are employed for the cantilever beam (i.e., by dividing the beam into smaller segments), the finite element solution approaches the exact solution.

2.3.2 Finite Element Formulation

For a structure of arbitrary geometry, such as that shown in Figure 2.4, the equilibrium equations are as stated in Eqs. 2.2a, 2.2b, and 2.2c. In general, there is no analytical solution for these coupled differential equations except for structural components of simple geometry, such as simple beams. Over the past 40 years, finite element methods have become indispensable for solving general mechanics problems. The methods and their tools provide numerical solutions that are sufficiently accurate for support of design decision making regarding the structural integrity of a design.

As stated earlier, the finite element method starts with the principle of virtual work. For a general structural problem (refer to Figure 2.4), the virtual work of the boundary value problem can be defined, according to the principle of virtual work, as

$$\delta W^{\text{int}} = \delta W^{\text{ext}} \tag{2.21}$$

where δW^{int} is the virtual work done by the internal forces, or

$$\delta W^{\text{int}} = \int_\Omega \delta U d\Omega = \int_\Omega \delta\left(\frac{1}{2}\boldsymbol{\sigma}\boldsymbol{\cdot}\boldsymbol{\varepsilon}\right)d\Omega = \int_\Omega \boldsymbol{\sigma}\boldsymbol{\cdot}\delta\boldsymbol{\varepsilon} d\Omega \tag{2.22}$$

where U is the strain energy, and $\boldsymbol{\sigma}$ and $\boldsymbol{\varepsilon}$ are stress and strain tensors, respectively. Note that for a 2D structure of biaxial stress state, $\boldsymbol{\sigma} = [\sigma_x, \sigma_y, \tau_{xy}]^T$ and $\boldsymbol{\varepsilon} = [\varepsilon_x, \varepsilon_y, \gamma_{xy}]^T$.

δW^{ext} is the virtual work done by the externally applied forces, or

$$\delta W^{\text{ext}} = \int_\Omega \boldsymbol{f}\boldsymbol{\cdot}\delta\boldsymbol{u} d\Omega + \int_{\Gamma_1} \boldsymbol{T}\boldsymbol{\cdot}\delta\boldsymbol{u} d\Gamma \tag{2.23}$$

where f and T are the body force and external traction vectors, respectively, and δu is the vector of virtual displacements. Therefore, the principle of virtual work states

$$\int_{\Omega} \sigma \cdot \delta e d\Omega = \int_{\Omega} f \cdot \delta u d\Omega + \int_{\Gamma_1} T \cdot \delta u d\Gamma, \text{ for all } \delta u \in Z \tag{2.24}$$

where Z is a kinematically admissible virtual displacement space.

Equation 2.24 must be solved through two important discretizations, *domain* and *function*. The finite element method discretizes the structural domain into small pieces, where each piece is called a *finite element* (i.e., an element with finite size or finite number of the elements). The finite element boundary edges (for surface elements) or faces (for 3D solid elements) are usually straight or flat for linear elements. For higher-order elements the element boundary can be a curve that matches better with the structural boundary. End points of the finite element edges (or corner points of the element faces) are called *nodes*. The structural domain discretized into finite elements is called a *mesh*. The finite element mesh approximates the structural domain, as illustrated in Figure 2.9. Note that forces, such as the traction force T, are also discretized. The linearly distributed force T becomes equivalent to point forces applied at nodes.

Material properties, section properties, and distributed loads, in addition to the physical domain, are discretized in accordance with the finite element mesh. For example, a circular cross-sectional beam, as shown in Figure 2.10, is discretized into four finite elements. As a result, the cross-section and distributed load (and material property if applicable) are also discretized, most likely as average values within each respective element.

The second discretization is function discretization. Structural responses, such as displacements (or stresses), within the finite element are interpolated using the responses at the finite element nodes and element shape functions (or interpolation functions), which are usually polynomial functions. As illustrated in Figure 2.11, the interpolated solutions usually do not match with the exact solutions. The discrepancy is called *interpolation error*.

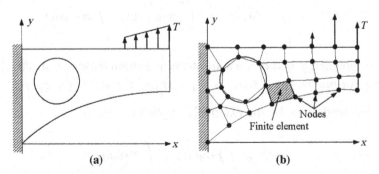

(a)   (b)

Figure 2.9: Domain discretization: (a) original structural domain and (b) finite element mesh.

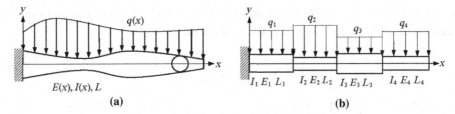

Figure 2.10: Domain discretization using a beam example: (a) physical problem and (b) finite element discretization.

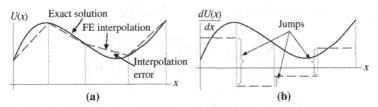

Figure 2.11: Function interpolation: (a) a piecewise linear interpolation for $U(x)$ and (b) derivative of $U(x)$ and jumps across the element boundary.

Displacements at the finite element nodes are called *degrees of freedom* (dof). For a 2D planar problem, there are two dof at each node, u_x and u_y. Consider a domain in a state of equilibrium discretized by a four-node quadrilateral finite element mesh, as depicted in Figure 2.12. According to the finite element method, the coordinates $x = [x,y]^T$ are interpolated from the nodal values $X_j = [X_j,Y_j]^T$:

$$x = \sum_{j=1}^{4} N_j X_j = \sum_{j=1}^{4} \begin{bmatrix} N_j & 0 \\ 0 & N_j \end{bmatrix} \begin{bmatrix} X_j \\ Y_j \end{bmatrix} \qquad (2.25)$$

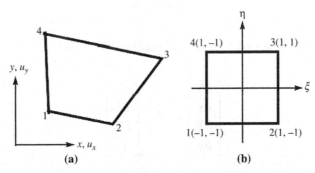

Figure 2.12: Two-dimensional isoparametric finite element of four nodes, (a) finite element, and (b) natural coordinates.

where N_j is the matrix of finite element shape functions for the jth node. For an isoparametric linear element, as shown in Figure 2.12, the shape functions are

$$N_1 = \frac{1}{4}(1 - \xi)(1 - \eta), \quad N_2 = \frac{1}{4}(1 + \xi)(1 - \eta)$$

$$N_3 = \frac{1}{4}(1 + \xi)(1 + \eta), \quad N_4 = \frac{1}{4}(1 - \xi)(1 + \eta)$$

(2.26)

in which ξ and η are the natural coordinates of the element. For an isoparametric finite element, displacement fields $u = [u_x, u_y]^T$ are similarly interpolated using the element shape functions and the nodal displacement values $U = [U_x, U_y]^T$:

$$u = \sum_{j=1}^{4} N_j U_j = \sum_{j=1}^{4} \begin{bmatrix} N_j & 0 \\ 0 & N_j \end{bmatrix} \begin{bmatrix} U_{xj} \\ U_{yj} \end{bmatrix}$$

The strain field is computed directly from Eq. 2.3:

$$\varepsilon = \sum_{j=1}^{4} B_j U_j$$

(2.28)

where the matrix B_j is defined in terms of derivatives of the shape functions N_j:

$$B_j = \begin{bmatrix} \dfrac{\partial N_j}{\partial x} & 0 \\ 0 & \dfrac{\partial N_j}{\partial y} \\ \dfrac{\partial N_j}{\partial y} & \dfrac{\partial N_j}{\partial x} \end{bmatrix}$$

(2.29)

and the chain rule is invoked to determine the coefficients of B_j:

$$\begin{bmatrix} \dfrac{\partial N}{\partial x} \\ \dfrac{\partial N}{\partial y} \end{bmatrix} = J^{-1} \begin{bmatrix} \dfrac{\partial N}{\partial \xi} \\ \dfrac{\partial N}{\partial \eta} \end{bmatrix}$$

(2.30)

where J is the Jacobian matrix,

$$J = \begin{bmatrix} \dfrac{\partial x}{\partial \xi} & \dfrac{\partial y}{\partial \xi} \\ \dfrac{\partial x}{\partial \eta} & \dfrac{\partial y}{\partial \eta} \end{bmatrix} \tag{2.31}$$

Also, $\boldsymbol{\sigma} = \boldsymbol{D}\boldsymbol{\varepsilon}$, where $\boldsymbol{D}$ is the material stress-strain matrix or constitutive matrix. For a 2D plane stress problem, the matrix $\boldsymbol{D}$ is

$$\boldsymbol{D} = \frac{E}{1 - v^2} \begin{bmatrix} 1 & v & 0 \\ v & 1 & 0 \\ 0 & 0 & \dfrac{1 - v}{2} \end{bmatrix} \tag{2.32}$$

If we insert the stress $\boldsymbol{\sigma}$ and strain $\boldsymbol{\varepsilon}$ into left side of Eq. 2.24 and integrate over an element domain Ω_e, we have

$$\int_{\Omega_e} \boldsymbol{\sigma} \cdot \delta \boldsymbol{\varepsilon} d\Omega = \int_{\Omega_e} \delta \boldsymbol{\varepsilon}^T \boldsymbol{D} \boldsymbol{\varepsilon} d\Omega = \delta \boldsymbol{U}^T \int_{\Omega_e} \boldsymbol{B}^T \boldsymbol{D} \boldsymbol{B} d\Omega \boldsymbol{U} = \delta \boldsymbol{U}^T \boldsymbol{K}_e \boldsymbol{U} \tag{2.33}$$

where $\boldsymbol{K}_e$ is the stiffness matrix of an element Ω_e:

$$\boldsymbol{K}_e = \int_{\Omega_e} \boldsymbol{B}^T \boldsymbol{D} \boldsymbol{B} d\Omega \tag{2.34}$$

Similarly, the right side of Eq. 2.24 can be discretized using element shape functions and expressed in an element as

$$\int_{\Omega_e} \boldsymbol{f}^b \cdot \delta \boldsymbol{u} d\Omega + \int_{\Gamma_{el}} \boldsymbol{f}^t \cdot \delta \boldsymbol{u} d\Gamma = \delta \boldsymbol{U}^T \boldsymbol{F}_e \tag{2.35}$$

where $\boldsymbol{F}_e$ is the element force vector.

The element stiffness matrix $\boldsymbol{K}_e$ and the force vector $\boldsymbol{F}_e$ are assembled with those of all the other elements to form the global stiffness matrix $\boldsymbol{K}$ and the force vector $\boldsymbol{F}$. After imposing the displacement boundary conditions, the reduced matrix equations can be obtained as

$$\boldsymbol{K}\boldsymbol{U} = \boldsymbol{F} \tag{2.36}$$

This matrix equation is then solved numerically for nodal displacements and later for stresses.

As can be seen in the previous discussion, by going through the domain discretization and response interpolations, the partial differential equations that govern the continuum mechanics problem are converted into a system of linear (matrix) equations. The size of a matrix equation is determined by the degrees of freedom at the finite element node and the number of nodes created in the finite element mesh.

Interpolating displacement fields guarantees that the finite element solutions of the displacement responses are C^0-continuous, as depicted in Figure 2.11a, which is desirable. However, one major deficiency of this displacement-based formulation is that the stress (and strain) solutions, obtained by taking the derivatives of the displacement solutions, result in jumps at the element boundary. This jump can be seen clearly in Figure 2.11b if we go back to take the derivative of $U(x)$ in Figure 2.11a. In the finite element method, stress jumps appear at element boundary points (for 1D elements), element boundary edges (for 2D elements), and element boundary faces for 3D elements.

Assuming the displacement solution shown in Figure 2.13a, the difference between the true solution u and the finite element solution is apparent. For a linear element (one with linear shape functions), the stress solution from FEA is $\hat{\sigma} = E\dfrac{du}{dx}$, as depicted in Figure 2.13b. To obtain a better stress field, a nodal averaging process is often employed to obtain the averaged nodal stresses $\bar{\sigma}$; the displacement shape functions N are used again to interpolate a new stress field $\sigma* = N\bar{\sigma}$. It is apparent that $\sigma*$ provides a continuous stress solution across element boundaries, which is obviously a better approximation than $\hat{\sigma}$. The difference between $\sigma*$ and $\hat{\sigma}$ can be a good error estimate in stress.

A stress error e_σ can be estimated by

$$e_\sigma = \sigma^* - \hat{\sigma} \tag{2.37}$$

Note that for 1D elements there is only one stress component: axial stress σ_x. Therefore, the stress error e_σ is a scalar function. For 2D or 3D elements, there are more stress components,

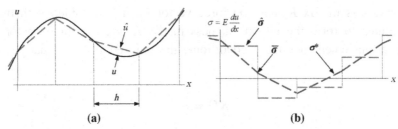

Figure 2.13: Stress jumps: (a) displacement fields and (b) stress functions.

where e_σ is a vector function. In general, the error estimate of the ith element can be evaluated in the energy norm as

$$\|e_\sigma\|_i = \left(\int_{\Omega_{e_i}} e_\sigma^T D^{-1} e_\sigma d\Omega \right)^{1/2} = \left(\int_{\Omega_{e_i}} e_\varepsilon^T D e_\varepsilon d\Omega \right)^{1/2} \tag{2.38}$$

where

D is the constitutive matrix

e_ε is the vector of strain component error

e_σ is the vector of stress component error interpolated using element shape functions

The global error estimate can be computed by summing $\|e_\sigma\|_i$ over the entire finite element domain Ω, using

$$\|e_\sigma\| = \left(\sum_{i=1}^{NE} \|e_\sigma\|_i^2 \right)^{1/2} \tag{2.39}$$

where NE is the total number of elements. This estimate can be an effective indicator in determining if the current mesh is adequate. If $\|e_\sigma\|$ is greater than a prescribed limit, the mesh needs to be refined. In that case, the element error estimate $\|e_\sigma\|_i$ provides an excellent guide in terms of which area or elements need refinement.

It is obvious from Figure 2.13 that the error estimate of Eq. 2.39 is reduced if more elements are employed for analysis. In general, a finite element model with smaller element sizes generates more elements and therefore more nodes. As a result, the model has a larger set of equations to solve, which usually leads to more accurate solutions but longer computation time. Increasing the number of elements or refining the finite element mesh to achieve more accurate solutions is referred to as the *h-adaptation* ("h" denotes element physical size). In h-version FEA, a solution is converged by refining the finite element mesh and solving a larger set of equations, thus consuming more computation resources. Therefore, we have to invest wisely. If there is no error analysis capability such as Eq. 2.39 offers in the finite element code, as a rule smaller elements at high-stress areas, and larger elements at low-stress areas, may be used.

Several general-purpose h-version FEA codes are commercially available, such as ANSYS®, MSC/Nastran®, and ABAQUS®. They all provide excellent finite element modeling and solution capabilities and are capable of solving general structural problems. For example, a 2D engine connecting rod, shown in Figure 2.14a (Edke and Chang 2010), was modeled and solved using ANSYS, where about 500 linear isoparametric elements were employed to yield about 6,000 degrees of freedom. A set of firing loads were applied at the inner boundary of the

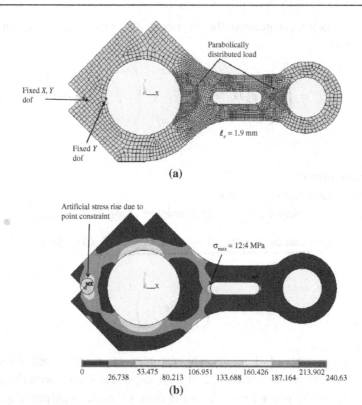

Figure 2.14: Two-dimensional engine connecting rod: (a) analysis problem with finite element mesh and (b) stress results obtained from ANSYS using linear isoparametric finite elements.

connecting rod where the crank shaft and piston pin are located. Boundary conditions were applied as suggested by Hwang et al. (1997). The maximum principal stress distribution in the connecting rod is shown in Figure 2.14b. Although the maximum stress appears to be at the fixed node on the left side, it is merely an artificial stress concentration due to displacement constraints. The real maximum stress of $\sigma_{max} = 124$ MPa occurs on the left semicircular edge of the slot.

In addition to the 2D isoparametric finite elements employed for the connecting rod example, there are many more element types available in commercial FEA codes. In general for h-version FEA, element shape functions are usually up to second order. A list of common finite element types supported in these codes is given in Figure 2.15.

2.3.3 p-Version FEA

The h-FEA achieves solution convergence by refining element size while retaining the polynomial order of the element shape functions (usually at lower order: $p = 1$ or $p = 2$); the

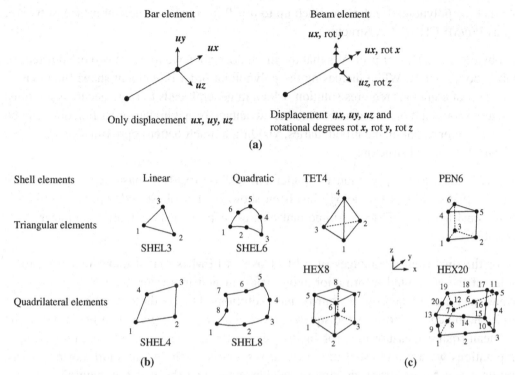

Figure 2.15: Typical finite element types: (a) bars and beams; (b) 2D plane, plate, and 3D shell elements; (c) 3D solid elements.

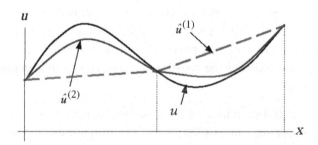

Figure 2.16: Displacement interpolations for $p = 1$ and $p = 2$.

p-FEA increases the polynomial order of the element shape functions to achieve solution convergence while maintaining the same finite element mesh. This concept is illustrated in Figure 2.16, where the exact solution u is approximated using a piecewise linear function $\hat{u}^{(1)}$ ($p = 1$) and a piecewise quadratic function $\hat{u}^{(2)}$ ($p = 2$). It is obvious that increasing the polynomial order of the interpolation function improves the accuracy of the approximation.

Note that the polynomial order can reach up to $p = 9$ in some commercial p-FEA software, such as Pro/MECHANICA Structure.

One obvious advantage of p-FEA is that the mesh does not have to be refined or adjusted for solution convergence. While increasing the polynomial order of element shape functions, a larger set of equations requires solution, which in general leads to more accurate solutions using more computational resources. Another advantage is that higher-order functions can be employed to represent element boundaries, yielding a much better approximation in geometry for general structures.

Research related to the p-version finite element method (p-FEM) has been ongoing since the early 1970s. p-FEM technology has been shown to be robust and superior to the conventional h-version FEM for important classes of problems, including nonlinear applications.

One of the important advantages of p-FEM over h-FEM is that it makes ensuring the quality of the computed information more efficient and more convenient. There are various measures of quality of approximate solutions. One is the energy norm. By definition, the energy norm is the square root of the strain energy, which is similar to the root-mean-square measure of error in stresses, as defined in Eq. 2.39. In engineering computations we are interested in structural responses, such as maximum normal stress, maximum von Mises stress, maximum displacement, and the first few natural frequencies. These are typical results offered by finite element solutions. It is important to know whether the finite element solutions are sufficiently close to the true (or exact) solution. Since generally we do not know the exact solution, this appears to be an unsolvable problem. However, we can always increase the polynomial order and rerun the analysis to achieve a better accuracy. While we are increasing the polynomial order for a series of analyses, if the difference in solutions obtained from the current polynomial order ($p = n$) and previous polynomial order ($p = n - 1$) is smaller than a prescribed tolerance, we consider the solution to converge to an acceptable value and so the solution process stops.

In p-version FEA, the stiffness matrix of the structure is formed by a set of hierarchical shape functions. "Hierarchical" means that when the polynomial degree increases from p to $p + 1$, the shape functions used in the polynomial degree p are not altered; in other words, the shape function set of polynomial degree p is a subset of that of polynomial degree $p + 1$. Methods for proper selection of the shape function set for p-version finite elements were addressed by Babuška et al. (1989). Since p-version FEA uses hierarchical shape functions, the data of the original computation in the lower polynomial element—portions of stiffness matrices, loading vectors, and so forth—can still be used for the higher-order element.

A human maxillary second molar is presented to demonstrate the advantages of using p-FEA to capture the geometry of the critical dentino-enamel junction (DEJ) and reach solution convergence. Creating an accurate geometry model for the tooth is challenging. The geometric modeling started with a histological section preparation of a human tooth. By tracing outlines of the tooth on the sections, discrete points were obtained and employed to construct B-spline curves that represent the exterior contours and DEJ of the tooth using least-square curve-fitting technique. The surface-skinning technique was then employed to quilt the B-spline curves to create a smooth boundary and DEJ of the tooth using B-spline surfaces. These surfaces were respectively imported into SolidWorks via its application protocol interface (API) to create solid models, as shown in Figure 2.17 (Chang et al. 2003).

The solid models were then imported into Pro/MECHANICA Structure for finite element meshing and analysis. The finite element mesh shown in Figure 2.18a was created manually in Pro/MECHANICA Structure. The automatic mesh generation capability provided in Pro/MECHANICA Structure was not employed because of its limitation on maintaining mesh continuity across the DEJ. The finite element mesh was created by splitting and merging surfaces and curves into desired segments. Then smaller solid volumes were created by choosing wrapping surfaces that form individual airtight cavities. Note that C^0-continuity had to be maintained to ensure validity of the finite element mesh since neither gap nor penetration between elements was allowed.

One finite element was assigned to each solid volume. All the elements in the model were 3D solid elements. This was the semiautomatic meshing capability offered by Pro/MECHANICA Structure, which is an independent module of Pro/ENGINEER.

Note that the DEJ was well preserved in the finite element model, as shown in Figure 2.18b. There are 192 solid elements in this model. According to the literature (e.g., Middleton et al. 1990), a uniformly distributed vertical load of 170 N was applied on top of the tooth model. The entire exterior boundary 2 mm below the DEJ was fixed. The isotropic material properties, Young's modulus, and Poisson's ratio, assigned to the enamel and dentin were based on literature designating 8.5×10^4 MPa and 0.33 for enamel, and 1.98×10^4 MPa and 0.31 for dentin, respectively.

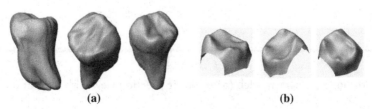

(a)   (b)

Figure 2.17: Tooth model in three views: (a) solid model and (b) surface for DEJ.

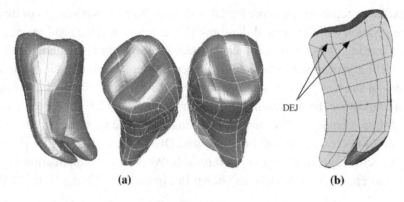

(a)   (b)

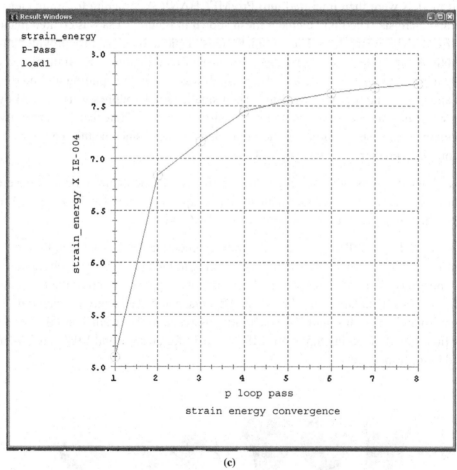

(c)

Figure 2.18: Tooth finite element model: (a) mesh, (b) section view with DEJ, and (c) convergence graph.

A finite element analysis of the solid tooth model was conducted using Pro/MECHANICA Structure. A linear static analysis was carried out for the loading, material, and constraint conditions described earlier. A 0.5% convergence in strain energy was defined as the criterion for ensuring solution convergence. The analysis took eight passes to reach convergent solutions, where the maximum polynomial order of element shape function was 8, and 40,059 linear equations were solved simultaneously in FEA. The convergence study ensured the accuracy of the FEA solutions. The convergence graph for strain energy is shown in Figure 2.18c. The strain energy curve became almost flat at passes 7 and 8, indicating that no significant improvement would be achieved by further increasing polynomial order. From the mathematical theory of the finite element method, the strain energy of the finite element solution converged to the exact solution from below, as revealed in the convergence graph in Figure 2.18c. The stress fringe plots shown in Figure 2.19a indicate that a stress concentration of 24 MPa occurs at the left exterior surface of the tooth. This is due to the asymmetry of the tooth geometry. The stress concentration derives mainly from the combination of axial and bending stresses in compression. In addition, the stress distribution along the DEJ, which largely determines the strength of the tooth, can be identified in Figure 2.19b (Chang et al. 2003).

In Pro/MECHANICA Structure, the "cut-plane" option is available for visualizing the FEA results interior to the object. Using this option, the interior stress distributions are revealed. Note that in Figure 2.20 the cut-planes were created for sections with 5% depth apart. A total of 18 sections were cut for visualizing the stress distribution. As shown in Figure 2.20, no significant stress jump appears in any of the sections, which demonstrates that the p-FEA is more accurate and suitable for tooth mechanics study.

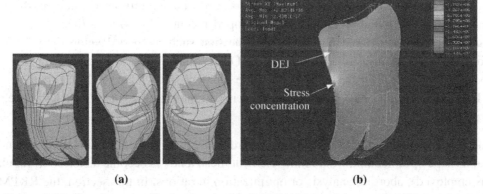

(a) **(b)**

Figure 2.19: Tooth stress distribution in Pro/MECHANICA: (a) complete fringe plot and (b) sectional stress distribution.

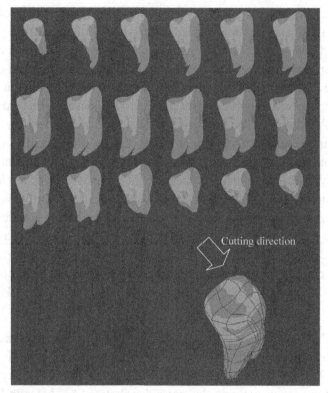

Figure 2.20: Stress fringe plots on cut-planes (in depth increments of 5%).

2.3.4 The Meshless Method

A number of meshless (or mesh-free) methods that do not require an explicit mesh for finite element formulation have been developed recently. Smooth particle hydrodynamics (SPH) (Monaghan 1988) was the first such method developed for handling infinite-domain astronomy problems. Others include the diffuse-element method (DEM) (e.g., Nayroles et al. 1992), the element-free Galerkin (EFG) method (e.g., Belytschko et al. 1994), and the reproducing kernel particle method (RKPM) (e.g., Chen et al. 1997).

The meshless method is appealing for solving problems in large-deformation analysis and structural shape optimization, where finite element mesh tends to become distorted (if regular FEA is employed), aborting analysis or optimization iterations. In this section, the RKPM will be briefly described, focusing only on the displacement interpolation using kernel functions.

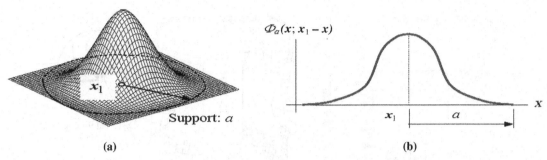

Figure 2.21: B-spline kernel function: (a) bi-cubic for planar problems and (b) cubic for 1D problems.

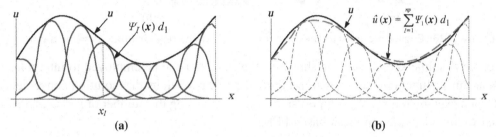

Figure 2.22: Displacement interpolations: (a) kernel functions and (b) summation of individual functions.

In RKPM, the displacement field is interpolated globally as

$$u(x) = \Sigma_{I=1}^{np} \, \Psi_I(x) \, d_I \qquad (2.40)$$

where

x represents a point in the structural domain

np is the total number of particles of the entire structure

$\Psi_I(x)$ is the RKPM interpolation function

d_I is the generalized displacement at the Ith particle

Equation 2.40 reveals the very basic feature of RKPM (and the meshless method in general): The displacement field is interpolated in the entire structural domain (instead of a finite element) through the RKPM interpolation function $\Psi_I(x)$. In theory, then, no mesh is required.

Because of the global interpolation function $\Psi_I(x)$, more computational effort than that from FEM is expected. This function consists of a kernel function $\Phi_a(x; x_I - x)$ with a support measure of a (and a correction function for meeting the reproducing conditions). Note that x_I is the position vector of the Ith particle. A typical kernel function for planar problems, a bi-cubic B-spline surface, is shown in Figure 2.21a. The kernel function must reveal a property $\int_\Omega \Phi(x_I - x)dx = 1$.

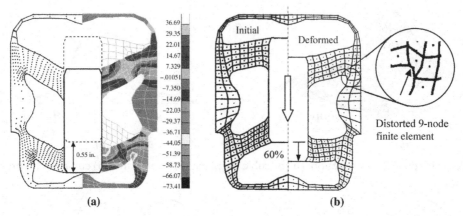

Figure 2.23: Engine mount: (a) meshless method and (b) regular finite element method.

The cubic B-spline kernel function shown in Figure 2.21b can be employed to interpolate a displacement function $u(x)$, shown in Figure 2.22. It is obvious that solution convergence can be achieved by refining the support size a or by increasing the number of particles; this is similar to h-p adaptation in mesh-based FEM.

The superiority of the RKPM in solving large-deformation problems has been demonstrated and widely recognized. Figure 2.23a demonstrates an application of the meshless method for calculating the displacement and stress of an engine mount. The core metal bar can move down to the end (total travel is 0.55 in.) without any mesh-related problem. However, with the regular finite element method, the element mesh is distorted, aborting the analysis at 60% of the metal bar movement, as shown in Figure 2.23b, where ABAQUS was employed for analysis (Grindeanu et al. 1998).

The RKPM method has been demonstrated to be feasible for solving large-deformation problems, for example, the engine mount problem shown in Figure 2.23. However, as illustrated in Figure 2.23a (see grids), a background mesh is still required for Gaussian integration in constructing a stiffness matrix. When the Jacobian of the background mesh becomes close to zero, Gaussian integration gives inaccurate results. Note that in general the geometry of the background mesh is much less restrictive than that of the shape of the true finite elements. The computational time for constructing the stiffness matrix and the load vector is longer than that of regular FEM. The meshless method is an active research topic and is continuously being developed and enhanced. Another newly developed FEM, among others, is extended FEM (XFEM), which supports discontinuity problems, interface problems, and crack propagation problems, among others.

2.3.5 Using Finite Element Method

In general, there are three major steps in using finite element method or FEA codes for solving structural problems. They are preprocessing, solutions, and postprocessing, as depicted in Figure 2.24.

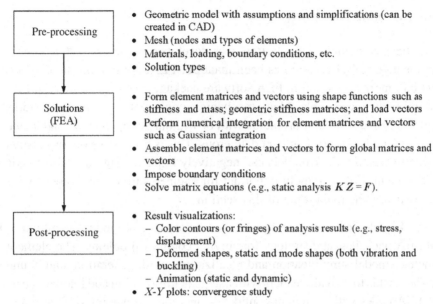

Figure 2.24: Overall process of structural analysis using finite element method.

The pre-processing step starts with creation of a geometric model that represents the structural domain. The geometric model can be directly created using FEA codes with relatively limited modeling capability or, preferably, created in a CAD system and then exported to FEA codes. Once the geometric model is available, the finite element mesh, including nodes and elements, must be generated automatically, using the mesh generation capability in the FEA code, or manually, which is often less desirable, especially for structures of complex geometry. In addition, material properties and loading and displacement boundary conditions must be defined. Finally, a solution type (e.g., static, vibration, buckling, nonlinear) must be specified.

The solutions step is dealt with by the FEA codes, which take the finite element model, formulate element matrices, assemble them for global matrices, impose boundary conditions, and solve the system of equations using numerical algorithms such as Gaussian elimination or LU decomposition (Atkinson 1989), in which the stiffness matrix is decomposed and is the multiplication of two matrices, L and U (L is a lower triangular matrix and U is an upper triangular matrix).

The analysis results can be visualized in various forms in the post-processing step. Color fringe or contour plots of the solutions, such as displacement or stress, may be displayed; a deformed shape or animation may be requested to better visualize structural deformation. In addition, graphs for various results may be generated, such as the convergence study graph that was shown in Figure 2.18c if a convergence study was carried out using a p-version FEA.

2.4 Finite Element Modeling

Since the first FEA computer code, Nastran, was developed in the late 1960s as a NASA initiative, tremendous advancement has been made for FEA, not only in theoretical development but, more important, in FEA software packages. Today commercial FEA codes are popular and widely accessible. Many mechanical engineers or engineering students can use FEA codes to solve structural problems with or without knowledge of fundamental FEA theory. FEA software is powerful and yet dangerous. This is because erroneous results due to mistakes made in creating FEA models can negatively impact design decision making in product development. It is extremely important for engineers to use FEA codes with competence, which is the main topic of this section.

We discuss the key elements of creating adequate finite element models, common pitfalls, commercial FEA packages, and strategy for solving complex problems. The elements to be discussed include model simplification and idealization, mesh generation, and boundary conditions. The goal is to provide readers with adequate information and knowledge to enable them to use FEA codes with sufficient confidence and competence for solving practical structural problems.

2.4.1 General Process and Potential Pitfalls

As shown in Figure 2.25, the first step in creating an FEA model is to fully understand the physical problem being considered. This includes the operating condition of the mechanical component or system, material properties, geometric shape and dimensions, boundary conditions, possible failure scenarios, and, most important, the questions FEA is to answer. Is it the maximum stress that could lead to material yield, the vibration frequency that could induce resonance, the load that poses a buckling hazard? We have to identify the structural performance or potential failure modes we want to monitor. This should determine the analysis type chosen in FEA.

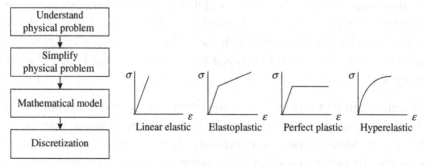

Figure 2.25: General process of creating FEA models.

The next step is about simplification and idealization of the structural problem. This is often necessary since the physical problems are usually too complex to solve as they are. A physical problem can be idealized by making adequate assumptions to reduce the complexity level to one that FEA is capable of solving, and yet with the numerical solutions closely resembling the behavior of the physical problem. These assumptions can be made by removing nonessential geometric features, reducing 3D structures to 2D or 1D components, converting impacts to equivalent static loads if feasible, fixing the rotating end of the structure for proper boundary conditions, and so forth. Moreover, the geometry of the structure may be simplified by taking advantage of the symmetric conditions, where only a portion of the structure is sufficient for solving the problem with the same accuracy. Idealization and simplification are two of the most challenging steps in conducting FEA. We will discuss this topic further in Section 2.4.2.

It is important to understand the physics of the structural problem to be solved and to identify the proper mathematical model that can be applied to solve the problem. There are at least three aspects to consider. The first is to understand in general how the deformation will mostly occur in the structure. Is it going to be a small deformation within the linear elastic range? Or, if the deformation is hyperelastic or plastic, is it beyond the linear elastic range? The second aspect is the load: Is it static or time dependent? Third, is it a multiphysics problem, such as aero-structure, fluid structure, acoustics, or thermal structure? All of these must be well understood before creating finite element models and choosing an adequate FEA code for solution.

The final step involves finite element mesh generation. The questions to ask are these: Does the problem require a very refined mesh, are the FEA software and computers able to handle a large finite element model? What type of elements will be used, linear or quadratic? Is p-version FEA a better choice for the problem at hand? If so, what will be the convergence criteria? Also, does the problem involve discontinuity in responses? These are essential questions to go over before creating the model. The more thought that is given to these questions, the better the preparation, which usually means a better FEA model and eventually time savings. Among the most important issues is what questions the FEA is to answer. These questions have to be very specific to avoid unnecessary or repeated effort in carrying out FEA.

2.4.2 Idealization and Simplification

In engineering analysis, a physical problem often has to be idealized and simplified before it can be solved, especially with FEA. Idealization is about making proper assumptions so that the problem can fit into an existing mathematical model for solutions. Simplification is about reducing the sizing of the FEA models yet maintaining solution accuracy. Idealization and simplification are probably the most important and yet challenging steps in creating FEA models. It takes time and experience to become competent in making these decisions. We will mention a few cases and draw some conclusions on this subject.

When working in CAD, especially in creating detailed design, the components being designed usually come with subtle geometry details. Very often it is difficult to mesh the entire model with the presence of small geometric features such as holes and fillets. Most of these small features can be suppressed or removed for finite element analysis since very often they do not affect the analysis results. Figure 2.26 illustrates a few such cases. However, features that affect the solutions cannot be ignored. For example, the small hole at the center of a plate under axial or bending load cannot be removed because the presence of the hole induces stress concentrations around it. In general, small features should be removed except for those that cause stress concentrations.

Another category in geometric idealization is reducing 3D solid components to surface or even line models to save computational effort. In general, a thin-shelled structure in 3D can be reduced to a shell surface model with its thickness entered as an element property in FEA. We may compress the outer and inner surfaces of the 3D thin-shelled solid model and use the compressed mid-surface as the shell surface for analysis, as shown in Figure 2.26c. Such a capability is available in both Pro/MECHANICA and SolidWorks Simulation. A 3D solid beam like that shown in Figure 2.26d can also be reduced to a 1D line model with element section properties, such as cross-sectional area and moment of inertia, and beam orientation, such as the third node, entered for the beam elements in FEA.

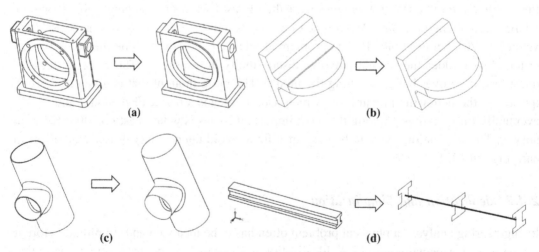

Figure 2.26: Geometric idealizations: (a) removing unnecessary features; (b) aligning surfaces; (c) reducing a 3D solid model to a surface or shell model (mid-surface); (d) reducing a 3D solid problem (beam) to a line model or frame structure (1D beam with cross-section properties, such as area, moment of inertia).

The purpose of geometric idealization is mainly to cut down the size of the finite element model. Results obtained from FEA for the original and reduced models differ, but the difference is insignificant if the idealization is properly made. Geometric simplification takes advantage of structural symmetry to reduce model size. It is different from idealization since simplification does not affect the analysis results but cuts down computation time. To take advantage of symmetry in geometric simplification, the structural geometry must be symmetric. The loading, boundary conditions, material properties, and geometric properties (thickness, beam section properties) must also be symmetric. Moreover, additional boundary conditions may have to be imposed on the simplified geometry to ensure symmetry of the structure in responses, such as the deformed shape.

For example, the rectangular plate shown in Figure 2.27a is loaded with two point loads horizontally, constrained at the left edge. This structure is symmetric with respect to the x-axis, assuming that the material is isotropic and homogeneous, that the two point forces F and P are identical, and that the thickness of the plate is uniform. Therefore, the plate can be split into two halves and only the upper (or lower) half analyzed, as shown. To impose the symmetric conditions, roller boundary conditions are added to the split boundary. The roller conditions ensure that the material on the middle axis only moves horizontally and that the material will not deform to cross the middle axis. The condition imposed is simply $u_y = 0$ along the x-axis. Solving half of the structure is much more efficient than solving the whole model. More important, the solutions for the half-structure are in theory identical to those of the full structure. The numerical solutions from FEA show very slight differences due to truncations in numerical computation.

Similarly, an axisymmetric problem, shown in Figure 2.27b, can be reduced to a smaller domain (in this case 1/3 of the whole structure) if load, boundary conditions, material, and section properties are also axisymmetric. In this case, it is much easier to impose the

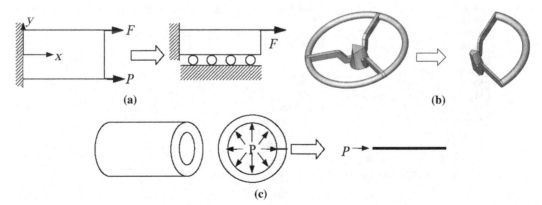

Figure 2.27: Geometry simplifications: (a) a rectangular plate, (b) an axisymmetric disk reduced to symmetrical portion, and (c) a long tube.

symmetric boundary conditions at the split faces using polar coordinate systems. Moreover, sufficiently long tubing with internal pressure *P* may be reduced to a 1D problem, as shown in Figure 2.27c.

The thin-walled tank example in Figure 2.28 is employed to further illustrate the idea of idealization and simplification. A full 3D solid model was first created and meshed for FEA, where 3,585 tetrahedron elements were generated using AutoGen (automatic mesh generator) in Pro/MECHANICA. A 1% strain energy convergence criterion was employed for solution convergence. As shown in Figure 2.28c, the number of equations solved in the final pass of the p-solution was 143,343. The maximum principal stress and displacement magnitude were 41.6 MPa and 0.427 mm, respectively.

The tank was cut in half due to symmetry. With half of the load and symmetric boundary conditions imposed, the half-tank was further reduced to the surface model, shown in Figure 2.28b. This surface model was meshed into 12 shell elements. The same convergence criterion was defined. The solution in strain energy converged in 9 passes. There were 2,733

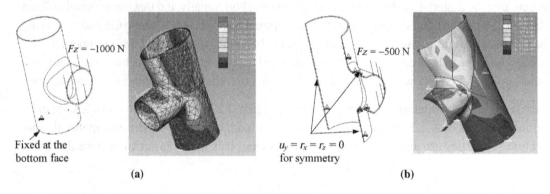

$F_z = -1000$ N

Fixed at the bottom face

(a)

$F_z = -500$ N

$u_y = r_x = r_z = 0$ for symmetry

(b)

Model	Model A: Full solid model	Model B: Half-surface model
Geometry	Solid	Surface
Number of finite elements	3,585 solid elements	12 shell elements
Number of equations in last pass	143,343	2,733
Number of passes	4	9
Maximum edge order	5	9
Strain energy convergence	<1%	<1%
Maximum displacement magnitude	0.4274 mm	0.4277 mm
Maximum principal stress	41.62 MPa	40.22 MPa
Total elapsed time*	826.02 seconds	11.89 seconds
Total CPU time*	106.02 seconds	3.25 seconds
Maximum memory used	346,574 KB	166,614 KB
Working directory disk usage	649,216 KB	4,096 KB

*Dell Precision PWS670, Intel Xeon 2.80 GHz 2.00 GB RAM

(c)

Figure 2.28: Thin-walled tank problem: (a) full 3D solid model, (b) half-surface model, and (c) performance comparison between these two models.

equations solved in the final pass (versus 143,343 equations in the full 3D solid model). The maximum principal stress and displacement magnitude were 40.2 MPa and 0.428 mm, respectively, which were very close to the principal stress and displacement of the full 3D solid model. However, as shown in Figure 2.28c, the CPU time required for the simplified model was only about 3% that of the full 3D model (3.25 out of 106 seconds). The savings in computation time due to model simplification and idealization were significant, as this example demonstrates.

Some FEA codes, including SolidWorks Simulation, offer geometric idealization to treat structural joints, such as pins, bolts, and rigid joints. Special mathematical models serve to replace the detailed geometry and finite element mesh of such connectors, greatly simplifying the analysis of structural assemblies. In Simulation, a 3D beam assembly can be easily reduced to frame structures with truss or beam elements, in which joints are automatically created to replace the connection in solid form between components. In addition to treating joints, Simulation provides contact capabilities, where components in an assembly are allowed to come into contact without penetrating each other.

2.4.3 Mesh Generation and Refinement

Mesh generation generates a polygonal or polyhedral mesh that approximates the geometric domain of the structure. This is a key step in finite element analysis. A given structural domain must be partitioned into simple elements meeting in well-defined ways. First, elements must be continuous or conformal, which means that they all have to share common element edges and corner nodes, as shown in Figure 2.29a. No elements should penetrate or overlap with any other. Second, all elements should be well shaped, which generally involves restrictions on the edge angles or aspect ratio of the elements (Figure 2.29b). For example, the best-shaped quadrilateral (or quad) element is square, and the best-shaped triangular element is equilateral triangle. However, in practice, square and equilateral elements are difficult to retain because of irregular structural boundaries. The mesh quality directly impacts the

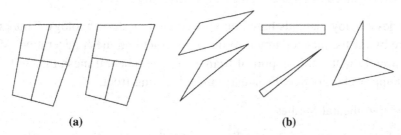

(a)   **(b)**

Figure 2.29: Finite element mesh: (a) conformal (*left*) and nonconformal (*right*) meshes; (b) poorly shaped elements: excessive angles (*left*), excessive aspect ratio (*center*), and negative Jacobian (*right*).

accuracy of the FEA as well as the computational resources involved. Finally, there should be as few elements as possible, but some portions of the domain may need smaller elements so that the computation is more accurate there.

Most FEA codes provide automatic mesh generation capabilities. For surface structures, triangular mesh is most common. Some codes also provide mesh with a mix of triangular and quadrilateral elements or even completely quadrilateral elements. For solid structures, only tetrahedral elements are currently supported by almost all commercial FEA software. Most provide some mesh control and refinement options. Certain ranges for angle and aspect ratio may be specified to ensure mesh quality. In some cases, global and local element sizes can be specified. In addition, a face, an edge, or points for local mesh refinement can be chosen.

Some FEA pre-processors, such as HyperMesh® and MSC/Patran®, offer semiautomatic mesh generation capabilities. These cases call for manually splitting the structural domain into continuous decomposed domains (the splits are called patches or volumes for 2D and 3D domains, respectively), and then creating mesh in these decomposed domains according to a mapping method to be discussed later. This approach provides flexibility in decomposing the structural domain and allows the user to create all quad or all hexahedral elements. The semiautomatic approach offers more flexibility in choosing types of finite elements, but requires much more manual effort.

Automatic Mesh Generation

In general, mesh generation technology deals with structured and unstructured meshing. Structured meshing is commonly referred to as "grid generation." Strictly speaking, a structured mesh can be recognized by all interior nodes of the mesh having an equal number of adjacent elements. The mesh generated by a structured grid generator is typically all quad or all hexahedral.

Unstructured mesh generation, on the other hand, relaxes the node valence requirement, allowing any number of elements to meet at a single node. Triangular and tetrahedral meshes are most commonly thought of when referring to unstructured meshing, although the quadrilateral and hexahedral meshes can be unstructured as well.

Most FEA codes employ methods based on unstructured mesh technology. Triangular and tetrahedral are by far the most common forms of unstructured mesh generation. Most techniques currently in use for support of triangular or tetrahedral meshing can fit into one of three main approaches: octree, Delaunay, and advancing front.

Triangular and Tetrahedral Meshes

The octree technique was primarily developed by Mark Shephard's group at Rensselaer Polytechnic Institute in the 1980s. In this method, cubes containing the geometric model are recursively subdivided until the desired resolution is reached. Figure 2.30a shows the

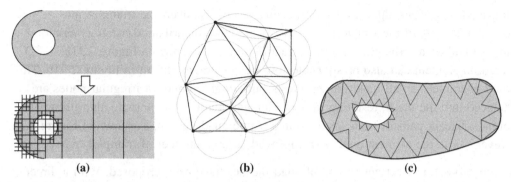

Figure 2.30: Meshing technology: (a) octree decomposition of a simple 2D object; (b) Delaunay triangulation in the plane with circumcircles shown; (c) advancing front, where one layer of triangles has been placed.

equivalent two-dimensional octree model decomposition. Irregular cells are then created where cubes intersect the boundary, often requiring a significant number of geometric intersection calculations. Tetrahedra are generated from both the irregular cells on the boundary and the internal regular cells.

By far the most popular mesh generation techniques for triangular and tetrahedral meshing are those utilizing the Delaunay criterion, as illustrated in Figure 2.30b. The Delaunay criterion, sometimes called the "empty sphere" property, states that a triangle net is a *Delaunay triangulation* if all the circumcircles of all the triangles in the net are *empty*. This is the original definition for a two-dimensional space. It is possible to use the criterion in a three-dimensional space if a circumscribed sphere replaces the circumcircle. A circumsphere can be defined as the sphere passing through all four vertices of a tetrahedron.

Another very popular family of triangular and tetrahedral mesh generation algorithms is represented by the advancing, or moving, front method. An active front is maintained where new tetrahedra are formed. Figure 2.30c is a simple two-dimensional example of the advancing front, where triangles have been formed at the boundary. As the algorithm progresses, the front advances to fill the remainder of the area with triangles. In three dimensions, for each triangular facet on the front, an ideal location for a new fourth node is computed. Also determined are any existing nodes on the front that may form a well-shaped tetrahedron with the facet. The algorithm selects either the new fourth node or an existing node to form the new tetrahedron based on which forms the best one. Also required are intersection checks to ensure that tetrahedron does not overlap as opposing fronts advance toward each other.

Quadrilateral Meshes

Unstructured quadrilateral (or quad) meshing algorithms can, in general, be grouped into two main approaches: direct and indirect. With an indirect approach, the domain is first meshed with triangles. Various algorithms are then employed to convert the triangles into quadrilaterals. One of the simplest methods for indirect quadrilateral mesh generation includes dividing all triangles into three quadrilaterals, as shown in Figure 2.31a. A 3D tetrahedron element can also be split into four hexahedra, as shown in Figure 2.31b. This method guarantees an all-quadrilateral mesh, but a high number of irregular nodes are introduced into the mesh, resulting in poor element quality. An alternate algorithm is to combine adjacent pairs of triangles to form a single quadrilateral, as shown in Figure 2.31c. However, while the element quality is improved, a large number of triangles may be left.

Many methods for direct generation of quad meshes have been proposed. With a direct approach, quadrilaterals are placed on the surface directly, without first going through the process of trianglar meshing. Among these methods, there appear to be two types. The first comprises methods that rely on some form of decomposition of the domain into simpler regions that can then be meshed following standard approaches, such as mapping method. The second comprises those that utilize a moving front for direct placement of nodes and elements.

One of the most advanced decomposition methods is *medial axis transformation* (MAT). The medial axis (MA) is defined as the locus of centers of locally maximal circles inside an object. The locus with the radii of the associated locally maximal circles is defined as the object's medial axis transform (MAT). The boundary and the corresponding MA of a 2D object are shown in Figure 2.32a. The points where three or more line entities of the MA meet are called branch points. Associated with the MA is a radius function R, which defines for each point on the axis its shortest distance to the object's boundary. At each branch point,

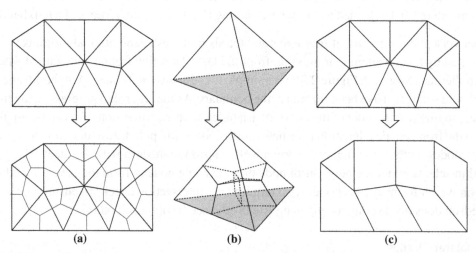

Figure 2.31: Indirect quad mesh: (a) a quad mesh generated by splitting each triangle into three quads through mid-point subdivision; (b) decomposition of a tetrahedron into four hexahedra; (c) a quad-dominant mesh generated by combining triangles.

line segments are created to connect the branch point to the tangent points of the circle with the boundary. These line segments, which decompose the physical domain of the structure, are called corridors, as illustrated in Figure 2.32b. Having decomposed the area into simpler regions, sets of templates are then employed to insert quadrilaterals into the domain, as shown in Figure 2.32c. A more recent method, called LayTrack, creates uniform spacing inside the corridors to generate uniform, isotropic quad-elements of good aspect ratio (for example, Quadros et al. 2004). A similar concept has been applied to 3D objects, where a medial surface is created by rolling maximal spheres inside an object, and then hexagon mesh is created, as depicted in Figure 2.32d.

A representative method in the second kind of direct quad mesh generation is plastering, which is an extension of the advancing front algorithm to 3D objects. In this method, elements are first placed starting with the boundaries and advancing toward the center of the volume, as shown in Figure 2.32e. A heuristic set of procedures for determining the order of element formation is defined. Similar to other advancing front algorithms, a current front is defined consisting of all quadrilaterals. Individual quads are projected toward the interior of the volume to form hexahedra. In addition, plastering must detect intersecting faces and determine when and how to connect to preexisting nodes. As the algorithm advances, complex interior voids may result, which in some cases are

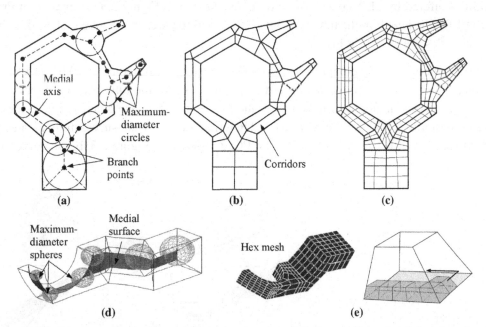

Figure 2.32: Direct methods for quad and hex meshes: (a) the medial axis of the domain; (b) decomposing domain into small corridors; (c) a quad mesh created in each corridors; (d) the medial surface and hex mesh for a 3D object; (e) the plastering process forming elements at the boundary.

impossible to fill with all-hex elements. Existing elements, already placed by the plastering algorithm, must sometimes be modified to facilitate placement of hexes toward the interior. The plastering algorithm has yet to be proven reliable on a large class of problems.

Mesh Improvement

It is uncommon for a mesh generation algorithm to be able to define a mesh that is optimal without some form of post-processing to improve the overall quality of the elements. The two main categories of mesh improvement include smoothing and clean-up. Most smoothing procedures involve an iterative process that repositions individual nodes to improve the local quality of the elements, such as edge angles and aspect ratios. Clean-up generally refers to any process that changes element connectivity.

From a practical perspective, almost all popular FEA codes provide automatic mesh generation capabilities that support tri- and quad-elements for shell and tetra-elements for solid structures. Ranges of critical mesh parameters, such as aspect ratio and edge angles, can be defined up front. Also, local refinements are supported in most codes.

Here we present a 3D torque tube example to provide a better picture of the mesh generation capabilities offered by FEA codes. The torque tube shown in Figure 2.33 is created in Pro/ENGINEER. The length of the tube is about 70 in. Small features in the tube such as thin fins and small holes are shown in Figure 2.33a.

The tube geometry is fairly complex. The AutoGen capability offered by Pro/MECHANICA Structure, a p-FEA program, was employed to create mesh for the tube. Default parameters for mesh generation, including allowed edge and face angles between 5 and 175 deg. and an allowed aspect ratio of 30, were employed; 8,246 tetra-elements were successfully created in about 90 seconds CPU time on a Pentium IV PC. As shown in Figure 2.33b, the curve

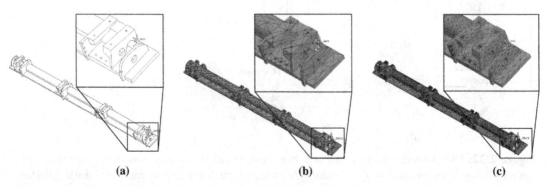

(a)   (b)   (c)

Figure 2.33: Three-dimensional torque tube: (a) CAD model, (b) first mesh of 8246 tetrahedral elements, and (c) second mesh of 63,816 tetrahedral elements.

boundary was largely matched with the mesh; however, the mesh quality was less desirable because of the large aspect ratio (the resulting maximum aspect ratio was 11.3).

The default parameters were adjusted by reducing the aspect ratio to 10 and the angle ranges to 25 and 155 deg.; AutoGen was again employed. This time, 63,816 tetra-elements were successfully created in 201 minutes CPU time on a Pentium IV PC, 130 times more than the first case. As shown in Figure 2.33c, the mesh quality was much improved (the resulting maximum aspect ratio was 4.40). The second mesh also captured the geometry around the holes much more accurately. Note that AutoGen does not always generate mesh successfully. With a smaller aspect ratio, it often fails. The best approach is to use default parameters to create a baseline mesh and then adjust mesh parameters for an improved mesh whenever possible.

Semiautomatic Mesh Generation

Although automated mesh generation methods for two- and three-dimensional structures have been studied intensively, many engineers still craft meshes manually for a certain class of analysis problem. One of the missing capabilities of current commercial mesh generators is the versatile control of mesh anisotropy and directionality.

In general, quad- and hex-elements are more desirable than tri- and tetra-elements. First, quad- and hex-elements use bilinear interpolation functions for the displacement field, providing more accurate solutions than those of tri- and tetra-elements in general. As a result, it takes fewer quad- or hex-elements to achieve accurate solutions. For example, if the triangular element is used in shell analysis, its membrane behavior is very poor and inaccurate results will be obtained for many problems (Wilson 2000, Braucr 1993, Zienkiewicz 1989). Also, it is reported in the literature that certain problems, such as plasticity and rubber analysis (Finney 2001) and eigenvalues problems (Benzley et al. 1995) converge better with quad- or hex-elements.

When the automatic mesh generator at hand is not able to create the desired mesh or fails to generate one, the mesh has to be created manually. The key idea for manual mesh generation is to decompose the structural domain into smaller continuous regions and then use a mapping method to create the mesh. This is the semiautomatic approach. CAD and CAE packages, such as MSC/Patran, HyperMesh, and ANSYS provide a mapping capability for mesh generation.

Two examples, a 2D connecting rod, shown in Figure 2.34, and a 3D turbine blade (see Figure 2.36), are employed to illustrate the semiautomatic mesh generation method. The domain of the connecting rod was first decomposed into 25 geometric patches in Patran, as shown in Figure 2.34a. In Patran, a geometric patch is represented mathematically as a bi-cubic parametric surface, as depicted in Figure 2.35. Obviously, one geometric patch is not able to represent a complicated geometric boundary. Usually, a structural domain is

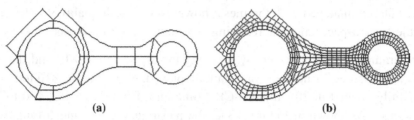

Figure 2.34: Two-dimensional connecting rod: (a) a Patran patch model and (b) a finite element mesh.

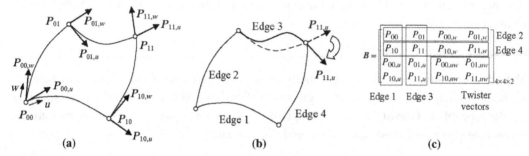

Figure 2.35: Bi-cubic parametric patch: (a) corner points and tangent vectors; (b) adjusting the tangent vector $P_{11,u}$; (c) data that define the patch formulated in a 4×4×2 matrix.

decomposed into several smaller patches in the modeling process. The decomposed patches must be at least C^0-continuous to generate a conformal mesh. Also, to avoid sharp edge angles in finite element mesh, the patch boundary edges must be properly created by adjusting the tangent vectors at the corner points.

A typical patch in Patran, shown in Figure 2.35, is a bi-cubic parametric surface in terms of parameters u and w, where u and $w \in [0,1]$. The patch edges are cubic curves in u or w. The shape of the cubic curve can be controlled by adjusting tangent vectors in both direction and magnitude. For example, changing the direction of the tangent vector $P_{11,u}$ alters the geometry of Edge 3 and therefore the geometry of the patch, as illustrated in Figure 2.35b. Aligning tangent vectors at the joint between patches ensures geometric smoothness across the patch boundary. A mesh of quad-elements can then be generated by specifying the number of elements along the u and w parametric directions, respectively, as shown in Figure 2.34b. This is essentially the mapping method.

Physically, a turbine blade is inserted into a slot of a disc mounted on a rotating shaft, as shown in Figure 2.36a. The blade can be divided into four parts: airfoil, platform, shank, and dovetail. When the turbine is operating, fluid pressure is applied to the surface of the airfoil and an inertia body force is applied to the whole blade structure because of rotation. To

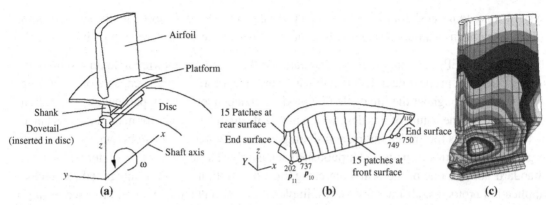

Figure 2.36: Three-dimensional turbine blade: (a) the physical model, (b) boundary patches created for the shank, and (c) the finite element model with a stress fringe plot.

generate the mesh, the shank was first decomposed into 15 smaller pieces, represented by C^1-continuous patches at the front and rear surfaces as shown in Figure 2.36b. Furthermore, the shank was decomposed in to 15 hyperpatches (volumes) using the respective front and rear patches. Hex-elements were then created in individual hyperpatches using the mapping method. The same procedure was employed to create mesh for the airfoil and dovetail, as shown in Figure 2.36c.

2.4.4 CAD Model Translations

Creating a geometric model using finite element codes—or even an FEA pre-processor such as MSC/Patran—by dealing with points, curves, and patches is often tedious and time consuming. It is much more convenient and productive to create a geometric model in CAD using feature-based solid modeling capabilities. Creating mesh using an automatic mesh generator in CAD is straightforward. However, from time to time it may be necessary to bring the CAD model into finite element code for a more desired mesh, such as quad- or hex-elements.

Some FEA provides a direct connection to popular CAD software. For example, in ANSYS, you may directly import the Pro/ENGINEER solid model for mesh. Very often such direct connection is not available. In that case, two options are commonly employed: IGES (initial graphics exchange system) (see Stokes 1995) and STEP (standard for the exchange of product model data).

The IGES standard was conceived at a meeting of the Society of Manufacturing Engineers (SME) in the fall of 1979. The initial committee that developed it comprised engineers from General Electric and Boeing (representing industry) and Computervision, Gerber, and Applicon (representing CAD vendors). IGES was seen as a strategy to decrease the number of system-to-system translators. IGES was first published in 1980 and subsequently adopted as an ANSI (American National Standards Institute) standard. The current version of IGES is 5.3.

ISO 10303 (STEP) is an international standard for the computer-interpretable representation and exchange of product data. Its objective is to provide a format that is capable of describing product data throughout the life cycle of a product, independent of any particular organization or technology. The nature of this description makes it suitable not only for neutral file exchange but also as a basis for implementing and sharing product databases and archiving. ISO 10303 is organized as a series of parts, each published separately. These parts of this International Standard fall into one of the following catagories: description methods, integrated resources, application protocols, abstract test suites, implementation methods, and conformance testing.

While there are more than 40 different documents called "Parts" and "Application Protocols" (APs) in STEP, individual industry sectors have identified the APs that most closely correlate to their standard work process. The two that are well supported by the CAD community for mechanical and aerospace engineering are AP203 and AP214.

We will take a look at a tracked vehicle roadarm to understand some pitfalls in CAD model translation for mesh generation. Starting with engineering drawings, the solid model of the roadarm was created in NX. As shown in Figure 2.37, its cross-section profile was composed of straight lines and circular arcs, with the arm tapered along the curve path. Moreover, there are sharp corners in the intersections of the arm segment and cylinders. This complicated geometry makes it difficult to model the roadarm and to generate regular finite element mesh.

In NX, cross-section profiles of the ends of the arm were first created using straight lines and arcs. Then wrapping surfaces were created by tapering the profile from one end to the other along the curve path. Sewing the wrapping surfaces of the arm segment to create the solid feature failed because of excessive gaps between the adjacent wrapping surfaces. To overcome this difficulty, a single closed spline was used to approximate the arm cross-section

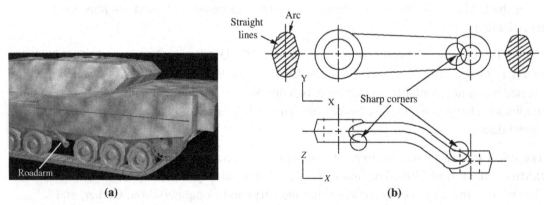

Figure 2.37: Tracked vehicle roadarm: (a) computer model and (b) schematic drawing.

contour, tapering along the curve profile. The two cylinders that intersect with the arm segment were created using standard extrusion features in NX.

The NX roadarm model was then translated into Patran using the NX–Patran. As shown in Figure 2.38a, the geometric shape and dimension of the roadarm model translated smoothly and accurately. However, because about 20 points were used to form the closed spline to approximate the cross-section profile, the interface created more patches in the arm segment than were necessary for maintaining accurate arm geometry and which could potentially produce a large finite element model in Patran. Moreover, the patches were not continuous, which caused difficulty in creating continuous hyperpatches and a conformal finite element mesh. Therefore, although the roadarm geometry was translated accurately, only a portion of the geometric entities that describe its key geometric data were valuable, such as the arm segment profile and cross-section contour. Continuous patches that wrap the arm segment and cylinders needed to be recreated in Patran.

An IGES translator was also used to produce a Patran model of the roadarm, as shown in Figure 2.38b. The IGES translation was not as accurate as the NX–Patran interface, however. In particular, the shape of the second cylinder was distorted and the profile of the arm segment was inaccurate. Moreover, the translated model contained discontinuous patches and areas with missing patches. Comparison of Figures 2.38a and 2.38b shows that the NX–Patran interface translated the model much better than the IGES translator did. More effort would be required to create usable Patran models based on the results of the IGES translator.

2.4.5 Loads and Boundary Conditions

One of the most challenging tasks in creating finite element models is dealing with boundary and loading conditions. The finite element model must be properly restrained by displacement constraints to eliminate rigid-body motion and, more important, to accurately capture physical

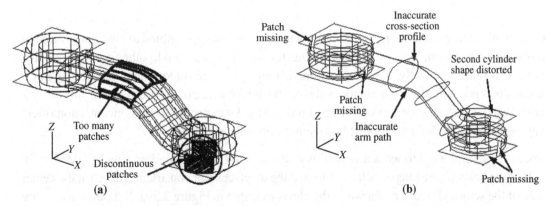

Figure 2.38: Roadarm model translation: (a) NX to Patran directly; (b) NX to Patran via IGES.

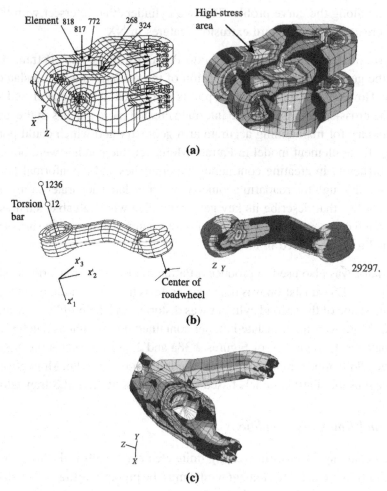

Figure 2.39: Creating proper boundary and loading conditions: (a) tracked vehicle clevis, (b) tracked vehicle roadarm, and (c) lower control arm of an HMMWV.

conditions. In many cases, the components being modeled are assembled to other parts through bolts, pins, and so on. Very often these connectors link the structure to other components that are constrained physically. These constrained components are seldom modeled. It may not always be a good idea to constrain the nodes on the inner surface of the holes where pins or bolts are located, since they are not constrained physically. Unexpected stress concentration often appears around the hole where displacement constraints are placed.

There are several ways to work around this problem. First, semi-cylinders may be added to fill half of the holes where pins or bolts reside, and the displacement constraints placed at the center axis of the semi-cylinders, as shown in the clevis example in Figure 2.39a. Similarly, forces are applied at the center axis of the semi-cylinders at the other end. As depicted in Figure 2.39a, there are stress concentrations close to the center axes where displacement constraints and forces are placed. However, these high stresses can be ignored since they are artificial. The real stress concentrations are properly captured at the top concave surfaces, as shown in Figure 2.39a.

Another approach is to create beams to connect the nodes at the inner surface of the holes and constrain the other end of the beams instead of constraining the nodes on the inner hole surface, as shown in Figure 2.39b. In the roadarm example, two elastic beams are added to the left hole that represents the torsion bar to which the roadarm is mounted on. The torsion bar is constrained to better resemble physical conditions. At the lower end of the torsion bar, numerous beams of greater stiffness (or even rigid beams) are added to connect to the nodes at the inner surface of the hole. As a result, there is no stress concentration around the hole, which again better resembles physical situations.

Similar to displacement constraints, application of loads to finite element models is critical. In many cases, loads are not stationary. Either a structural dynamic analysis must be conducted or the load must be converted to stationary for static analysis. Also, an isolated point load is fine for a beam structure but not for 2D or 3D solids since an isolated point load tends to create stress concentrations. Physically, applying a point load to a structure is like pushing a needle into an elastic object, which yields high stress concentration and is not physically meaningful. In general, a distributed pressure load is more realistic.

If an isolated point load is unavoidable, a better approach is to add beams to diffuse the force flow, similar to that employed for displacement constraints, as shown in the roadarm example. As shown in Figure 2.39b, an elastic beam is added to the hole on the right to simulate the axis of the roadwheel. At the lower end of the beam, a point force is applied that simulates the force acting on the roadarm from the roadwheel. At the upper end of the beam, numerous rigid beams are added to connect the end to the nodes at the inner surface of the hole. As shown in Figure 2.39b, high stress concentration is properly captured at the top surface closer to the left hole.

A similar example, shown in Figure 2.39c, involves a lower control arm of a high-mobility multipurpose wheeled vehicle (HMMWV). Again, numerous rigid beams are added to connect the nodes at the inner surface of the large hole. At the other end of the beams, a point load is applied to simulate the force delivered by the shock absorber to the control arm. Similar treatment can be seen at the left end in order to add a point load that simulates the force acting from the ball joint. As shown in Figure 2.39c, stress concentration is properly captured. No artificial stress concentrations are revealed.

2.4.6 Results Checking

Finite element codes are powerful and capable of solving general structural problems. However, these codes are dangerous. Mistakes made in the finite element model yield erroneous results, which may put the product design in jeopardy if they are not carefully checked and corrected. It is important to note that no software tools are error-proof. Only limited warnings and error checking are provided, so it is absolutely critical for engineers to find every possible way to verify the FEA results, either by hand calculations (which are not always possible) or by reviewing the FEA results and exercising educated engineering judgment. Good engineers make sound judgments; incompetent engineers live with fuzzy guesses.

It is a good practice to predict, or at least have some rough idea about the FEA results beforehand. After running FEA, if the FEA results deviate significantly from prediction, it is either that the prediction is wrong or that the FEA results are incorrect (due to mostly modeling errors). In either case, this is a golden opportunity to learn and improve our design engineering skill.

The first thing to check is the unit system. CAD software usually offers choices in unit systems. Once chosen, the system must be followed consistently in creating the FEA model. For those who are familiar with Pro/ENGINEER, the default unit system is in.-lb_m-sec—that is, inch in length, pound mass in mass, and second in time. The question is, what is the difference between pound mass and pound force (the latter being more familiar)? This can be a pitfall if care is not taken.[1]

A very simple cantilever beam is discussed to further illustrate the danger of ignorance, especially regarding the unit system. In this example the default unit system in.-lb_m-sec in Pro/ENGINEER is chosen. The beam has a 1 in. by 1 in. square cross-section and a 10 in. length. The left end of the beam is fixed and a distributed load of 1,000 is added to the top edge of the front end face, as shown in Figure 2.40a. The material is aluminum 2014 with properties 4.09E+09 and 0.33 for Young's modulus and Poisson's ratio, respectively, given by Pro/MECHANICA. The beam is meshed into 12 tetra-elements (Figure 2.40a).

The FEA results obtained by a single-pass run in Pro/MECHANICA show that the maximum bending stresses of 71,188 occur at the top and bottom faces (tension and compression, respectively, close to the left end), as shown in Figure 2.40c. The deformed shape in Figure 2.40b looks fine. The issue is whether the FEA results are correct and what these numbers mean. The best way to resolve the issue is to calculate the bending stress using classical beam theory from strength of materials. According to the theory, the maximum bending stress is

$$\sigma_b = \frac{Mc}{I} = \frac{(F \times L)(h/2)}{bh^3/12} = \frac{(1000 \times 10)(1/2)}{1 \times 1^3/12} = 60,000 \tag{2.41}$$

---

[1]CNN reported on September 30, 1999, that NASA lost a $125 million Mars orbiter because a Lockheed Martin engineering team used English imperial units of measurement while the agency's team used the more conventional metric system for a key spacecraft operation.

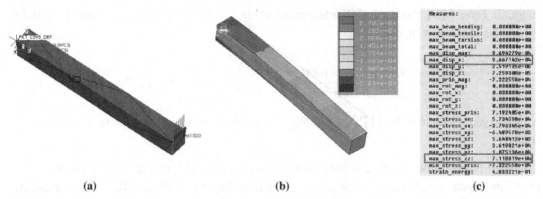

Figure 2.40: Cantilever beam: (a) boundary and load conditions with mesh, (b) deformed shape with bending stress fringe plot, and (c) results of default measures.

where *F*, *L*, *b*, and *h* are force, length, width, and height of the beam cross-section, respectively. The theory says that the bending stress is 60,000, but the FEA results indicate a value of 71,188—a 30% difference, which is significant for this simple example. Which one is correct? Which one should be trusted?

In fact, both are correct in their respective context. The difference is due to Poisson's ratio, which classical beam theory does not take into consideration. In FEA, we entered 0.33 for the ratio and expected to see lateral shrinkage when the transverse load was applied, which contributed partially to the stress along the longitudinal direction. That is why the bending stress was higher than calculated. FEA certainly provides more detailed and realistic results. If Poisson's ratio is set to zero and the analysis is rerun, the bending stress becomes very close to 60,000. Note that in using other software (for example, SolidWorks Simulation), there may be slightly higher stresses at the corners of the beam cross-section at the root that are due to stress concentration. The question now is, what is the stress unit? Pro/MECHANICA indicates that the stress unit is $lb_m/(in.sec^2)$ instead of psi (lb per $in.^2$) that we are more familiar with. What is $lb_m/(in.sec^2)$? Will the beam yield or fracture at 60,000 $lb_m/(in.sec^2)$?

According to Newton's second law, the force unit in the default in.-lb_m-sec system is $lb_m in./sec^2$. As discussed in the chapter appendix, 1 lb_f = 386 $lb_m in./sec^2$, and the force in $lb_m in./sec^2$ units is 386 times smaller than that in lb_f, which we are more familiar with. When you apply a 1 lb_f force to a 1 lb_m mass block, it accelerates 386 in/sec^2, whereas a 1 lb_m $in./sec^2$ force only accelerates the same block 1 $in./sec^2$.

In what units is the 1,000 force applied? In fact, the force applied is 1,000 $lb_m in./sec^2$, which is only about 2.5 lb_f. Such a small force causes a very small stress in the beam—that is, 60,000 $lb_m/(in. sec^2)$, which is only about 155 psi. This also explains why the vertical displacement is so small. From FEA, the vertical displacement is 9.667×10^{-4} in. This result can be verified using classical beam theory:

$$\delta = \frac{PL^3}{3EI} = \frac{1000 \times 10^3}{3(1.06 \times 10^9)(1 \times 1^3/12)} = 9.780 \times 10^{-4} \text{in.} \tag{2.42}$$

which is very close to that of FEA. Certainly this small displacement is due to a small 2.5 lb$_f$ force.

Classical beam theory is in fact very useful in verifying FEA results. Many practical problems, even complex ones such as the roadarm in Figure 2.39b, can be verified using it. Stress and displacement result verification for the roadarm is left as an exercise. However, there are still large portions of the physical problems that are beyond beam theory. When analytical solutions are out of the question, what can be done? First, the FEA model must be checked to make sure that there is no modeling error, by reviewing the deformed shape of the structure to make sure all displacement boundary conditions, including symmetry conditions if applicable, are satisfied. Then how the loads are applied must be reviewed. Do they resemble the physical conditions? Are there any potential artificial stress concentrations? Also, it must be ascertained that the material properties are correctly defined.

After the model is verified, the stress pattern should be reviewed to see if it is consistent with the load and boundary conditions defined. A convergence study may be necessary. Finally, educated engineering judgment must be exercised to check the numerical results. For example, in the cantilever beam case, if the force unit was mistakenly understood as 1,000 lb$_f$, the question arises as to why such a large force produced only 9.667×10^{-4} in. of displacement on a 1 in. $\times$1 in. $\times$10 in. aluminum bar. Engineer must not be afraid to ask questions. When all are answered, the FEA results are most likely to be accurate, which creates greater confidence.

2.4.7 Strategy for Complex Problems

Structural analysis problems are usually very complex. This complexity can be attributed to structural geometry, loading conditions, material behavior, and problem type. The most challenging part of dealing with a complex problem is not creating an FEA model for analysis but creating a model that correctly answers the questions asked at the outset.

The best strategy for tackling a complex problem is to start with a simple one. The problem must be simplified by making proper assumptions, hopefully to a point that the solutions of the simplified FEA model can be verified analytically and yet reasonably resemble the physical problem. Solutions of the simplified FEA model serve as a baseline for any improved results that follow.

The fidelity or efficacy of the simplified FEA model must be improved by removing or relaxing the assumptions. Any improvement increases the complexity of the problem and increases FEA computation time. One general strategy is to relax the assumptions that cause the least computation increment first. For example, we want to stay in a linear elastic range initially and gradually extend to nonlinear or plasticity, if applicable, which takes more analysis time. The FEA results obtained from a linear elastic model are most likely not what can be used. Very often this model serves well for model verification. Solutions to time-dependent problems, such as dynamics, impact load, or viscoelasticity, are probably the most time consuming. Also, it is more efficient to start with a simplified geometry by suppressing small and insignificant features and with a relatively coarse mesh. The mesh may be gradually refined and the suppressed features brought back that can impact the results.

In any improvement, it is critical to check the FEA results and compare them with the baseline result before moving to the next improvement. A baseline can be stress or displacement, calculated using analytical equations or FEA models that are verified beforehand. Any significant deviations should raise a warning flag and merit careful rechecking. Mesh refinement and convergence studies are usually the final activities carried out to improve the accuracy of results.

Basically, we want to start with a simple problem, relax assumptions one or two at a time, check results using baseline or engineering judgment, and then iteratively increase the fidelity level of the FEA models until they closely reassemble the physical problem at hand and produce FEA results of acceptable accuracy and confidence.

2.5 Commercial FEA Software

A large number of commercial FEA software tools are currently available. General-purpose codes support general applications, such as static, buckling, dynamic, and multiphysics. Some have strong ties with CAD through designated interfaces; some are even embedded in CAD. There are also special-purpose codes, such as LS-DYNA®, for nonlinear structural dynamic simulation for problems such as crashworthiness. In this section, a brief overview of commercially available FEA codes is presented, including their strengths and weaknesses.

2.5.1 General-Purpose Codes

There are several general-purpose FEA codes embedded in their respective CAD systems. These are SolidWorks Simulation (called COSMOSWorks® before 2009) in SolidWorks; Pro/MECHANICA Structure in Pro/ENGINEER; CATIA® FEA in CATIA V5; and FEA in NX. All provide a seamless connection between CAD and FEA without the need for a geometric translator. All user interactions are performed within the CAD environment, including pre-process, analysis, and post-process. The learning curve for these software tools is usually less steep if the user has prior experience with the respective CAD systems.

All of the codes support basic FEA problems, including static, vibration, and buckling. All are primarily h-version, except for Pro/MECHANICA, which is a p-version FEA. SolidWorks Simulation supports nonlinear and fatigue analysis. Pro/MESH, an h-FEA mesh generator in Pro/ENGINEER, supports both pre- and post-processing of commercial FEA codes. Pro/MESH automatically meshes Pro/ENGINEER parts, supports defining FEA models in Pro/ENGINEER, and supports a loose connection to major FEA codes (through ASCII input data files). Pro/ENGINEER also reads and displays FEA results. Meshing capability in Pro/MESH is somewhat limited, supporting only tetrahedral elements for 3D solid structures, and it sometimes produces undesirable meshes.

In addition to Pro/MECHANICA, another popular p-FEA is StressCheck™, probably the most theoretically sound p-FEA code, which was developed by Barna Szabó at Washington University in St. Louis; it was later commercialized by Engineering Software Research and Development, Inc. StressCheck supports 2D and 3D models, linear and nonlinear problems, and fracture mechanics problems.

Three well-known and popular FEA codes—ANSYS, MSC/Nastran, and ABAQUS—offer more engineering capabilities and more options in finite element types. ANSYS was one of the earliest FEA codes and is probably the most popular code in academia. It is h-FEA and recently began offering p-elements. ANSYS supports fairly complete analysis capabilities for solving general engineering problems. Nastran is one of the best general-purpose FEA codes, providing probably the most complete set of capabilities for general engineering problems, including linear, nonlinear, elastic, plastic, aero-structure, acoustic, transient dynamics, and others. Nastran's quadrilateral plate element is considered to be the most robust in the field. MSC/Nastran supports mainly h-elements. It supports p-elements as well, but they are not its main capability. ABAQUS is one of the best commercial FEA codes for nonlinear analysis; its nonlinear hyperelastic and plastic capabilities are among the best. Recently ABAQUS incorporated XFEM for crack propagation simulation.

In terms of CAD connections, ANSYS offers direct connections to most major CAD packages, including Pro/ENGINEER, SolidWorks, CATIA, UG, Solid Edge®, and Autodesk® Inventor®, in addition to neutral formats such as IGES. On the other hand, MSC/Nastran ABAQUS interface with MSC/Patran for pre- and post-processing, especially in importing CAD geometry. In addition to MSC/Patran, HyperMesh from Altair® Engineering provides excellent CAD interfaces as well as semiautomated mesh capabilities. It also offers excellent post-processing capabilities similar to those of Patran.

2.5.2 Specialized Codes

Specialized codes usually offer analysis capabilities geared to much focused engineering fields in addition to standard capabilities. One typical software considered by industry and academia to be the tool for crashworthiness analysis is LS-DYNA, which also offers excellent capabilities for solving nonlinear problems, including nonlinear dynamics, contact, crack propagation, and coupling problems, such as structural, fluid, thermal, and acoustics.

Another specialized code for nonlinear FEA is MSC.Marc®. Marc and Mentat combined deliver a complete solution (pre-processing, solution, and post-processing) for implicit nonlinear FEA. Marc is capable of reliably solving problems that involve changing contact conditions between components and/or large strain (plasticity or elastomeric behavior, for example), especially for rubber-related applications such as tires and seals. Mentat is the premier environment for understanding, exploring, and interacting with nonlinear analysis. It is tightly integrated with the MSC.Marc FEA program, allowing data to be defined interactively through a graphic user interface.

Another code worth mentioning is COMSOL Multiphysics®, a finite element analysis software package for various physics and engineering applications, especially coupled phenomena and multiphysics. COMSOL Multiphysics also offers an extensive interface to MATLAB® and its toolboxes for a large variety of programming, pre-processing, and post-processing possibilities.

2.6 Case Study and Tutorial Examples

Before wrapping up this chapter, we will go through a case study in which scanned anatomy images are used to create finite element models. In addition, tutorial examples, including a simple cantilever beam and a thin-shelled tube, are provided. Step-by-step instructions in creating these tutorial examples are given in the tutorial projects P3 or S3 of this book. Model files can be found on the companion website (http://booksite.elsevier.com/9780123984609).

2.6.1 Case Study

In this case study, we present a practical and systematic method for constructing an accurate finite element model for a human middle ear (Sun et al. 2002). The middle ear is the portion of the ear internal to the eardrum and external to the oval window of the cochlea, as shown in Figure 2.41. It contains three ossicles (malleus, incus, and stapes), which transform vibrations of the eardrum into waves in the fluid and membranes of the inner ear. The hollow space of the middle ear is the tympanic cavity, or *cavum tympani*. The eustachian tube joins the tympanic cavity with the nasal cavity (nasopharynx), allowing pressure to equalize between the middle ear and the throat.

Middle ear components have a tiny and complex geometry, and they play an essential role in sound transmission. Sounds collected and conveyed in the external ear canal are first transformed into mechanical vibrations of the eardrum and ossicular chain, and then transformed into traveling waves in the fluid-filled cochlea (inner ear). Devices for hearing

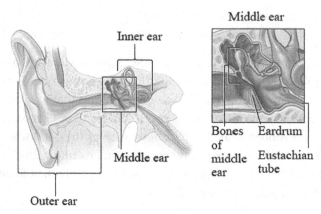

Figure 2.41: Human middle ear anatomy. *(Source: (left) http://upload.wikimedia.org/wikipedia/commons/ d/d2/Anatomy_of_the_Human_Ear.svg & (right) http://en.wikipedia.org/wiki/Ear#Middle_ear)*

restoration in cases of middle ear conductive impairments (sensorineural loss) are often mechanical in nature and thus can be studied using mechanical computational models.

From a mechanics perspective, the middle ear is essentially an impedance match between the external ear and the inner ear. Its transfer function, related to the dynamic behavior of the ossicular chain, is complex because of its anatomy. The objective of this case study was to create an accurate three-dimensional (3D) geometric model of the human middle ear that allows observation of its 3D structure for morphological studies and for developing finite element models for mechanics analysis.

The modeling process started with extracting a normal fresh human temporal bone that was fixed, decalcified, and embedded in celloidin. The resulting block was removed from the chloroform, trimmed, mounted on a microtome plate, and stored in 70% alcohol. Before the celloidin block was sectioned, four parallel fiducial holes (perpendicular to the cutting plane) were made in it with a drill press. Permanent ink was injected into the fiducial holes to stain the marks. Sections were then cut on a sliding microtome at a thickness of 20 μm, as shown in Figure 2.42a.

Next, in sequence, every tenth section was stained and mounted on glass slides. Finally, each histology slide was scanned into a computer using a flatbed scanner and saved as an image file (Figure 2.42b). The images were aligned with a template constructed by a typical section image using the fiducial marks. Aligned images were then trimmed as standard-sized images, brought onto a sketch plane in CAD software (in this case SolidWorks), and fitted in a reference rectangle of the same size as the image so that the actual size could be represented accurately. The images were then digitized by marking points along the outlines of the middle ear structures as identified by an otologic surgeon. These structures included eardrum, ossicles, attached ligaments, and muscle tendons (Figure 2.42c). The B-spline curves were constructed to best fit these points using the curve-fitting technique (Figure 2.42d) and were quilted to form the closed boundary

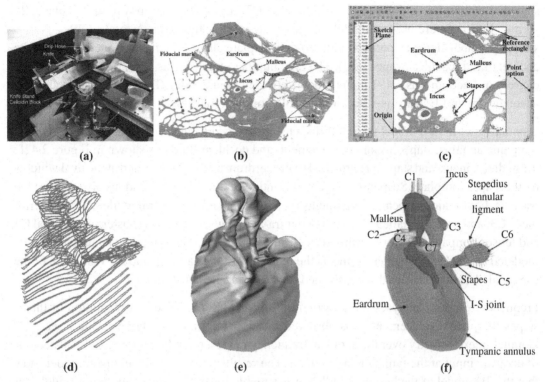

Figure 2.42: Construction of the middle ear finite element model: (a) temporal bone slicing and slide preparations; (b) sample section image with fiducial marks for alignment; (c) image digitization; (d) section curves; (e) smooth solid model; (f) finite element model. (Sun et al. 2002)

surfaces for individual middle ear components using the surface-skinning technique (Tang and Chang 2001). The solid models of the middle ear components were constructed using these surfaces and then assembled to create the entire middle ear model, as shown in Figure 2.42e.

All surfaces of the geometric model were translated into HyperMesh for finite element modeling. Based on these surfaces, the FE mesh of the middle ear components and attached ligaments/tendons was created with the HyperMesh meshing capabilities. Figure 2.42f shows the anterior-medial view of the middle ear FE model with attached ligaments/tendons and cochlear fluid constraints. A total of 1,746 three-node triangular and four-node quadrilateral shell elements were created to mesh the eardrum. Surrounding the eardrum periphery, the tympanic annulus was modeled using 113 three-node triangular and four-node quadrilateral shell elements. A total of 812 eight-node hexahedral, six-node pentahedral, and four-node tetrahedral solid elements were created to mesh three ossicles, two joints (incudomalleolar and incudostapedial joints), and six ligaments/tendons (superior mallear ligament C1, lateral mallear ligament C2, posterior incudal ligament C3, anterior mallear ligament C4, posterior stapedial tendon C5, and tensor tympani tendon C7). Lying in the stapes footplate plane, 25 spring elements were used to model the stapedius annular ligament. A total of 49 spring–dashpot elements perpendicular to the footplate plane were incorporated to model the cochlear fluid (cochlear impedance). The total number of nodes was 1,497, and the number of degrees of freedom was 4,491 in the model shown in Figure 2.42f. The size of the model was adequate for an FE analysis with a reasonable computation time.

Boundaries of the FE model included suspensory ligaments, intra-aural muscle tendons, tympanic annulus, stapedius annular ligament, and cochlear fluid. As shown in Figure 2.42f, the malleus, incus, and stapes are attached to the eardrum by the malleus and at the oval window by the stapes footplate. Suspensory ligaments and intra-aural muscle tendons also support the ossicles. Four major suspensory ligaments (superior malleus C1, lateral malleus C2, posterior incus C3, and anterior malleus C4) and two intra-aural muscle tendons (posterior stapedial C5 and tensor tympani C7) were regarded as elastic constraints. The tympanic annulus was modeled as an elastic ring that connects the periphery of the eardrum and the bony wall of the ear canal. The cochlear fluid was assumed as a viscoelastic constraint, C6.

Frequency response analyses were carried out using ANSYS. Excellent resolution of the stapes footplate displacements was observed in six human temporal bones using laser Doppler interferometry over the acoustic frequency ranges, including 80, 90, and 100 dB SPL of acoustic input at the tympanic membrane. The results from the finite element model show that the 3D model of the human middle ear is suitable for further investigation of middle ear mechanics and transfer functions.

2.6.2 Tutorial Examples

The cantilever beam and the thin-walled tube examples discussed in Section 2.4 are included in the tutorial lessons. This beam model was simply prepared for an easy start with both SolidWorks Simulation and Pro/MECHANICA Structure. Default options and values were mostly used in this example. We discuss reducing the 3D beam to a 1D model in Simulation. Note that no such conversion capability is yet available in Pro/MECHANICA. Once the reader is more familiar with one of these two software tools, he or she may move to the second tutorial example, the thin-walled tube, and other examples in the tutorial.

Cantilever Beam

A cantilever beam of 1 in. × 1 in. × 10 in. was modeled in both Simulation and Pro/MECHANICA. As discussed in Section 2.4.6, Pro/MECHANICA created 12 tetra-elements and employed a p-version solver to create results. Instead of using the default unit system in.-lb_m-sec, a more commonly employed system, in.-lb_f-sec, was chosen. The force added at the front top edge was 1000 lb_f; the boundary conditions (restrained at the rear end face) and material properties (with modulus $E = 1.06 \times 10^7$ psi and Poisson's ratio $\nu = 0.33$) remained the same. A multipass analysis with 1% strain energy was defined as the convergence criterion. The analysis took 5 passes (with maximum p-order = 5) and solved 1,017 equations to achieve a 0.8% convergence in strain energy. Essential solution-related information can be found in the Run Status window shown in Figure 2.43a. The convergence graph shown in Figure 2.43b indicates that an excellent convergence was achieved in 5 passes. The maximum von Mises stress was 5,776 psi at the constrained end, as shown in Figure 2.43c. Note that the maximum bending stress and vertical displacement were 7,353 psi and 0.03742 in., respectively (not shown in the figure). As before, the displacement result can be verified using classical beam theory. The bending stress would be close to 6,000 psi if Poisson's ratio were set to 0.

The same example was analyzed in SolidWorks Simulation. About 7,500 tetra-elements, as shown in Figure 2.44a, were created using the default mesh setting, where the global size of the element was 0.2155 in. There were about 36,000 dof in this model—about 35 times more than in the Pro/MECHANICA model. The maximum vertical displacement was 0.0382 in. (Figure 2.44b), which was very close to that of Pro/MECHANICA (0.03742 in.) and the analytical solution. The maximum bending stress shown in Figure 2.44c was 6,088 psi (with Poisson's ratio set to 0), which was also very close to the analytical solution.

The 3D beam was idealized to a 1D finite element model in Simulation, as shown in Figure 2.45a. The same load and boundary conditions were applied at respective ends where joints were created when the solid beam was converted to a 1D beam. The auto mesh generator created 46 beam elements (Figure 2.45b) with 276 dof (as opposed to the 36,000 employed in the 3D model of Simulation). Note that for this simple example, one beam element gave exact solutions since the exact solution of the beam displacement was a cubic function and cubic shape functions were employed in element formulation. The bending stress and maximum

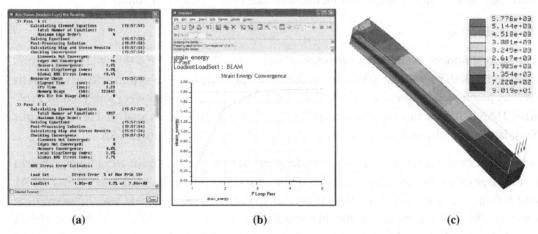

(a)   (b)   (c)

Figure 2.43: Simple beam modeled with Pro/MECHANICA: (a) the Run Status window, (b) the strain energy convergence graph, and (c) the von Mises stress fringe plot.

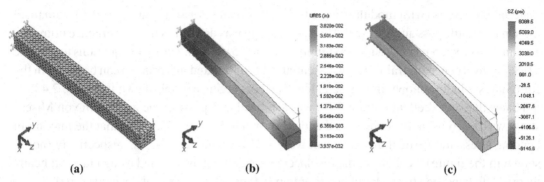

(a)   **(b)**   **(c)**

Figure 2.44: Simple beam modeled with SolidWorks Simulation: (a) the finite element mesh, (b) the von Mises stress fringe plot, and (c) the displacement Ux (vertical) fringe plot.

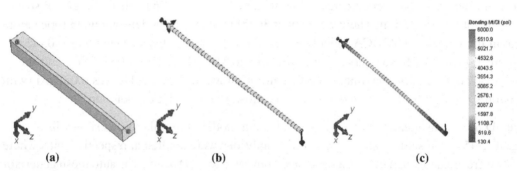

(a)   **(b)**   **(c)**

Figure 2.45: One-dimensional beam modeled in SolidWorks Simulation: (a) the beam with two end joints, (b) the finite element mesh with load and boundary conditions, and (c) the bending stress fringe plot.

vertical displacement were 6,000 psi (Figure 2.45c) and 0.03778 in., respectively. These results were excellent. Bending stress was identical to that of the analytical solution.

Thin-Walled Tube

The second tutorial example was mentioned in Section 2.4.2, where a thin-walled tube was analyzed in both its original 3D form and in half, taking advantage of symmetric conditions as well as a compressed surface model. Both solid and surface models were analyzed in Pro/MECHANICA Structure. In this section, we discuss the same models analyzed in SolidWorks Simulation. Step-by-step instructions can be found in the tutorial project S3.

A half-tube model was created in SolidWorks in the SI unit system. The outer radius and height of the large tube (vertical) were 51 and 240 mm, respectively; for the small tube, they were 41 and 100 mm, respectively. The fillet radius was 15 mm, and the thickness was 2 mm over the entire model. The bottom face of the tube was fixed just as in the Pro/MECHANICA model. A symmetry boundary condition was imposed on the faces of the middle section where the model was split in half. A 500-N load was applied at the front face of the small tube, as shown in

Figure 2.46a. The default mesh setting was chosen (with a global size of 4.383 mm), yielding 14,433 tetra-elements (Figure 2.46b). The analysis results shown in Figures 2.46c and 2.46d indicate that the maximum von Mises stress and displacement magnitude were 46.7 MPa and 0.421 mm, respectively—very close to those of Pro/MECHANICA.

An attempt was made to insert mid-surfaces between respective pairs of the outer and inner surfaces to create the surface model, similar to what was done in Pro/MECHANICA Structure. Two mid-surfaces, compressed from the outer and inner surfaces of the large and small tubes, were created successfully. However, SolidWorks failed to create the mid-surface for the fillet because the inner face of the fillet pair was not the exact offset of the outer face.

A surface model, shown in Figure 2.47a, was then created from scratch. There were about 2,200 triangular elements, as shown in Figure 2.47b, with about 27,000 dof using a mesh setting with a global element size of 6.153 mm. Note that this model, in terms of degrees of freedom (dof), was about 10 times larger than the Pro/MECHANICA model. The maximum displacement was 0.6937 mm (Figure 2.47c), which was larger than that of Pro/MECHANICA (0.42 mm). The maximum von Mises stress, shown in Figure 2.4.7d, was 55.2 MPa, also larger than that of Pro/MECHANICA (42 MPa). Note that a convergent study may be necessary to verify the accuracy of the FEA results obtained from SolidWorks Simulation. Serious design engineers may use FEA codes, such as ANSYS, MSC/Nastran, or ABAQUS to acquire more dependable FEA results.

2.7 Summary

In this chapter, we discussed methods for structural analysis, both analytical and numerical. Analytical methods often support structural analysis of problems with simple geometry. Usually solutions can be obtained in closed form and solved by hand calculation. However, these methods are not general enough to support problems encountered in engineering design.

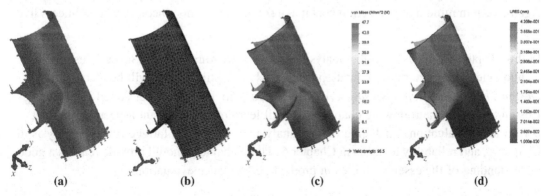

Figure 2.46: Thin-walled tubing modeled with SolidWorks Simulation: (a) the solid model with load and boundary conditions, (b) the finite element mesh, (c) the von Mises stress fringe in deformed shape, and (d) the displacement magnitude fringe plot.

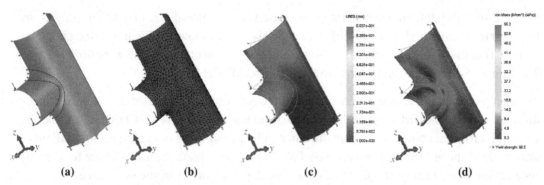

Figure 2.47: Tube surface model in SolidWorks Simulation: (a) the surface model with load and boundary conditions, (b) the finite element mesh, (c) the displacement magnitude fringe plot, and (d) the von Mises stress fringe in deformed shape.

Realistically, design engineers must depend on FEA for making design decisions that ensure structural integrity and satisfy performance requirements. Although analytical methods are not general, they are indispensable for mechanical engineers to understand the fundamentals in structural analysis and to develop references for verifying FEA results.

We devoted much discussion to finite element methods. We focused our discussion on the essential aspects of using FEA software for modeling and analysis from a practical perspective. We hope this chapter helps readers become more familiar with FEA. Practice should lead to more confidence and competence in using FEA tools for creating adequate models and obtaining reasonable results to support product design.

Although FEA software is powerful, as discussed in this chapter, it is also dangerous. Any mistakes made in modeling, analysis, or even results interpretation can lead to a wrong design decision. FEA users must be very careful in using the software for solving structural problems, learning how to check or verify FEA results before presenting them to others. Again, keep in mind that good engineers make sound judgments; incompetent engineers live with fuzzy guesses.

In this chapter, we focused on the analysis aspect of structural design. Several design technologies for structural optimization, such as shape optimization, will be discussed in *Design Theory and Methods using CAD/CAE*, a book in The Computer Aided Engineering Design Series. In the following chapters, we will learn other important aspects of product performance evaluations, including motion analysis in Chapter 3, fatigue fracture analysis in Chapter 4, and reliability analysis in Chapter 5. These chapters should provide readers a good understanding of the essential topics in product performance evaluation.

Questions and Exercises

2.1. Calculate the maximum stress and its location for the inclined cantilever beam in Example 2.1 using an analytical method.

2.2. Calculate the rotation angle at the tip of the inclined cantilever beam in Example 2.1 using an analytical method.

2.3. Calculate the displacement functions of the cantilever beam example in Example 2.4 using Eqs. 2.3 and 2.5.

2.4. Repeat Example 2.4 using the following stress functions: $\sigma_x = C_1 x^2 y$ and $\sigma_x = C_1 x^2$. Do any of the two functions provide correct stress results? Is one better than the other?

2.5. The simple pin-connected bars in the figure below are loaded with a force P. If all bars are of equal rigidity EA, what are the horizontal and vertical displacements at Point D? Hint: use Castigliano's theorem

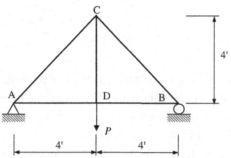

2.6. Create an FEA model to calculate the maximum displacement and stress for the curve beam shown in Example 2.3. The dimensions of the beam are $R = 10$ in. and $d = 1$ in.; the material is steel; $E = 30 \times 10^6$ psi; $F = 10$ lb (see figure below).

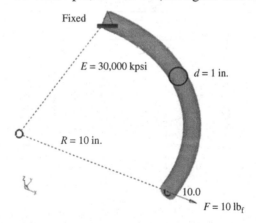

2.7. An L-shaped circular bar of diameter 1.25 in. is loaded with an evenly distributed force of 400 lbs at its end face (see figure on the next page). Note that the elbow radius (the corner of the L-shape) is 1 in. and the material is 1060 alloy.

(a) Create an FEA model using FEA software to calculate the maximum principal stress and the maximum shear stress in the bar.

(b) Calculate the maximum principal stress and maximum shear stress using analytical beam theory. Where are the maximum stresses located? Compare your calculations with those obtained from FEA, both values and locations. Are they close? Why or why not? Comment on your comparison.

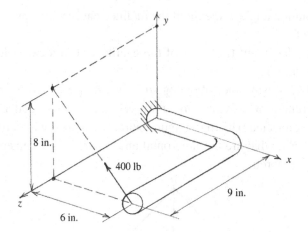

2.8. Create and analyze stress for a thin-shelled model of thickness 2 mm (see figure below) using FEA. Assume 1060 steel as the material. Refine the mesh around high-stress areas, and show a screen capture for the mesh and axial stress. Calculate the maximum stresses at both Sections A and B using the stress concentration theory that you learned in a course on strength of materials or design of mechanical components. Check your calculations with those obtained from FEA. Do they agree? Why or why not? Comment on your conclusion.

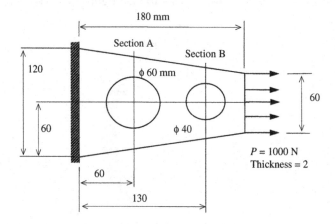

2.9. **Mini-Project**: The figures that follow show a tracked-vehicle roadarm that connects the roadwheel to the torsion bar of the chassis. The geometric shape of the roadarm is shown in the figure and its cross-sectional dimensions are given in the table. Displacement constraints are defined at the inner surface of the hole where the torsion bar is attached. A point load is applied at a point shown in the second figure that simulates the shaft of the roadwheel. Note that the force consists of two components:

$F_{x3} = 0.8F$ and $F_{x2} = -0.6F$. The roadarm is made of AISI 1030 HR steel, with material properties of Young's modulus $E = 3.0 \times 10^7$ psi and Poisson's ratio $\nu = 0.3$.

(a) If the safety factor is 2, calculate the allowable force F using distortion energy theory.

(b) Locate the hot spot where maximum stress occurs. Draw a stress element at the spot identified and show all stress components.

(c) If the actual force applied to the roadarm is greater than F as calculated in (a), suggest a design change from the given 8 cross-sectional dimensions that will reduce the highest stress most effectively.

Note that you will have to make assumptions in order to simplify the problem for a possible solution. You will need to write down all of your assumptions and justify them with reasoning based on engineering judgment. In (a) and (b), you will need to give a quantitative estimate on the possible error in your calculations resulting from these assumptions.

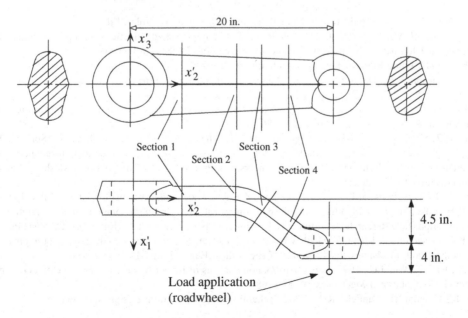

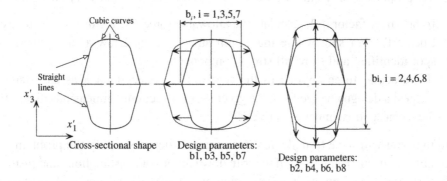

Geometric dimension	Description	Initial design
b1	Width of cross-section 1	1.968 in.
b2	Height of cross-section 1	3.250 in.
b3	Width of cross-section 2	1.968 in.
b4	Height of cross-section 2	3.170 in.
b5	Width of cross-section 3	2.635 in.
b6	Height of cross-section 3	3.170 in.
b7	Width of cross-section 4	2.635 in.
b8	Height of cross-section 4	3.170 in.

References

Atkinson, K.E., 1989. An Introduction to Numerical Analysis, second ed. John Wiley.

Babuška, I., Griebel, M., Pitkaranta, J., 1989. Problem of selecting the shape functions for a p-Type finite element. International Journal for Numerical Methods in Engineering 28 (8), 1891−1908.

Bathe, K.J., 2007. Finite Element Procedures. Prentice Hall, 1996.

Beer, F.P., Johnston, E.R., DeWolf, J.T., 2002. Mechanics of Materials, third ed. McGraw-Hill.

Belytschko, T., Chen, J.S., 2007. Meshfree and Particle Methods. John Wiley.

Belytschko, T., Lu, Y.Y., Gu, L., 1994. "Element free Galerkin methods". International Journal for Numerical Methods in Engineering vol. 37, pp. 229−256.

Bendsoe, M.P., Sigmund, O., 2003. Topology Optimization, Theory, Methods, and Applications. Springer-Verlag.

Benzley, S.E., Perry, E., Merkley, K., Clark, B., 1995. A comparison of all hexagonal and all tetrahedral finite element meshes for elastic and elasto-plastic analysis, In Proceedings, 4th International Meshing Roundtable. Sandia National Laboratories.

Brauer, J.R. (Ed.), 1993. What Every Engineer Should Know About Finite Element Analysis. Marcel Dekker.

Chang, K.H., Edke, M., 2010. Manufacturing in shape optimization of structural components, Computational Optimization: New Research Developments. In: Linton, R.F., Carroll Jr., T.B. (Eds.), Nova Science Publishers.

Chang, K.H., Magdum, S., Khera, S., Goel, V.K., 2003. An advanced computer modeling and prototyping method for human tooth mechanics study. Annals of Biomedical Engineering 31 (5), 621−631.

Chen, J.S., Pan, C., Wu, T.C., 1997. Large deformation analysis of rubber based on a reproducing kernel particle method. Computational Mechanics 19, 53−168.

Choi, S.-K., Grandhi, R., Canfield, R.A., 2007. Reliability-Based Structural Design Springer.

Edke, M. and Chang, K.H.,"Shape Sensitivity Analysis For 2-D Mixed Mode Fractures Using Extended FEM (XFEM) And Level Set Method (LSM)," Mechanics Based Design of Structures and Machines, 38(03), pp. 328–347, 2010

Finney R.H., Finite element analysis. In: Gent, A.N. (Ed.), Elasticity in Engineering with Rubber: How to Design Rubber Components, second ed. Alan N. Gent. Hanser Verlag. 2001.

Grindeanu I., Chang K.H., Choi K.K., Chen J.S., Design sensitivity analysis of hyperelastic structures using a meshless method, AIAA Journal 36 (4): 618–627, 1998.

Hamrock, B.J., Schmid, B., Jacobson, O., 2005. Fundamentals of Machine Elements, second ed. McGraw-Hill.

Hwang, H.Y., Choi, K.K., Chang, K.H., 1997. Second-order shape design sensitivity analysis using a p-version finite element tool. Journal of Structural Optimization 14, 91–99.

Juvinall, R.C., Marshek, K.M., 2000. Fundamentals of Machine Component Design, third ed. John Wiley.

Li, S., Liu, W.K., 2004. Meshfree Particle Methods. Springer Verlag.

Middleton, J., Jones, M.L., Wilson, A.N., 1990. Three-dimensional analysis of orthodontic tooth movement. Journal of Biomedical Engineering 12, 319–327.

Moës, N., Dolbow, J., Belytschko, T., 1999. A finite element method for crack growth without remeshing. International Journal for Numerical Methods in Engineering 46 (1), 131–150.

Monaghan, J.J., 1988. An introduction to SPH. Computer Physics Communications 48, 88–96.

Nayroles, B., Touzot, G., Villon, P., 1992. Generalizing the finite element method: diffuse approximation and diffuse elements. Computational Mechanics 10, 301–318.

Norton, R.L., 2011. Machine Design: An Integrated Approach, fourth ed. Prentice Hall.

Pilkey, W.D., Wunderlich, W., 2002. Mechanics of Structures: Variational and Computational Methods, second ed. CRC Press.

Quadros, W.R., Ramaswami, K., Prinz, F.B., Gurumoorthy, B., LayTracks: a new approach to automated geometry adaptive quadrilateral mesh generation using medial axis transform, International Journal for Numerical Methods in Engineering, 09/2004; 61(2): 209–237.

Szabó, B., Babuška, I., 1991. Finite Element Analysis. Wiley-Interscience.

Stokes, H., 1995. Solid modeling and the initial graphics exchange specification (IGES). In: LaCourse, D.E. (Ed.), Handbook of Solid Modeling. McGraw-Hill.

Strang, G., Fix, G., 1973. An Analysis of the Finite Element Method. Prentice Hall.

Sun, Q., Chang, K.H., Dormer, K., Dyer, R., Gan, R.Z., 2002. An advanced computer-aided geometric modeling and fabrication method for the human middle ear. Medical Engineering and Physics 24 (9), 595–606.

Tang, P-S. and Chang, K.H., Integration of Topology and Shape Optimizations for Design of Structural components, Journal of Structural Optimization, vol. 22 (1), pp. 65–82, 2001.

Wilson, E.L., 2000. Three Dimensional Static and Dynamic Analysis of Structures: A Physical Approach with Emphasis on Earthquake Engineering. Computers and Structures, Inc.

Zienkiewicz, O.C., Taylor, R.L., 1989. The Finite Element Method. McGraw-Hill.

Sources

Abaqus, CATIA: www.3ds.com

ANSYS: www.ansys.com

COMSOL Multiphysics: www.comsol.com

HyperMesh: www.altairhyperworks.com

MATLAB: www.mathworks.com

MSC/Nastran, MSC/Patran: www.mscsoftware.com

Pro/ENGINEER, Pro/MECHANICA, Pro/MFG, Pro/SHEETMETAL, Pro/WELDING, etc.: www.ptc.com

SolidWorks Motion, SolidWorks Simulation: www.solidworks.com

Solid Edge, UG: www.plm.automation.siemens.com

StressCheck: www.esrd.com

Appendix: The Default in.-lb$_m$-sec Unit System

The default unit system employed by Pro/ENGINEER, and therefore the Pro/MECHANICA Structure is in.-lb$_m$-sec (inch-pound mass-second). This system is not common to many engineers. The basic physical quantities involved in determining a unit system are length, time, mass, and force. These four basic quantities are related through Newton's second law:

$$F = ma \qquad (A.1)$$

where F, m, and a are force, mass, and acceleration (length per second squared), respectively.

In the default in.-lb$_m$-sec system, the force unit is determined by length (in.), mass (lb$_m$), and second (sec) through

$$1 \text{ lb}_m \text{ in./sec}^2 (\text{force}) = 1 \text{ lb}_m (\text{mass}) \times 1 \text{ in./sec}^2 (\text{acceleration}) \qquad (A.2)$$

where the force unit, lb$_m$ in./sec^2, is a derived unit.

From Eq. A.2, a force of 1 lb$_m$ in./sec^2 generates an acceleration of 1 in./sec^2 when applied to a 1 lb$_m$ mass block, as shown in Figure A.1a. The same block weighs 1 lb$_f$ on earth (see Figure A.1b), where the gravitational acceleration is assumed to be 386 in./sec^2; that is

$$1 \text{ lb}_f (\text{force}) = 1 \text{ lb}_m (\text{mass}) \times 386 \text{ in./sec}^2 (\text{acceleration}) \qquad (A.3)$$

Therefore, from Eqs. A.2 and A.3, we have 1 lb$_f$ = 386 lb$_m$ in./sec^2. In other words, the force quantity entered into Pro/MECHANICA Structure is in lb$_m$ in./sec^2 by default, which is 386 times smaller than the 1 lb$_f$ that we are more used to. When a 1 lb$_f$ force is applied to the same mass block, it accelerates 386 in./sec^2, as shown in figure A.1c. Therefore, care is essential in

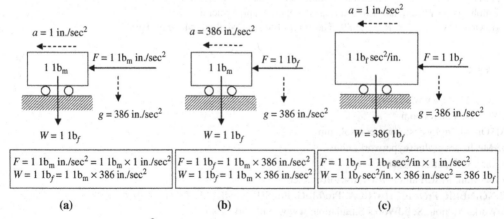

Figure A.1: (a) A 1 lb$_m$ in./sec^2 force applied to a 1 lb$_m$ mass block, (b) a 1 lb$_f$ force applied to a 1 lb$_m$ mass block, and (c) a 1 lb$_f$ force applied to a 1 lb$_f$ sec^2/in. mass block.

entering numerical figures when defining analysis models. For example, if a 1,000 unit force is applied to a mechanical component in the default unit system, it is in fact $1,000/386 = 2.59$ lb_f, which is very small.

On the other hand, the mass unit, 1 $lb_m = 1/386$ lb_f $sec^2/in.$, means that a 1 lb_m mass block is 386 times smaller than a 1 lb_f sec^2/in. Therefore, block. a 1 lb_f sec^2/in. block weighs 386 lb_f on earth. A 1 lb_f force applied to the mass block accelerates at a 1 $in./sec^2$ rate, as illustrated in Figure A.1c.

Motion Analysis

Chapter Outline

Many products or mechanical systems involve moving parts. Parts must move in a certain way to perform required functionality and achieve desired performance. Components must not collide or interfere with each other during normal operations. Also, the system must be efficient (usually lightweight) yet durable. Essentially, the product must be well designed, and designers must understand the kinematic and dynamic behavior of the system and be sure beforehand that the components of the mechanical system move according to the design intent. Discovering problems after the product is manufactured means it is usually too late and too costly to correct them. Motion analysis offers viable alternatives to support designers in simulating and analyzing the movement of parts in mechanical systems in the early design stages.

Traditionally, motion analysis has been divided into two major categories: kinematics and dynamics. Kinematics is the study of motion without regard to the forces that cause it. Dynamics, on the other hand, is the study of motion that results from forces. Except for simple ones, such as particle dynamics, very few of these problems can be solved by hand using analytical equations. The majority must be solved using numerical methods. Also, most problems must be formulated in a certain way so that they can be generalized and handled by computer software—that is, so-called computer-aided kinematic and dynamic analyses. Although computer-aided approaches and software tools are powerful, understanding analytical methods is crucial. Their underlying physics helps us understand how mechanisms behave in certain ways in a given motion scenario. Understanding the underlying physics also helps us make good modeling decisions and choose adequate simulation parameters that lead to faster and more reliable simulations.

Furthermore, it is critical that we verify numerical results obtained from computer tools whenever possible. Very often, analytical solutions for complex problems are not available. However, it is good practice, especially for new users, to start with something simple so that the solutions obtained from the motion analysis software can be verified analytically. The bottom line is that we must have adequate knowledge to use the computer tools correctly and effectively.

In addition to analytical methods, we devote most of the discussion in this chapter to computer-aided approaches for kinematic and dynamic analyses. Note that our focus is not on computational theory but rather on the essential elements for motion modeling and analysis from a practical perspective. The goal of this chapter is to help the reader become confident and competent in using motion software tools for creating adequate models and obtaining reasonable results to support design. Those who are interested in kinematics and dynamics theory can refer to other excellent books, such as *Computer Aided Kinematics and Dynamics of Mechanical Systems*, by Haug, or *Dynamics: Theory and Applications*, by Kane and Levinson. (Both are listed in the references at the end of the chapter.) Major motion simulation software packages, both general-purpose and specialized, are presented to provide a general understanding of tool availability and sufficient information to make adequate choices.

Several practical examples are introduced. Two vehicle examples—a Formula SAE racecar and a high-mobility multipurpose wheeled vehicle (HMMWV)—illustrate vehicle suspension simulation and design. Driving simulators that employ real-time simulation for vehicle dynamics illustrate some of the most advanced motion simulation applications. Also, a theme park water slide application shows the variety of engineering problems that motion analysis supports. Finally, two tutorial examples deal respectively with a sliding block and a single-piston engine using both Pro/MECHANICAL Mechanism Design and SolidWorks® Motion. These practice examples can also be found in the tutorial lessons. Detailed instructions for bringing up these models and the steps for carrying out motion simulations can be found in Projects P2 and S2 of this book. Example models are available for download at the book's companion website (http://booksite.elsevier.com/9780123984609).

Overall, the objectives of this chapter are the following:

- To provide basic kinematics and dynamics theory using simple examples to further understanding of the underlying computation methods for motion analysis.
- To familiarize readers with motion modeling and analysis for effective use of motion analysis tools to support product design.
- To familiarize readers with existing commercial motion analysis software.
- To instruct readers in the use of either Pro/MECHANICAL Mechanism Design or SolidWorks Motion for basic applications, with the help of the tutorial lessons.

3.1 Introduction

Motion analysis comprises the physical laws and mathematics required to study and predict the performance and behavior of mechanical systems. When a product design is first created in a CAD environment, it is unclear if the system will behave as intended, if the components will move in the designated manner, or collide or interfere. Also uncertain is the system's functionality and overall performance. Certainly these questions can be answered by a functional prototype of the product. However, since it is usually too late and too costly to correct design problems by then, it is crucial that design engineers understand how the system behaves and are able to predict if the design meets the functional requirements. Instead of building a functional prototype, motion analysis offers a viable alternative to answer some of these design questions with accuracy that is adequate to support design decision making. In practice, motion analysis supports engineering design by simulating the kinematic and dynamic performance of the product, and it proves the soundness of a design without dependence on physical testing.

A broad range of products and mechanical systems involve moving parts. Some of these products, as shown in Figure 3.1, include mechatronics devices that we use on a daily

FIGURE 3.1: Complex mechanical systems: (a) mechatronic device—hard disk drive (*Source: http:// upload.wikimedia.org/wikipedia/commons/1/1e/Hard_disk_dismantled.jpg*); (b) heavy equipment—backhoe (*Source: http://www.docstoc.com/docs/85268831/Transportation-backhoe*); (c) ground vehicle—HMMWV (*Source: http://upload.wikimedia.org/wikipedia/commons/7/72/Iraqi_ Humvees.jpg*); (d) theme park ride—waterslide.

basis such as computer hard disk drives, heavy construction equipment (e.g., backhoes), ground vehicles (e.g., the Army's high-mobility multipurpose wheeled vehicle), and theme park rides such as a waterslide. Design objectives and functional requirements vary for different products. For a hard disk drive, increasing the spindle speed, for example, from 7,200 rpm to 10,000 rpm, significantly increases the data-accessing speed in our computers. For a backhoe, power and torque must be sufficient to generate adequate digging forces while minimizing the impact force and the mechanical vibration felt by the operator. In the case of the HMMWV, maneuverability is critical since the vehicle is intended to be driven on rough roads. As for the waterslides, safety is always number one. Besides functionality, structural integrity must be ensured. Reaction forces between moving parts may cause part failures due to yield, buckle, or fatigue.

Most products involve multiple moving parts. From a motion analysis perspective, such products are often referred to as multibody systems, which have to be modeled or formulated by incorporating mass and inertia properties of individual parts as well as the connections (also called joints) between parts. In most cases, these parts are assumed rigid and so are referred to as rigid-body systems. In some cases, certain parts have to be considered flexible, such as the arm of the read/write head in a hard drive, which deforms substantially during motion. The arm that carries the read/write head flies over the surface of the disk and must reach precise locations in a very short time. The arm deforms mainly as a result of inertia. Position and velocity of the read/write head must be predicted in high precision, considering the deformation. However, not all motion problems are complex. Some applications can be simplified to particle dynamics. For example, the rider on the waterslide can be assumed a particle with concentrated mass when the engineer is carrying out motion simulation for safety analysis. Certainly for a more detailed simulation the rider itself must be modeled as a multibody system.

The complexity of a motion simulation model varies. An HMMWV suspension model can have as little as 14 bodies. A body is an entity in a motion analysis model and can be stationary (as a ground body) or in motion; however, no relative motion is allowed between parts within a body. The number of bodies in a mechanical system can rise easily, which increases the size of the motion model and therefore requires more computational time to solve the problem. A motion model can be simple and also very complex, depending on the level of fidelity retained in it. Determination of the level of model fidelity is often driven by the questions we want the motion simulation to answer.

Some questions require knowledge of position, velocity, and acceleration of individual parts in the system, and they can often be answered by conducting a kinematic analysis. Kinematic analysis provides the physical positions, velocity, and acceleration of all parts in an assembly with respect to time, without consideration of the forces leading to the motion. This analysis is also useful for evaluating motion for interference of complex mechanical systems. On the other hand, questions involving forces must be answered by dynamic analysis, which is the study of motion in response to externally applied loads. This analysis is necessary when torque or force is involved in motion analysis—for example, the torque required to generate sufficient digging force for a backhoe. In those cases, the dynamic behavior of a mechanism is governed by Newton's laws of motion.

Even though design questions vary and they often must be determined for specific applications of certain fidelity levels, there are a few basic questions that are common and typical for all motion applications. A single-piston engine, shown in Figure 3.2, is used to illustrate some typical questions:

- Will the components of the mechanism collide or interfere in operation? For example, will the connecting rod collide with the inner surface of the piston or the inner surface of the engine case during operation?
- Will the components in the mechanism move according to our intent? For example, will the piston stay entirely in the piston sleeve? Will the system lock up when the firing force aligns vertically with the connecting rod?

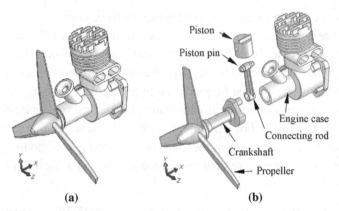

FIGURE 3.2: A single-piston engine: (a) unexploded view and (b) exploded view.

- How much torque or force does it take to drive the mechanism? For example, what is the minimum firing load to move the piston? Note that in this case, proper friction forces must be added to simulate the resistance between moving parts before a realistic firing force can be calculated.
- How fast will the components move (e.g., what is the linear velocity of the piston)?
- What is the reaction force or torque generated at a connection between components (or bodies) during motion? For example, what is the reaction force at the joint between the connecting rod and the piston pin? This reaction force is critical since the structural integrity of the piston pin and the connecting rod must be ensured—that is, they must be strong and durable enough to sustain the load in operation.

Motion simulation helps us answer these common questions accurately and realistically, as long as the motion model is properly created. Some motion analysis tools also help us search for better design alternatives. A better design alternative is very much problem dependent. It is critical that a design problem be clearly defined by the designer up front before searching for better design alternatives. For the engine example, a better alternative can be a design that meets the functional requirements and reveals, for example,

- A smaller reaction force applied to the connecting rod
- No collisions or interference between components

A common approach to searching for design alternatives is to vary the component sizes of the mechanical system. Vary component sizes to explore better design alternatives, the parts and assembly must be properly parameterized to capture design intents. At the parts level, design parameterization implies creating solid features and relating dimensions so that when a dimension value is changed the part can be properly rebuilt. At the assembly level, design parameterization involves defining assembly mates and relating dimensions across parts. When an assembly is fully parameterized, a change in dimension value can be automatically

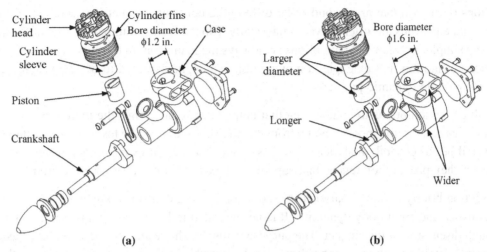

FIGURE 3.3: Single-piston engine—exploded view: (a) bore diameter 1.2 in. and (b) bore diameter 1.6 in.

propagated to all parts affected. Parts affected must be rebuilt successfully; at the same time, they must maintain proper position and orientation with respect to one another without violating any assembly mates or revealing part penetration or excessive gaps. For example, in the single-piston engine (see Figure 3.3), a change in the bore diameter of the engine case alters not only the geometry of the engine case itself but also all other associated parts, such as the piston, the piston sleeve, and even the crankshaft, as illustrated. Moreover, all parts have to be properly rebuilt and the entire assembly must stay intact through assembly mates.

In this chapter, we start by briefly reviewing analytical methods and then computer-aided approaches for kinematic and dynamic analyses. Essential elements in using computer-aided software for modeling and analysis of motion problems, such as joints, force elements, and the like, are discussed in detail. We touch a bit on the topic of real-time simulation, which realizes the vision of operator-in-the-loop simulation and which has led to several driving simulators. We also discuss commercial motion software, both general-purpose and specialized. Example problems modeled and solved using some of the commercial codes are presented as well.

3.2 Analytical Methods

Analytical methods employ analytical formulations to solve motion problems. They are applicable to simple problems, such as particle dynamics, kinematics, and single-body dynamics. Even though the formulations are based on physical laws and principles, they usually have to be derived and solved for individual problems. The formulations lead to equations of motion that are second-order differential equations, sometimes coupled. These differential equations are often solved numerically. Closed-form solutions are available only for some simpler problems. Assumptions are always made up front in formulating or deriving equations of motion. For example, a moving object must be assumed to be of concentrated mass for particle dynamic problem.

As with any assumptions in engineering that help simplify problems, the more the assumptions of a motion model deviate from the physical problem, the less useful the analysis result will be. In general, analytical solutions cannot directly support product design. However, they provide references that support complex motion problem verification.

This section briefly reviews analytical methods for particle dynamics, kinematics of mechanisms, and rigid-body dynamics. It is not intended to be an in-depth presentation and thorough discussion of the subject. The purpose is just for the reader to understand the basics that support problem solving using software tools. Both Newton's law and the energy method are discussed for simple examples. Analytical methods can only go so far. General multibody dynamics analysis must rely on computer-aided methods and computer software for solutions.

3.2.1 Particle Motion

Particle motion is the simplest motion problem to solve. In general, moving objects are assumed to be particles with concentrated mass. A particle moves in only three directions (three translational degrees of freedom). No rotation is considered.

There are two main approaches for deriving equations of motion for particle motion problems: Newton's law and the energy method. Newton's law is straightforward; however, it is limited to very simple problems. A free-body diagram is usually drawn for the particle in motion; then Newton's second law, $F = ma$, is employed to create an equation of motion as a differential equation. In terms of the energy method, kinetic energy and potential energy of the moving particle are first stated. Either the energy conservation law or the more advanced Lagrange equations of motion can be employed to derive differential equations that govern particle motion.

A simple pendulum example is discussed next to show how the equations of motion can be derived using Newton's method, and how the differential equations can be solved analytically.

Example 3.1

A simple pendulum shown in the figure is released from a position slightly off the vertical line. Once released, it rotates freely due to gravity. Calculate the position, velocity, and acceleration of the pendulum.

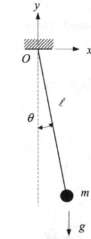

Simple pendulum

Solution

There are four assumptions that we have to make to apply particle dynamics theory to this simple problem:

- The mass of the rod is negligible.
- The sphere is of a concentrated mass.
- The rotation angle is small.
- No friction is present.

From the free-body diagram shown, the moment equation at the origin (point O) about the z-axis (normal to the paper) can be written as

$$\sum M = -mg\ell \ \sin \theta = I\ddot{\theta} = m\ell^2\ddot{\theta} \tag{3.1}$$

Thus,

$$\ddot{\theta} + \frac{g}{\ell} \sin \theta = 0 \tag{3.2}$$

and

$$\ddot{\theta} + \frac{g}{\ell}\theta = 0 \tag{3.3}$$

when $\theta \approx 0$.

It is well known that the solution of the differential equation is

$$\theta = A_1 \cos \omega_n t + A_2 \sin \omega_n t \tag{3.4}$$

where $\omega_n = \sqrt{\dfrac{g}{\ell}}$, and A_1 and A_2 are constants to be determined by initial conditions.

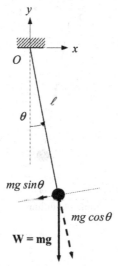

Free-body diagram

The initial conditions for the pendulum are $\theta(0) = \theta_0$ deg., and $\dot{\theta}(0) = 0$ deg./sec. Plugging the initial conditions into the solution, we obtain $A_1 = \theta_0$ and $A_2 = 0$. Therefore, the solutions are

$$\theta = \theta_0 \cos \omega_n t \tag{3.5a}$$

$$\dot{\theta} = -\theta_0 \omega_n \sin \omega_n t \tag{3.5b}$$

$$\ddot{\theta} = -\theta_0 \omega_n^2 \cos \omega_n t \tag{3.5c}$$

Equations (3.5a) through (3.5c) represent angular position, velocity, and acceleration of the pendulum or, more precisely, the revolute joint at point O that allows the rotation motion of the rod and sphere.

Note that the four assumptions stated up front must be incorporated into the motion simulation models if the solutions obtained in the example are to be employed to verify the motion analysis using computer tools. In general, the simulation model and the conditions of the analytical solution must be consistent when the solutions are employed for the verification of the simulation model.

For simple problems, such as the pendulum in Example 3.1, Newton's method works well. However, it is somewhat limited. The energy method is often more powerful in formulating equations of motion for more complex problems. One of the simplest forms of the energy method is the conservation of mechanical energy, which states that the total mechanical energy, which is the sum of the kinetic energy T and the potential energy U, is a constant with respect to time:

$$\frac{d}{dt}(T + U) = 0 \tag{3.6}$$

For the simple pendulum example, the kinetic and potential energy are, respectively, $T = \frac{1}{2}J\dot{\theta}^2$, where J is the mass moment of inertia; that is, $J = m\ell^2$ and $U = mg\ell(1 - \cos\theta)$. Hence, from Eq. 3.6,

$$\frac{d}{dt}\left(\frac{1}{2}m\ell^2\dot{\theta}^2 + mg\ell(1 - \cos\theta)\right) = m\ell^2\ddot{\theta} + mg\ell\sin\theta = 0 \tag{3.7}$$

Therefore,

$$\ddot{\theta} + \frac{g}{\ell}\sin\theta = 0 \quad \text{and} \quad \ddot{\theta} + \frac{g}{\ell}\theta = 0 \quad \text{where } \theta \approx 0$$

which is identical to Eqs. 3.2 and 3.3. Note that the same equation of motion has been derived with two different approaches.

A more general form of the energy method is based on Lagrange's equations. In a dynamic system with n degrees of freedom it is usually possible to choose n geometric quantities, $q = [q_1, q_2, \dots, q_n]^T$, which uniquely specify the position of all moving components in the system. These quantities are known as generalized coordinates. For example, in the case of the simple pendulum the position of the mass is completely determined by the angle θ. The angle θ is a generalized coordinate of the simple pendulum problem. Lagrange's equation of motion based on Hamilton's principle for particle dynamics (Greenwood 1987) can be stated as

$$\frac{d}{dt}\left(\frac{\partial L}{\partial \dot{q}}\right) - \frac{\partial L}{\partial q} = Q \tag{3.8}$$

where the Lagrangian L is defined as $L \equiv T - U$, $\dot{q} = \partial q/\partial t$. When the system is conservative, $Q = 0$. For a nonconservative system (for example, if friction is present), $Q = F$, where F is the vector of generalized forces.

For the simple pendulum example, $q = [\theta]$, $L = T - U = \frac{1}{2}m\ell^2\dot{\theta}^2 - mg\ell(1 - \cos\theta)$, and $Q = 0$. Thus, using Eq. 3.8, we have

$$\frac{d}{dt}(m\ell^2\dot{\theta}) - (-mg\ell\sin\theta) = m\ell^2\ddot{\theta} + mg\ell\sin\theta = 0 \tag{3.9}$$

Therefore, $\ddot{\theta} + \frac{g}{\ell}\sin\theta = 0$, which is the same as Eq. 3.2 (and Eq. 3.3 if $\theta \approx 0$).

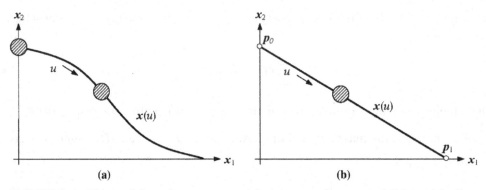

FIGURE 3.4: Object sliding along a curve $x(u)$: (a) general curve and (b) straight line.

Now consider a particle sliding along a slope defined by a parametric curve $x(u)$, shown in Figure 3.4a, with no friction. Note that the curve can be planar or spatial. The position, velocity, and acceleration of the particle can be obtained by solving the equation of motion derived from the Lagrange equation. The kinetic energy and potential energy are, respectively, $T = \frac{1}{2}m\dot{x}^2 = \frac{1}{2}m(x_{,u}\dot{u})^2$ and $U = mgx_2(u)$, where m is the particle mass and $x_{,u} = \partial x/\partial u$. The Lagrangian is $L = \frac{1}{2}m(x_{,u}\dot{u})^2 - mgx_2(u)$. Using Eq. 3.2, the equation of motion can be derived as

$$x_{,u}^2\ddot{u} + 2x_{,u}x_{,uu}\dot{u}^2 + gx_{2,u} = 0 \qquad (3.10)$$

This is an ordinary second-order differential equation of u. Note that the equation of motion shown in Eq. 3.10 can be used in applications such as rollercoasters, where position, velocity, and acceleration of riders modeled as concentrated mass are calculated. Equation 3.10 can be extended to support particle motion on a spatial parametric surface $x(u,w)$ for applications such as waterslides and bobsleds. More on this topic is given in Section 3.6.

Note that, in general, Eq. 3.10 must be solved using a numerical method. Only very simple cases—for example if $x(u)$ is a straight line as shown in Figure 3.4(b)—can be solved analytically. The following example illustrates the details.

Example 3.2

A particle slides along a straight line, as shown in Figure 3.4b, from p_0 to p_1, where $p_0 = [0,1]^T$ and $p_1 = [1,0]^T$. Calculate the position, velocity, and acceleration of the particle and the time required for it to reach point p_1. Assume the initial velocity to be zero.

Solution
The parametric equation of a straight line is

$$x(u) = p_0(1-u) + p_1 u = [u, 1-u]^T \tag{3.11}$$

Also,

$$x_{,u}(u) = [1,-1]^T, \; x_{,uu}(u) = 0, \; (x_{,u}(u))^2 = x_{,u}(u)^T x_{,u}(u) = 2, \text{and } x_{2,u} = -1$$

Therefore, the equation of motion becomes

$$\ddot{u} - \frac{1}{2}g = 0 \tag{3.12}$$

We integrate twice to obtain

$$\dot{u} = \frac{1}{2}gt + c_1 \tag{3.13a}$$

and

$$u = \frac{1}{4}gt^2 + c_1 t + c_2 \tag{3.13b}$$

Note that $c_1 = 0$ due to initial velocity, and $c_2 = 0$ due to initial position; that is, $u(0) = 0$, so

$$x(u) = \left[\frac{1}{4}gt^2, 1 - \frac{1}{4}gt^2\right]^T, \quad \dot{x}(u) = \left[\frac{1}{2}gt, -\frac{1}{2}gt\right]^T, \quad \text{and} \quad \ddot{x}(u) = \left[\frac{1}{2}g, -\frac{1}{2}g\right]^T \tag{3.14}$$

The time required to reach point p_1, where $u = 1$, is $t = 2/\sqrt{g}$. Note that all of these results can be verified using Newton's method as well. This is left as an exercise.

For a system of multiple particles, the forces acting on the ith particle, both external and internal, can be written as

$$m_i \ddot{r}_i = F_i + \sum_{j=1}^{n} f_{ij} \tag{3.15}$$

where

m_i is the mass of the ith particle.
r_i is the particle's position vector relative to the fixed reference.
F_i is the force acting on the ith particle.
f_{ij} is the interaction force of the jth particle acting on the ith particle.

In addition to Newton's method, shown in Eq. 3.15, the energy method is applicable and useful for multiparticle problems. For a multiparticle system, the kinetic energy is the sum of the energy of individual particles; so is the potential energy.

Example 3.3

This problem is slightly modified from Greenwood (1987). A particle of mass m moves on a frictionless plane. It is connected to a second identical particle by an inextensible string that passes through a small hole at point O, as shown in the figure. The second particle moves only vertically through O. The initial conditions are $r(0) = r_0$, $\dot{r}(0) = 0$, $\dot{\theta}(0) = \omega_0$.

The problem can be divided into three parts: (1) find the minimum initial angular velocity ω_{min}—that is, when $r(0) = r_0$ is given—so that the second particle is stationary. If the initial angular velocity is $\omega_0 = \dfrac{1}{\sqrt{3}}\omega_{min}$, (2) find the maximum and minimum r of the first particle and (3) find the tension force between the particles at both the maximum and minimum r. Note that the maximum and minimum r occur when $\dot{r} = 0$.

Solution
Part 1 is straightforward. The second particle is assumed stationary: $\dot{r} = \ddot{r} = 0$. Hence, from the free-body diagrams shown here, the following can be obtained by applying Eq. 3.15:

$$f = mg = mr\dot{\theta}^2 = mr_0\omega_0^2 \tag{3.16}$$

Thus,

$$\omega_0 = \sqrt{\frac{g}{r_0}} = \omega_{min} \tag{3.17}$$

That is, if $\omega_0 = \sqrt{\dfrac{g}{r_0}}$, the second particle is stationary.

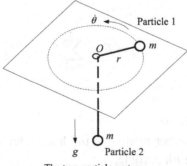

The two-particle system

For part 2, if the initial angular velocity is $\omega_0 < \sqrt{\dfrac{g}{r_0}} = \omega_{min}$, then $mg > mr_0\omega_0^2$. The first particle starts moving inward toward point O and the second particle moves downward. Since the angular momentum of the system is preserved, the angular velocity of the first particle is

$$H = mr^2\dot\theta = mr_0^2\omega_0 \tag{3.18}$$

Hence,

$$\dot\theta = \left(\frac{r_0}{r}\right)^2 \omega_0 \tag{3.19}$$

Free-body diagram of Particle 1

Free-body diagram of Particle 2

which provides important information about the relation between r and $\dot\theta$. The first particle rotates faster when it comes closer to point O. In the meantime, its kinetic energy increases. The first particle moves inward and reaches a minimum r; then it moves outward and reaches the maximum r. The particle continues the movement because no friction is present. The maximum and minimum rotating radii r_{max} and r_{min}, which happen when $\dot r = 0$, are to be found. Apparently, Eqs. 3.18 and 3.19 are not enough to solve the problem. We need more equations, such as energy conservation. The total mechanical energy of the system is

$$E = T + U = \left(\frac{1}{2}m\dot r^2 + \frac{1}{2}mr^2\dot\theta^2\right)_1 + \left(\frac{1}{2}m\dot r^2 + mgr\right)_2 = m\dot r^2 + \frac{1}{2}mr^2\dot\theta^2 + mgr \tag{3.20}$$

where the subscripts 1 and 2 represent the energy of the first and second particles, respectively. Also, mgr is the potential energy of the second particle, assuming $U_2 = 0$ when $r = 0$.

From the initial conditions, we have

$$E = \frac{1}{2}mr_0^2\omega_0^2 + mgr_0 \tag{3.21}$$

We bring Eq. 3.19 into Eq. 3.20 and equate it with Eq. 3.21, which yields

$$m\dot r^2 + m\frac{r_0^4\omega_0^2}{2r^2} + mgr = \frac{1}{2}mr_0^2\omega_0^2 + mgr_0 \tag{3.22}$$

To find the minimum r, we set $\dot{r} = 0$; therefore, Eq. 3.22 becomes

$$r^3 - \left(r_0 + \frac{r_0^2\omega_0^2}{2g}\right)r^2 + \frac{r_0^4\omega_0^2}{2g} = (r - r_0)\left(r^2 - \frac{r_0^2\omega_0^2}{2g}r - \frac{r_0^3\omega_0^2}{2g}\right) = 0 \qquad (3.23)$$

Solving Eq. 3.23, we have $r = r_0$ and

$$r = \frac{r_0^2\omega_0^2}{4g}\left(1 \pm \sqrt{1 + \frac{8g}{\omega_0^2 r_0}}\right)_0 \qquad (3.24)$$

Certainly, r cannot be less than zero, so only the positive solution is physically meaningful. If

$$\omega_0 = \sqrt{\frac{g}{3r_0}},$$

the minimum distance is

$$r = \frac{r_0^2\omega_0^2}{4g}\left(1 + \sqrt{1 + \frac{8g}{\omega_0^2 r_0}}\right) = \frac{1}{2}r_0.$$

So in this case, the maximum distance is $r = r_0$.

For part 3 of the problem, we have, from the free-body diagrams, the following two equations:

$$mr\dot{\theta}^2 - f = m\ddot{r} \qquad (3.25a)$$

$$f - mg = m\ddot{r} \qquad (3.25b)$$

By adding Eqs. 3.25a and 3.25b, we have

$$\ddot{r} = \frac{1}{2}r\dot{\theta}^2 - \frac{1}{2}g = \frac{1}{2}r\left[\left(\frac{r_0}{r}\right)^2\omega_0\right]^2 - \frac{1}{2}g \qquad (3.26)$$

For $r = r_0$, we have

$$\ddot{r} = \frac{1}{2}r_0\frac{g}{3r_0} - \frac{1}{2}g = -\frac{1}{3}g$$

For $r = 1/2r_0$, we have

$$\ddot{r} = \frac{r_0^3 g}{6r^3} - \frac{1}{2}g = \frac{4}{3}g - \frac{1}{2}g = \frac{5}{6}g$$

Bringing these into Eq. 3.25b, we have

$$f|_{r=r_{max}} = mg + m\ddot{r} = mg - \frac{1}{3} mg = \frac{2}{3} mg \qquad (3.27a)$$

and

$$f|_{r=r_{min}} = mg + m\ddot{r} = \frac{11}{6} mg \qquad (3.27b)$$

3.2.2 Rigid-Body Motion

In physics, rigid-body dynamics is the study of the motion of rigid bodies in response to external loads. Unlike particles, which move only in three degrees of freedom (translation in three directions), rigid bodies occupy space and have geometrical properties, such as center of mass, mass moment of inertia, and so forth, that characterize motion in six degrees of freedom (translation in three directions plus rotation in three directions). Rigid bodies are also nondeformable, as opposed to deformable bodies. As such, rigid-body dynamics is used heavily in analyses and computer simulations of physical systems and machinery where rotational motion is important but material deformation does not have a significant effect on the motion of the system.

Two equations are particularly important for a discussion of rigid-body motion using Newton's method. The first is the translation equation of motion given as

$$F = m\ddot{x}_c \qquad (3.28)$$

This equation states that the vector sum of the external forces acting on a rigid body is equal to the total mass of the body times the absolute acceleration of the center of mass $\ddot{x}_c$. The second equation is the rotational equation of motion given as

$$M = \dot{H} \qquad (3.29)$$

where the reference point for calculating the applied moment M and the angular momentum H is either fixed in an inertial frame or located at the mass center of the rigid body.

If a rigid body rotates about a reference point O, as shown in Figure 3.5, the angular momentum of a small element dV relative to point O is

$$dH_O = x \times dm\dot{x} \qquad (3.30)$$

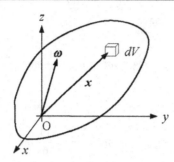

FIGURE 3.5: Typical volume element in a rotating rigid body.

where

x is the position vector of the small element relative to the reference point O.
dm is the mass of the small element dV.

Note that $\dot{x}$ is the velocity of dm as viewed by a nonrotating observer translating with O; the velocity is $\dot{x} = \omega \times x$. Therefore, the angular momentum of the rigid body is, with ρ as the mass density of the body,

$$H_O = \int_V \rho x \times (\omega \times x)dV \tag{3.31}$$

where

$\omega = \omega_x i + \omega_y j + \omega_z k$.
$x = xi + yj + zk$.
i, j, and k are the unit vectors along the x-, y-, and z-axes, respectively.

Note that

$$\omega \times x = \begin{vmatrix} i & j & k \\ \omega_x & \omega_y & \omega_z \\ x & y & z \end{vmatrix} = (z\omega_y - y\omega_z)i + (x\omega_z - z\omega_x)j + (y\omega_x - x\omega_y)k \tag{3.32}$$

and

$$x \times (\omega \times x) = \left[(y^2 + z^2)\omega_x - xy\omega_y - xz\omega_z\right]i$$
$$+ \left[-yx\omega_x + (x^2 + z^2)\omega_y - yz\omega_z\right]j \tag{3.33}$$
$$+ \left[-zx\omega_x - zy\omega_y + (x^2 + y^2)\omega_z\right]k$$

Thus, Eq. 3.31 becomes

$$H_O = H_x i + H_y j + H_z k$$
$$= (I_{xx}\omega_x + I_{xy}\omega_y + I_{xz}\omega_z)i + (I_{yx}\omega_x + I_{yy}\omega_y + I_{yz}\omega_z)j + (I_{zx}\omega_x + I_{zy}\omega_y + I_{zz}\omega_z)k \tag{3.34}$$

where I represents the mass moments of inertia; that is,

$$sI_{xx} = \int_V \rho(y^2 + z^2)dV, \quad I_{yy} = \int_V \rho(x^2 + z^2)dV, \quad I_{zz} = \int_V \rho(x^2 + y^2)dV$$

$$I_{xy} = I_{yx} = -\int_V \rho xy dV, \quad I_{xz} = I_{zx} = -\int_V \rho xz dV, \quad I_{yz} = I_{zy} = -\int_V \rho yz dV \qquad (3.35)$$

If the x-, y-, and z-axes align with the principal axes of the rigid body, $I_{xy} = I_{yz} = I_{xz} = 0$. Therefore, Eq. 3.34 gives

$$H_x = I_{xx}\omega_x, \ H_y = I_{yy}\omega_y, \ \text{and} \ H_z = I_{zz}\omega_z \qquad (3.36)$$

In this case, Eq. 3.29 gives

$$\begin{aligned} M_x &= \dot{H}_x = I_{xx}\dot{\omega}_x \\ M_y &= \dot{H}_y = I_{yy}\dot{\omega}_y \\ M_z &= \dot{H}_z = I_{zz}\dot{\omega}_z \end{aligned} \qquad (3.37)$$

A simple example is presented next to illustrate the use of angular momentum for solving rigid-body dynamics problems.

Example 3.4

A uniform circular cylinder of mass m, length L, and radius a rolls without slipping on a plane inclined at an angle α with the horizontal. Solve for the angular acceleration of the cylinder.

Solution
Taking the center of mass O as the reference point, we can use the rotational equations, Eqs. 3.29 and 3.31, to derive the equation of motion for this rigid body. From the free-body diagram the external moment applied to the body is $M = Fa$, where F is an unknown friction force ensuring pure rolling.

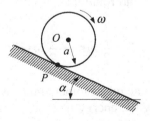

The cylinder rotates about its mass center O; hence, the angular momentum H_O, from Eq. 3.31, is

$$H_O = \int_V pr(\omega r)\, dV = \int_0^a pr(\omega r)(2\pi rL)\, dr = \frac{1}{2}\rho\omega\pi a^4 L$$

(3.38)

$$= \frac{1}{2}\rho a^2 \omega (\pi a^2 L) = \frac{1}{2}\, ma^2\omega = \frac{1}{2}I\dot\theta$$

where the position vector x becomes radius r. Hence, from Eq. 3.29 we have

$Fa = \frac{1}{2}\, I\ddot\theta = \frac{1}{2}\, ma^2\ddot\theta$; therefore,

$$F = \frac{1}{2}ma\ddot\theta$$

(3.39)

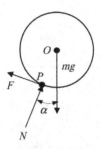

Since F is unknown, we need one more equation, which can be obtained from Newton's second law. Applying the law along the inclined plane, we have

$$mg \sin \alpha - F = ma\ddot\theta$$

(3.40)

If we bring $F = \frac{1}{2}\, ma\ddot\theta$ into Eq. 3.40, we have $mg \sin \alpha = \frac{3}{2}\, ma\ddot\theta$; hence,

$$\ddot\theta = \frac{2g \sin \alpha}{3a}$$

(3.41)

Solving Eq. 3.41 for angular position and velocity is straightforward.

The energy method is also applicable to rigid-body problems. For a rigid body in motion, the kinetic energy is both translational and rotational:

$$T = \frac{1}{2}\, m\dot{x}^2 + \frac{1}{2}\, \boldsymbol{\omega}\cdot\boldsymbol{H}$$

(3.42)

in which both the velocity $\dot{x}$ and the angular momentum $\boldsymbol{H}$ are calculated at the mass center of the body. Note that you may apply the energy method to Example 3.4 to obtain the same equation of motion. Here is how to proceed.

The total kinetic energy of the rolling cylinder is, according to Eq. 3.42,

$$T = \frac{1}{2}\, m(a\dot\theta)^2 + \frac{1}{2}\dot\theta\left(\frac{1}{2}\, ma^2\dot\theta\right) = \frac{3}{4}\, ma^2\dot\theta^2$$

On the other hand, the potential energy is $U = -mg(a \sin \alpha)$. The Lagrangian is then

$$L = T - U = \frac{3}{4} ma^2 \dot{\theta}^2 + mga \sin \alpha \qquad (3.43)$$

Applying Eq. 3.8, we have

$$\frac{d}{dt}\left(\frac{\partial L}{\partial \dot{\theta}}\right) - \frac{\partial L}{\partial \theta} = \frac{d}{dt}\left(\frac{3}{2}ma^2\dot{\theta}\right) - mga \sin \alpha = \frac{3}{2} ma^2 \ddot{\theta} - mga \sin \alpha = 0 \qquad (3.44)$$

Thus, the result of Example 3.4 is obtained.

The momentum equation and its derivative for the rolling cylinder in Example 3.4 are straightforward since the cylinder only rotates in one direction and rotates along its principal axis, which passes through its mass center.

3.2.3 Multibody Kinematic Analysis

Multibody kinematic analysis involves formulating equations of motion and solving them for position, velocity, and acceleration of individual bodies in the system in time. Such an analysis is important for general mechanism analysis and design, particularly for workspace analysis and robotics, where position, velocity, and acceleration of the moving components must be known in order to assess the functionality and performance of the mechanical system.

One of the most famous mechanisms commonly mentioned in spatial kinematic analysis is the Stewart platform, as shown in Figure 3.6a. The Stewart platform mechanism, originally suggested by Stewart (1965), is a parallel kinematic structure that can be used as a basis for controlled motion with six degrees of freedom (dof). The mechanism itself consists of a stationary platform (base platform) and a mobile platform that are connected via six legs (struts) mounted on universal joints. The legs have a built-in mechanism that allows changing the length of each individual leg. The desired position and orientation of the mobile platform is achieved by combining the lengths of the six legs, transforming the six transitional dof into three positional (displacement vector) dof and three orientation dof (angles of rotation of a rigid body in space). The advantages of the Stewart platform are six degrees of freedom and a split-hair accuracy of mobile platform positioning. This makes Stewart platform widely usable in robot-building infrastructures. For example, such a parallel kinematic structure is used in the National Advanced Driving Simulator (NADS) shown in Figure 3.6b.

In this section, instead of analyzing complex mechanisms such as the Stewart platform, which involves complex mathematical formulations, we discuss simple applications that can be solved analytically. One of the simplest and most useful is the four-bar linkage, among which

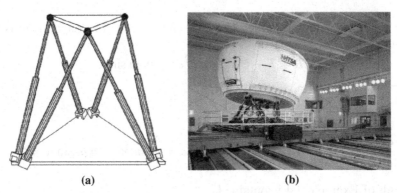

(a)   (b)

FIGURE 3.6: Stewart platform: (a) schematic view and (b) the University of Iowa's National Advanced Driving Simulator.

the slider-crank mechanism (Figure 3.7) can be seen in many applications such as internal combustion engines and oil well–drilling equipment. For the internal combustion engine, the mechanism is driven by a firing load that pushes the piston, converting the reciprocal motion into rotational motion at the crank. In drilling equipment, a torque is applied at the crank. The rotational motion is converted to reciprocal motion at the drill bit that digs into the ground. Note that in all cases the length of the crank must be smaller than that of the rod to allow the mechanism to operate. This is commonly referred to as Grashof's law. Also in this section we briefly review the analytical methods employed for multibody kinematic analysis using the slider-crank example. The methods reviewed should be applicable to more complex linkages.

The slider-crank mechanism consists of four bodies: crank, rod, slider, and the ground that is fixed to the reference frame. There are four joints—three revolute and one translation—defined among the bodies to restrain the relative motion between them. Assuming that the movable bodies only move on the plane, there are three degrees of freedom for each one: two translations (along the x- and y-coordinates) and one rotation (along the z-axis). Also, because a planar revolute joint constrains translational motion of the body, allowing only rotational degrees of freedom, it removes two dof. Similarly, a translation joint allows a body to move only along the translational direction, again removing two dof. Hence, the total degrees of freedom of the planar slider-crank mechanism, or the so-called Gruebler's

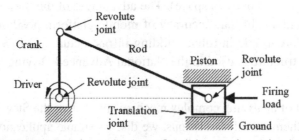

FIGURE 3.7: Schematic view of a slider-crank mechanism.

count, are

$$F = 3(nb - 1) - 3f_1 - 1f_2 = 3(4 - 1) - 3 \times 2 - 1 \times 2 = 1 \qquad (3.45)$$

where nb is the number of bodies in the system, and f_1 and f_2 are the dof that revolute and translation joints remove, respectively. The free degree of freedom, which is commonly assumed the rotation of the crank or the translation of the slider, can be driven by a rotation or translational motor. Once such a motor is added, the overall degrees of freedom of the system become zero. Note that for kinematic analysis, Gruebler's count of the mechanism must be equal to zero.

There are three methods for solving kinematics of mechanism, which are commonly found in a textbook on mechanism design. They are the relative velocity or graphical method, the instant center method, and the analytical method based on complex variables. The following example illustrates the position and velocity computation for the slider-crank mechanism using the analytical method or, more specifically, the complex variable method.

Example 3.5

Conduct a kinematic analysis for the slider-crank mechanism shown in Figure 3.7.

Solution

In kinematic analysis, forces and torque are not involved. Because all bodies are assumed massless, mass properties defined for them do not influence the analysis results.

The slider-crank mechanism is a planar kinematic analysis problem. A vector plot that represents the positions of joints of the planar mechanism is shown in the following figure. The vector plot serves as the first step in computing the mechanism's position, velocity, and accelerations.

The position equations of the system can be described by the following vector summation:

$$\mathbf{Z}_1 + \mathbf{Z}_2 = \mathbf{Z}_3 \qquad (3.46)$$

where

$$\mathbf{Z}_1 = Z_1 \cos \theta_A + iZ_1 \sin \theta_A = Z_1 \, e^{i\theta_A}$$
$$\mathbf{Z}_2 = Z_2 \cos \theta_B + iZ_2 \sin \theta_B = Z_2 \, e^{i\theta_B}$$
$$\mathbf{Z}_3 = Z_3, \text{ since } \theta_c \text{ is always } 0$$

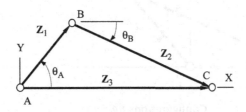

The real and imaginary parts of Eq. 3.46, corresponding to the X- and Y-components of the vectors, can be written as

$$Z_1 \cos \theta_A + Z_2 \cos \theta_B = Z_3 \tag{3.47a}$$

$$Z_1 \sin \theta_A + Z_2 \sin \theta_B = 0 \tag{3.47b}$$

In Eqs. 3.47a and 3.47b, Z_1, Z_2, and θ_A are given. We are solving for Z_3 and θ_B. These equations are nonlinear. Solving them directly for Z_2 and θ_B is not straightforward. Instead, we calculate Z_3 first, using trigonometric relations:

$$Z_2^2 = Z_1^2 + Z_3^2 - 2Z_1 Z_3 \cos \theta_A$$

Thus,

$$Z_3^2 - 2Z_1 \cos \theta_A Z_3 + Z_1^2 - Z_2^2 = 0$$

Solving Z_3 from the preceding quadratic equation, we have

$$Z_3 = \frac{2Z_1 \cos \theta_A \pm \sqrt{(2Z_1 \cos \theta_A)^2 - 4(Z_1^2 - Z_2^2)}}{2} \tag{3.48}$$

where two solutions of Z_3 represent the two possible configurations of the mechanism shown in the following figure. Note that point C can be either at C or C′ for any given Z_1 and θ_A.

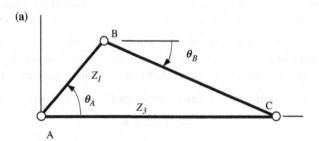

Configurations 1: $\theta_B < 180°$

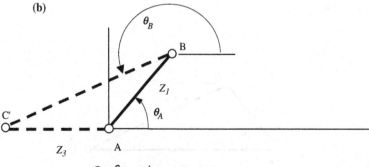

Configurations 2: $\theta_B > 180°$

From Eq. 3.47b, θ_B can be solved by

$$\theta_B = \sin^{-1}\left(\frac{-Z_1 \sin \theta_A}{Z_2}\right) \tag{3.49}$$

Similarly, θ_B has two possible solutions corresponding to vector $\mathbf{Z}_3$.

Taking derivatives of Eqs. 3.47a and 3.47b with respect to time, we have

$$-Z_1 \sin \theta_A \dot{\theta}_A - Z_2 \sin \theta_B \dot{\theta}_B = \dot{Z}_3 \tag{3.50a}$$

$$Z_1 \cos \theta_A \dot{\theta}_A + Z_2 \cos \theta_B \dot{\theta}_B = 0 \tag{3.50b}$$

where $\dot{\theta}_A = \dfrac{d\theta_A}{dt} = \omega_A$ is the angular velocity of the rotation driver, which is a constant. Note that Eqs. 3.50a and 3.50b are linear functions of $\dot{Z}_3$ and $\dot{\theta}_B$. Rewrite the equations in matrix form:

$$\begin{bmatrix} Z_2 \sin \theta_B & 1 \\ Z_2 \cos \theta_B & 0 \end{bmatrix} \begin{bmatrix} \dot{\theta}_B \\ \dot{Z}_3 \end{bmatrix} = \begin{bmatrix} -Z_1 \sin \theta_A \dot{\theta}_A \\ -Z_1 \cos \theta_A \dot{\theta}_A \end{bmatrix} \tag{3.51}$$

Equation 3.51 can be solved by

$$\begin{bmatrix} \dot{\theta}_B \\ \dot{Z}_3 \end{bmatrix} = \begin{bmatrix} Z_2 \sin \theta_B & 1 \\ Z_2 \cos \theta_B & 0 \end{bmatrix}^{-1} \begin{bmatrix} -Z_1 \sin \theta_A \dot{\theta}_A \\ -Z_1 \cos \theta_A \dot{\theta}_A \end{bmatrix}$$

$$= \begin{bmatrix} -\dfrac{Z_1 \cos \theta_A \dot{\theta}_A}{Z_2 \cos \theta_B} \\ -\dfrac{Z_1 \left(\cos \theta_B \sin \theta_A \dot{\theta}_A - \sin \theta_B \cos \theta_A \dot{\theta}_A \right)}{\cos \theta_B} \end{bmatrix} \tag{3.52}$$

Therefore,

$$\dot{\theta}_B = -\frac{Z_1 \cos \theta_A \dot{\theta}_A}{Z_2 \cos \theta_B} \tag{3.53}$$

and

$$\dot{Z}_3 = Z_1 \left(\tan \theta_B \cos \theta_A \dot{\theta}_A - \sin \theta_A \dot{\theta}_A \right) \tag{3.54}$$

3.2.4 Multibody Dynamic Analysis

A multibody system is used to model the dynamic behavior of interconnected rigid or flexible bodies, each of which may undergo large translational and rotational displacements. The systematic treatment of the dynamic behavior of interconnected bodies has led to a large number of important multibody formalisms in the field of mechanics.

The simplest bodies or elements of a multibody system were treated by Newton (free particle) and Euler (rigid body), as briefly discussed earlier. Basically, the motion of bodies is described by their kinematic behavior. The dynamic behavior results from the equilibrium of applied forces and the rate of change in momentum. Nowadays, the term *multibody system* relates to a large number of engineering research fields, especially robotics and vehicle dynamics. As an important feature, multibody system dynamics usually offers an algorithmic, computer-aided way to model, analyze, simulate, and optimize the arbitrary motion of possibly thousands of interconnected bodies.

The equations of motion are used to describe the dynamic behavior of a multibody system. Each multibody system formulation may lead to a different mathematical appearance of the equations of motion while the physics behind them remains the same. The motion of constrained bodies is described by equations that result basically from Newton's second law. These equations are written for the general motion of individual bodies with the addition of constraint conditions. Usually the equations of motion are derived from Newton and Euler equations or from Lagrange equations. We mentioned Newton's laws and Lagrange's equations; now we briefly discuss Euler equations, which govern the rotational motion of rigid bodies.

A rigid body rotates together with its body-fixed reference frame x-y-z at an angular velocity ω with respect to the inertia (fixed) frame X-Y-Z, as shown in Figure 3.8. Note that we assume that the body-fixed reference frame also originates at the mass center of the rigid body. For a rotating body, the angular momentum H_o is most conveniently expressed in the body reference frame x-y-z, where point o is the origin. H_o can be written as

$$\boldsymbol{H}_o = H_x \boldsymbol{i} + H_y \boldsymbol{j} + H_z \boldsymbol{k} \tag{3.55}$$

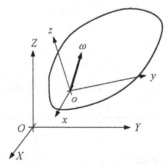

FIGURE 3.8: Rotating rigid body.

where the unit vectors i, j, and k align and rotate with the x-, y-, and z-axes, respectively. From Eq. 3.29, we have

$$M_o = \dot{H}_o \tag{3.56}$$

where M_o is the vector of external moment applied to the body referring to the body reference frame x-y-z.

The derivative of H_o is

$$
\begin{aligned}
\dot{H}_o &= \left(\dot{H}_x i + \dot{H}_y j + \dot{H}_z k\right) + \left(H_x \dot{i} + H_y \dot{j} + H_z \dot{k}\right) \\
&= \left(\dot{H}\right)_r + \left(H_x \omega \times i + H_y \omega \times j + H_z \omega \times k\right) = \left(\dot{H}\right)_r + \omega \times H
\end{aligned}
\tag{3.57}
$$

where $\left(\dot{H}\right)_r$ is the rate of change of the angular momentum with respect to o as viewed by an observer on the moving x-y-z coordinate system. We also assume that x, y, and z align with the principal axes of the body; hence, from Eq. 3.37,

$$\left(\dot{H}\right)_r = \dot{H}_x i + \dot{H}_y j + \dot{H}_z k = I_{xx}\dot{\omega}_x i + I_{yy}\dot{\omega}_y j + I_{zz}\dot{\omega}_z k \tag{3.58}$$

And the second term, after computing the cross-product and considering the principal axis assumption, is

$$\omega \times H = \left(I_{zz} - I_{yy}\right)\omega_y \omega_z i + \left(I_{xx} - I_{zz}\right)\omega_x \omega_z j + \left(I_{yy} - I_{xx}\right)\omega_y \omega_x k \tag{3.59}$$

Combining Eqs. 3.56 through 3.59, we have

$$
\begin{aligned}
M_{ox} &= I_{xx}\dot{\omega}_x + \left(I_{zz} - I_{yy}\right)\omega_y \omega_z \\
M_{oy} &= I_{yy}\dot{\omega}_y + \left(I_{xx} - I_{zz}\right)\omega_x \omega_z \\
M_{oz} &= I_{zz}\dot{\omega}_z + \left(I_{yy} - I_{xx}\right)\omega_y \omega_x
\end{aligned}
\tag{3.60}
$$

These are known as Euler's equations of motion. Example 3.6 illustrates a few details in solving a simple dynamic problem of a rigid-body system in rotation using the Euler equations.

Example 3.6

A two-body system connected by a revolute (or pin) joint at point P is shown in the following figure. Body 1 rotates along the Y-axis at angular velocity Ω, and Body 2 rotates along the z-axis at angular velocity $\dot{\phi}$. Assume no friction at the pin joint between the two bodies. Note that point o is the mass center of Body 2 and the x-, y-, and z-axes align with the principal axes of Body 2. Formulate equations of motion that need to be solved for the rotation angle of Body 2 as well as the reaction forces (f_1, f_2, and f_3) and torque (M_1, M_2, and M_3) between the two bodies at the pin joint.

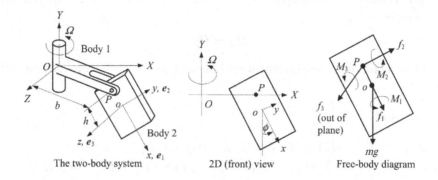

The two-body system   2D (front) view   Free-body diagram

Solution

We formulate rotation equations of motion using Euler's equations first. Then we formulate equations for calculating reaction forces.

First, the angular velocity of Body 2 at its mass center o is

$$\boldsymbol{\omega}_o = \dot{\phi}\boldsymbol{e}_3 + \Omega\boldsymbol{j} \tag{3.61}$$

Note that in this example, unit vectors $\boldsymbol{i}, \boldsymbol{j}$, and $\boldsymbol{k}$ align with the X-, Y-, and Z-axes, respectively; similarly, unit vectors $\boldsymbol{e}_1, \boldsymbol{e}_2$, and $\boldsymbol{e}_3$ align with the x-, y-, and z-axes, respectively.

To apply Euler's equations, the angular velocity must be represented in the body-fixed reference frame (i.e., the x-y-z coordinate system). Hence, Eq. 3.61 becomes

$$\boldsymbol{\omega}_o = \dot{\phi}\boldsymbol{e}_3 + \Omega(-\cos\phi\boldsymbol{e}_1 + \sin\phi\boldsymbol{e}_2) = -\Omega\cos\phi\boldsymbol{e}_1 + \Omega\sin\phi\boldsymbol{e}_2 + \dot{\phi}\boldsymbol{e}_3 \tag{3.62}$$

and

$$\begin{aligned}
\dot{\boldsymbol{\omega}}_o &= \Omega\sin\phi\dot{\phi}\boldsymbol{e}_1 - \Omega\cos\phi\dot{\boldsymbol{e}}_1 + \Omega\cos\phi\dot{\phi}\boldsymbol{e}_2 + \Omega\sin\phi\dot{\boldsymbol{e}}_2 + \ddot{\phi}\boldsymbol{e}_3 + \dot{\phi}\dot{\boldsymbol{e}}_3 \\
&= (\Omega\sin\phi\dot{\phi}\boldsymbol{e}_1 + \Omega\cos\phi\dot{\phi}\boldsymbol{e}_2 + \ddot{\phi}\boldsymbol{e}_3) - \Omega\cos\phi\boldsymbol{\omega}_o \times \boldsymbol{e}_1 + \Omega\sin\phi\boldsymbol{\omega}_o \times \boldsymbol{e}_2 + \dot{\phi}\boldsymbol{\omega}_o \times \boldsymbol{e}_3 \\
&= \Omega\sin\phi\dot{\phi}\boldsymbol{e}_1 + \Omega\cos\phi\dot{\phi}\boldsymbol{e}_2 + \ddot{\phi}\boldsymbol{e}_3
\end{aligned} \tag{3.63}$$

As shown in the free-body diagram, the reaction forces f_1, f_2, and f_3, as well moments M_1, M_2, and M_3 at the pin joint P, and the weight of Body 2 are the only external loads applied to the body. Hence, according to Euler's equations, the moment equations are

$$\sum M_x = I_{xx}\Omega\sin\phi\dot{\phi} + (I_{zz} - I_{yy})\Omega\sin\phi\dot{\phi} = M_1 \tag{3.64a}$$

$$\sum M_y = I_{yy}\Omega\cos\phi\dot{\phi} - (I_{xx} - I_{zz})\Omega\cos\phi\dot{\phi} = M_2 + hf_3 \tag{3.64b}$$

$$\sum M_z = I_{zz}\ddot{\phi} - (I_{yy} - I_{xx})\Omega^2\cos\phi\sin\phi\dot{\phi} = M_3 - hf_2 = -hf_2 \tag{3.64c}$$

where $M_3 = 0$, which is due the fact that Body 2 is free to rotate along the z-axis at the pin joint.

Now for the translational motion, we apply Newton's second law to Body 2. We calculate the acceleration a_o of Body 2 at its mass center first. From particle dynamics,

$$a_o = a_p + a_{o/p} = a_p + \alpha_o \times r + \omega_o \times \omega_o \times r \qquad (3.65)$$

where r is the position vector from point P to o (i.e., $r = he_1$), and

$$a_p = -b\Omega^2 i = -b\Omega^2 (\sin \phi e_1 + \cos \phi e_2) = -b\Omega^2 \sin \phi e_1 - b\Omega^2 \cos \phi e_2 \qquad (3.66)$$

After vector multiplications and arranging terms, we have

$$\begin{aligned} a_o &= a_p + \dot{\omega}_o \times r + \omega_o \times \omega_o \times r \\ &= \left(-b\Omega^2 \sin \phi - h\Omega^2 \sin^2 \phi - h\dot{\phi}^2 \right) e_1 \\ &\quad + \left(-b\Omega^2 \cos \phi - h\Omega^2 \cos \phi \sin \phi + h\ddot{\phi} \right) e_2 \\ &\quad + \left(-2h\Omega^2 \dot{\phi} \cos \phi \right) e_3 \end{aligned} \qquad (3.67)$$

Applying Newton's second law to Body 2, we have

$$\sum F_x = -m \left(-b\Omega^2 \sin \phi - h\Omega^2 \sin^2 \phi - h\dot{\phi}^2 \right) = f_1 + mg \cos \phi \qquad (3.68a)$$

$$\sum F_y = m \left(-b\Omega^2 \cos \phi - h\Omega^2 \cos \phi \sin \phi + h\ddot{\phi} \right) = f_2 - mg \sin \phi \qquad (3.68b)$$

$$\sum F_z = m \left(-2h\Omega^2 \dot{\phi} \cos \phi \right) = f_3 \qquad (3.68c)$$

These equations are too complex to solve by hand.

As shown in Example 3.6, multibody system dynamics problems are difficult to solve using analytical methods. In fact, the more critical issue in dynamics problems is the formulation of equations of motion. Certainly there are methods and principles such as Newton's law, conservation of energy, conservation of momentum, and so on. They must be applied to given problems in a case-by-case manner. Force and moment applied to bodies have to be identified, free-body diagrams have to be drawn, and then equations of motion can be formulated. Very often, we end up with equations of motion that are too complex to solve by hand, such as those in Example 3.6. There is no single method that is powerful and general enough to be consistently applied to formulate and solve general multibody dynamics problems.

A different approach that can be systematically applied to solve general problems must be employed for motion analysis of general mechanical systems. The computer-aided approach, which employs Newton's equations of motion governing the motion of rigid bodies, and constraint equations, governing the relative motion between bodies, is introduced in the next section. Formulation of the computer-aided method is uniform and consistent; therefore, it is more suitable for computer implementation. Equations of motion can be solved using numerical methods implemented on computers.

3.3 Computer-Aided Methods

Large displacements and rotations that occur in the kinematic and dynamic performance of mechanical systems lead to nonlinear mathematical models that must be formulated and analyzed. The analytical complexity of nonlinear algebraic equations of kinematics and nonlinear differential equations of dynamics makes it impossible to obtain closed-form solutions in most applications.

The main objective of computer-aided methods is to create a systematic approach for both formulating and solving equations of motion that can be implemented on a digital computer. However, only if such a systematic approach is adopted can the burden of extensive analytical derivation be taken away from the shoulders of the engineer and delegated to the computer. The basic idea in computer-aided methods is introducing constraints that define relative motion between bodies in a multibody mechanical system, taking derivatives of the constraint equations to obtain velocity and acceleration equations, and then solving these equations using numerical methods. For dynamic analysis, Newton's law of motion is coupled with the constraint equations using Lagrange multipliers.

A thorough and in-depth discussion of the theory and mathematical formulation of computer-aided kinematic and dynamic analyses is beyond the scope of this book. Only a brief treatment is provided in this section. To simplify the mathematical expressions, the formulation of constraint equations and equations of motion focuses on mostly planar motion problems. This section aims at providing a short discussion on the theory and formulation of multibody systems with simple examples to help the reader understand the behind-the-scenes operations and to create models and use the computer tools more effectively. Several excellent references, such as Haug (1989), treat this subject with great details.

3.3.1 Kinematic Analysis

We start with a basic formulation for planar kinematic analysis. Formulation of selected joints that are commonly found in motion analysis are included.

Any set of variables that uniquely specify the position and orientation of all bodies in the mechanism—that is, the configuration of the mechanism—is a set of generalized coordinates $q = [q_1, q_2, ..., q_{nc}]^T$, where nc is the number of these coordinates. If a planar mechanism is made up of nb rigid bodies, the number of planar Cartesian generalized coordinates is $nc = 3nb$, since each planar body moves in two directions and rotates along the direction that is normal to the plane. Note that for spatial mechanisms, $nc = 6nb$ since each spatial body moves and rotates in three directions. Generalized coordinates may be independent (i.e., free to vary) or dependent (i.e., required to satisfy constraint equations).

A kinematic constraint between bodies imposes conditions on the relative motion between a pair of bodies. When these conditions are expressed as algebraic equations in terms of generalized coordinates, they are called *holonomic* kinematic constraint equations. A system of nh holonomic kinematic constraint equations can be expressed as

$$\mathbf{\Phi}^K(\boldsymbol{q}, t) = [\mathbf{\Phi}_1^K(\boldsymbol{q}, t), \mathbf{\Phi}_2^K(\boldsymbol{q}, t), \dots \mathbf{\Phi}_{nh}^K(\boldsymbol{q}, t)]^T = \mathbf{0} \tag{3.69}$$

Note that some or all the constraints may be stationary; that is, $\mathbf{\Phi}^K(\boldsymbol{q}, t) = \mathbf{\Phi}^K(\boldsymbol{q}) = 0$.

If the constraints of Eq. 3.69 are independent, then the system is said to have $nc - nh$ degrees of freedom; in other words, dof $= nc - nh$. To determine the motion of the system, we must either define additional driving constraints for the dof so as to uniquely determine $\boldsymbol{q}(t)$ (kinematic analysis) or define forces that act on the system, in which case $\boldsymbol{q}(t)$ is the solution of equations of motion (dynamic analysis). If the independent driving constraints are specified for kinematic analysis, denoted

$$\mathbf{\Phi}^D(\boldsymbol{q}, t) = 0 \tag{3.70}$$

then the configuration of the system as a function of time can be determined. Thus, the combined constraints of Eqs 3.69 and 3.70,

$$\mathbf{\Phi}(\boldsymbol{q}, t) = \begin{bmatrix} \mathbf{\Phi}^K(\boldsymbol{q}, t) \\ \mathbf{\Phi}^D(\boldsymbol{q}, t) \end{bmatrix} = \mathbf{0} \tag{3.71}$$

can be solved for $\boldsymbol{q}(t)$. Such a system is called kinematically driven. Note that to guarantee solutions for Eq. 3.71, the size of $\mathbf{\Phi}$ (i.e., the number of constraint equations in it) must equal the number of generalized coordinates. In addition, all constraint equations in Eq. 3.71 must be independent. The very best way to test if the equation is solvable is that the Jacobian of $\mathbf{\Phi}_q$ must be nonsingular. Note that $\mathbf{\Phi}_q = \partial \mathbf{\Phi} / \partial \boldsymbol{q} = \mathbf{\Phi}_{,q}$. We use this shorthand notation throughout the chapter.

Example 3.7

The same pendulum discussed in Example 3.1 is presented again to illustrate the formulation of particle kinematics using a computer-aided kinematic analysis method.

Solution

For the simple pendulum shown in the next figure, the kinematic constraint is described by the pair of equations,

$$\mathbf{\Phi}^K(\boldsymbol{q}) \equiv \begin{bmatrix} x_1 - \ell \sin \theta \\ y_1 + \ell \cos \theta \end{bmatrix} = \mathbf{0}$$

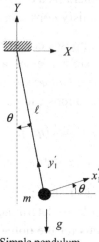

Simple pendulum

where $q = [x_1, y_1, \theta]^T$. Note that x_1 and y_1 represent the particle location (or the origin of the $x_1' - y_1'$ frame). Clearly, the above equations can be solved for x_1 and y_1 as a function of θ, so one independent variable can specify the pendulum's motion. Thus, the system has one degree of freedom (dof = 1).

To specify the motion of the pendulum, a driving constraint must be introduced, for example, specifying the time history of θ:

$$\Phi^D(\theta, t) \equiv \theta - f(t) = 0 \tag{3.73}$$

Combining the previous two equations yields the system kinematic constraint equation:

$$\Phi(q, t) = \begin{bmatrix} \Phi^K(q) \\ \Phi^D(q, t) \end{bmatrix} = \begin{bmatrix} x_1 - \ell \sin\theta \\ y_1 + \ell \cos\theta \\ \theta - f(t) \end{bmatrix} = 0 \tag{3.74}$$

This is a formulation for kinematic analysis. In reality, the pendulum rotates due to gravity, which falls into the scope of dynamic analysis. If we are interested in finding the reaction force or torque at the revolute joint, Eq. 3.74 is not sufficient. We need additional equations to calculate the reaction loads, in which dynamic analysis will be carried out.

If Φ_q is nonsingular—that is, $|\Phi_q| \neq 0$ at some value of q that satisfies Eq. 3.74—then Eq. 3.74 can be solved for q as a function of time. To test this condition, note that

$$|\Phi_q| = \begin{vmatrix} 1 & 0 & -\cos\theta \\ 0 & 1 & -\sin\theta \\ 0 & 0 & 1 \end{vmatrix} = 1 \tag{3.75}$$

Thus, from Eq. 3.74, once the rotation angle θ is given, the location of the mass center of the particle can be determined by solving for x_1 and y_1.

In general q is not known as an explicit function of time; it cannot be differentiated to obtain $\dot{q}$ and $\ddot{q}$. An alternative approach that is well suited to numerical computation is the chain rule of differentiation for evaluating derivatives of both sides of Eq. 3.71 with respect to time to obtain the velocity equation

$$\dot{\Phi} = \Phi_q \dot{q} + \Phi_t = 0$$

or

$$\Phi_q \dot{q} = -\Phi_t \equiv v \qquad (3.76)$$

If Φ_q is nonsingular, Eq. 3.76 can be solved for $\dot{q}$ at discrete instants of time. Similarly, both sides of Eq. 3.76 can be differentiated again with respect to time, using the chain rule of differentiation to obtain

$$\Phi_q \ddot{q} = -\left(\Phi_q \dot{q}\right)_q \dot{q} - 2\Phi_{qt}\dot{q} - \Phi_{tt} \equiv \gamma \qquad (3.77)$$

Since Φ_q is nonsingular, Eq. 3.77 can be solved for $\ddot{q}$ at discrete instants of time.

Example 3.8

A slider-crank mechanism similar to that in Example 3.2 is presented to illustrate the computer-aided formulation for kinematic analysis. To simplify the mathematical presentation, a two-body system, composed of crank and rod, is assumed. The crank is driven by a driver of constant angular velocity ω.

Solution
There are two planar bodies in this system and one driving constraint; hence, we need $3nb - 1 = 5$ geometric constraint equations to uniquely define the kinematics of the system. First, Point O_1 on the crank (Body 1) coincides with the origin O of the X-Y frame; that is, $x_1 = y_1 = 0$. These are the first two constraint equations. The third equation is $y_2 = 0$ since Point O_2 on the coupler (Body 2) must lie on the X-axis. The final two constraint equations are obtained by making Points P_1 and P_2 on the crank and rod, respectively, coincident:

$$r^{P_1} \equiv \begin{bmatrix} x_1 + \ell_1 \cos \theta_1 \\ y_1 + \ell_1 \sin \theta_1 \end{bmatrix} = \begin{bmatrix} x_2 - \ell_2 \sin \theta_2 \\ y_2 + \ell_2 \cos \theta_2 \end{bmatrix} \equiv r^{P_2} \qquad (3.78)$$

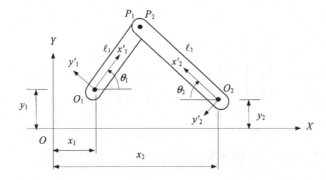

Therefore, in terms of $q = [x_1, y_1, \theta_1, x_2, y_2, \theta_2]^T$, the system of kinematic constraint equations, combining with the driver, is

$$\Phi(q,t) \equiv \begin{bmatrix} x_1 \\ y_1 \\ y_2 \\ x_1 - x_2 + \ell_1 \cos\theta_1 + \ell_2 \sin\theta_2 \\ y_1 - y_2 + \ell_1 \sin\theta_1 - \ell_2 \cos\theta_2 \\ \theta_1 - \omega t \end{bmatrix} = 0 \tag{3.79}$$

Solving for q yields

$$q \equiv \begin{bmatrix} x_1 \\ y_1 \\ \theta_1 \\ x_2 \\ y_2 \\ \theta_2 \end{bmatrix} = \begin{bmatrix} 0 \\ 0 \\ \omega t \\ \ell_1 \cos\theta_1 + \ell_2 \sin\theta_2 \\ 0 \\ \pm\cos^{-1}\left(\dfrac{\ell_1 \sin\theta_1}{\ell_2}\right) \end{bmatrix} \tag{3.80}$$

in which ℓ_1 and ℓ_2 are known and $\dot{\theta}$ is given; hence, θ_1 can be solved, then θ_2, and finally x_2.

Taking derivatives on both sides of Eq. 3.79 with respect to time t yields

$$\Phi_q(q,t)\dot{q} \equiv \begin{bmatrix} 1 & 0 & 0 & 0 & 0 & 0 \\ 0 & 1 & 0 & 0 & 0 & 0 \\ 0 & 0 & 0 & 0 & 1 & 0 \\ 1 & 0 & -\ell_1 \sin\theta_1 & -1 & 0 & \ell_2 \cos\theta_2 \\ 0 & 1 & \ell_1 \cos\theta_1 & 0 & -1 & \ell_2 \sin\theta_2 \\ 0 & 0 & 1 & 0 & 0 & 0 \end{bmatrix} \begin{bmatrix} \dot{x}_1 \\ \dot{y}_1 \\ \dot{\theta}_1 \\ \dot{x}_2 \\ \dot{y}_2 \\ \dot{\theta}_2 \end{bmatrix} = \begin{bmatrix} 0 \\ 0 \\ 0 \\ 0 \\ 0 \\ \omega \end{bmatrix} = -\Phi_t \tag{3.81}$$

Since $|\Phi_q| \neq 0$ (Exercise 3.6), $\dot{q}$ can be solved using Eq. 3.81 as

$$\dot{q} \equiv \begin{bmatrix} \dot{x}_1 \\ \dot{y}_1 \\ \dot{\theta}_1 \\ \dot{x}_2 \\ \dot{y}_2 \\ \dot{\theta}_2 \end{bmatrix} = \begin{bmatrix} 0 \\ 0 \\ \omega \\ -\ell_1\omega(\sin\theta_1 + \cos\theta_1 \cot\theta_2) \\ 0 \\ -\dfrac{\ell_1 \cos\theta_1 \omega}{\ell_2 \sin\theta_2} \end{bmatrix} \tag{3.82}$$

in which $\dot{\theta}$, θ_1, and θ_2 can be solved in Eq. 3.80, then $\dot{x}_2$, and finally $\dot{\theta}_2$.

Taking a derivative on both sides of Eq. 3.81 with respect to time t yields

$$\Phi_q(q,t)\ddot{q} \equiv \begin{bmatrix} 1 & 0 & 0 & 0 & 0 & 0 \\ 0 & 1 & 0 & 0 & 0 & 0 \\ 0 & 0 & 0 & 0 & 1 & 0 \\ 1 & 0 & -\ell_1 \sin\theta_1 & -1 & 0 & \ell_2 \cos\theta_2 \\ 0 & 1 & \ell_1 \cos\theta_1 & 0 & -1 & \ell_2 \sin\theta_2 \\ 0 & 0 & 1 & 0 & 0 & 0 \end{bmatrix} \begin{bmatrix} \ddot{x}_1 \\ \ddot{y}_1 \\ \ddot{\theta}_1 \\ \ddot{x}_2 \\ \ddot{y}_2 \\ \ddot{\theta}_2 \end{bmatrix}$$

$$= \begin{bmatrix} 0 \\ 0 \\ 0 \\ \dot{\theta}_1^2 \ell_1 \cos\theta_1 + \dot{\theta}_2^2 \ell_2 \sin\theta_2 \\ \dot{\theta}_1^2 \ell_1 \sin\theta_1 - \dot{\theta}_2^2 \ell_2 \cos\theta_2 \\ 0 \end{bmatrix} \tag{3.83}$$

$\ddot{q}$ can be solved using Eq. 3.83. This is left as an exercise (Exercise 3.7).

With the Cartesian generalized coordinates (or absolute coordinates)—that is, position and orientation of every single body in the system—a large system of equations is obtained. For example, for a slider-crank mechanism that consists of three moving bodies, shown in Figure 3.9a, the absolute coordinates are

$$q = [x_1, y_1, \theta_1, x_2, y_2, \theta_2, x_3, y_3, \theta_3]^T \tag{3.84}$$

However, using Cartesian generalized coordinates, the constraint equations can easily be formulated automatically, which is ideal for implementation on computers.

There is a simpler set of generalized coordinates that can be chosen for problem formulation. For the system shown in Figure 3.9b, the joint coordinates that correspond to kinematic joints in system can be chosen as

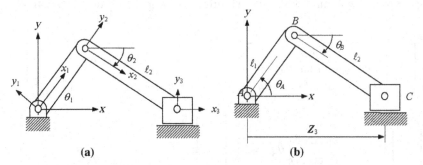

(a)   **(b)**

FIGURE 3.9: A slider-crank mechanism, (a) with Cartesian generalized coordinate system, and (b) with joint (relative) coordinates.

$$q = [\theta_A, \theta_B, Z_3]^T \tag{3.85}$$

where

θ_A and θ_B correspond to the rotation dof of the two revolute joints A and B, respectively. Z_3 represents the translation dof of the slider joint at point C.

Note that the parameters defined in Eq. 3.85 are consistent with those of Example 3.5.

Assuming the rotation angle of the crank is prescribed by a driving constraint $\theta = f(t) = \omega t$, the constraint equations of the system can be formulated in terms of $q = [\theta_A, \theta_B, Z_3]^T$ as

$$\Phi(q,t) = \begin{bmatrix} \ell_1 \cos \theta_A + \ell_2 \cos \theta_B - Z_3 \\ \ell_1 \sin \theta_A + \ell_2 \sin \theta_B \\ \theta_A - \omega t \end{bmatrix} = 0 \tag{3.86a}$$

The velocity and acceleration equations can be obtained, respectively, as

$$\Phi_q \dot{q} = \begin{bmatrix} -\ell_1 \sin \theta_A & -\ell_2 \sin \theta_B & -1 \\ \ell_1 \cos \theta_A & \ell_2 \cos \theta_B & 0 \\ 1 & 0 & 0 \end{bmatrix} \begin{bmatrix} \dot{\theta}_A \\ \dot{\theta}_B \\ \dot{Z}_3 \end{bmatrix} = \begin{bmatrix} \omega \\ 0 \\ 0 \end{bmatrix} = -\Phi_t \tag{3.86b}$$

and

$$\Phi_q(q,t)\ddot{q} = \begin{bmatrix} -\ell_1 \sin \theta_A & -\ell_2 \sin \theta_B & -1 \\ \ell_1 \cos \theta_A & \ell_2 \cos \theta_B & 0 \\ 1 & 0 & 0 \end{bmatrix} \begin{bmatrix} \ddot{\theta}_A \\ \ddot{\theta}_B \\ \ddot{Z}_3 \end{bmatrix}$$

$$= \begin{bmatrix} \ell_1 \cos \theta_A \dot{\theta}_A^2 + \ell_2 \cos \theta_B \dot{\theta}_B^2 \\ \ell_1 \sin \theta_A \dot{\theta}_A^2 + \ell_2 \sin \theta_B \dot{\theta}_B^2 \\ 0 \end{bmatrix} \equiv \gamma \tag{3.86c}$$

Solving Eqs. 3.86a, 3.86b, and 3.86c, the solutions of q, $\dot{q}$, and $\ddot{q}$ are, respectively,

$$\begin{bmatrix} \theta_A \\ \theta_B \\ Z_3 \end{bmatrix} = \begin{bmatrix} \omega t \\ \sin^{-1}(-\ell_1 \sin \theta_A/\ell_2) \\ \ell_1 \cos \theta_A + \ell_2 \cos \theta_B \end{bmatrix} = \begin{bmatrix} \omega t \\ \sin^{-1}(-\ell_1 \sin \theta_A/\ell_2) \\ \ell_1 \cos \theta_A \pm \sqrt{\ell_2^2 - (\ell_1 \sin \theta_A)^2} \end{bmatrix} \tag{3.87a}$$

$$\begin{bmatrix} \dot{\theta}_A \\ \dot{\theta}_B \\ \dot{Z}_3 \end{bmatrix} = \begin{bmatrix} \omega \\ -\dfrac{\ell_1 \cos \theta_A}{\ell_2 \cos \theta_B} \dot{\theta}_A \\ -\ell_1 \sin \theta_A \dot{\theta}_A + \ell_2 \sin \theta_B \dot{\theta}_B \end{bmatrix} = \begin{bmatrix} \omega \\ -\dfrac{\ell_1 \cos \theta_A}{\ell_2 \cos \theta_B} \dot{\theta}_A \\ \ell_1 \dot{\theta}_A (\tan \theta_B \cos \theta_A - \sin \theta_A) \end{bmatrix} \tag{3.87b}$$

$$\begin{bmatrix} \ddot{\theta}_A \\ \ddot{\theta}_B \\ \ddot{Z}_3 \end{bmatrix} = \begin{bmatrix} 0 \\ \left(\ell_1 \sin \theta_A \dot{\theta}_A^2 + \ell_2 \sin \theta_B \dot{\theta}_B^2 \right) \Big/ \left(\ell_2 \cos \theta_B \right) \\ -\ell_1 \cos \theta_A \dot{\theta}_A^2 - \ell_2 \cos \theta_B \dot{\theta}_B^2 - \ell_2 \sin \theta_B \ddot{\theta}_B \end{bmatrix} \qquad (3.87c)$$

Note that the results in Eq. 3.87a and 3.87b are identical to those in Example 3.5. Exercise 3.7 (at the end of the chapter) will indicate if the results match with Eq. 3.87c for accelerations.

3.3.2 Kinematic Joints

Kinematic joints (or simply *joints*) are critical parts of a mechanism. The resultant motion in operating a mechanism is largely determined by the kinematic joints connecting the members of the mechanism. The kinematic joints allow motion in some directions and constrain it in others. The types of motion allowed and constrained are related to the characteristics and intended use of the joint, which can be usually characterized by the degrees of freedom it allows.

For a planar mechanism, there are two kinds of joints: planar J1 joints that allow one dof (restrict two dof); and planar J2 joints that allow two dof (restrict one dof). The revolute and slider joints discussed before are J1 joints and the pin-in-slot is a J2 joint, as shown in Figure 3.10. Three-dimensional or spatial joints are classified into two categories based on the type of contact between the two members making a joint: lower pair joint and higher pair joint. The contact can be point, line, or area. A third category of kinematic joint comprises the joints formed by combining two or more lower pair and/or higher pair joints. Such joints are termed compound joints.

The two members forming a lower pair joint have area contact between the two mating surfaces. The contact stress is thus small for lower pair joints as compared to higher pair

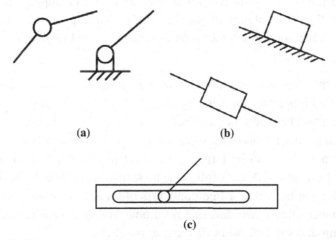

FIGURE 3.10: Planar joints. (a) J1: revolute/hinge/pin joint, (b) J1: prismatic/slider, and (c) J2: pin-in-slot.

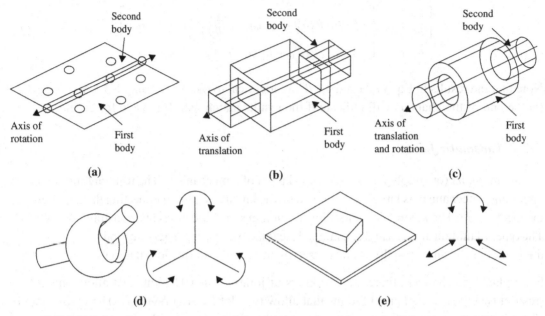

FIGURE 3.11: Common lower pair joints: (a) revolute/hinge/pin, (b) prismatic/slider, (c) cylindrical, (d) spherical/ball, and (e) planar.

joints. Lower pair joints have a long service life because the wear and stress are spread over a larger surface area of contact, and they allow better lubrication. The degrees of freedom for a lower pair joint are usually fewer, as the requirement for area contact between the members constrains the joint geometry.

Some of the lower pair joints we commonly see in mechanical systems are shown in Figure 3.11. Different joints allow different kinds of motion. For example, a revolute (or hinge or pin) joint allows the rotation of one rigid body with respect to another rigid body about a common axis. Therefore, a revolute joint eliminates a total of five dof: three translational and two rotational.

The contact between the two members of a higher pair joint has point or line geometry. The contact stress for a higher pair joint is large because the contact area is very small. If there is pure rolling contact between the members, then there is no relative sliding between the contact surfaces and thus friction and wear are minimal. The number of degrees of freedom for a higher pair joint can be large as the point or line contact allows for less constrained motion of members. A cam-follower, as shown in Figure 3.12a, is a good example of higher pair joints, where a line contact is observed between the cam and the follower. A cam-follower allows two dof, one rotational and one translational, along the center axes of the cam and the followers that are in parallel.

Lower pair and/or higher pair joints are combined as per the design requirement to obtain compound joints. Compound joints composed of higher pair joints can be kinematically

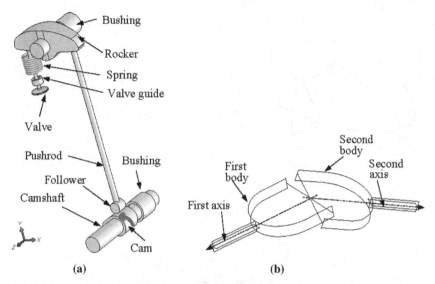

FIGURE 3.12: Higher pair joints: (a) a cam-follower in the mechanism of an engine inlet or outlet valve; (b) universal joint.

equivalent to lower pair joints and vice versa. By such combinations desirable features from the combining joints are retained to achieve robust joints.

A good example of a composite joint is a ball or roller bearing. The actual members in contact are balls or rollers with the inner and outer race. This is a rolling contact, which is a higher pair. However, the overall joint has the motion geometry of a revolute joint, a lower pair. A ball bearing has the low friction properties of rolling contacts and the high load capacity of revolute joints. Ball or roller bearings are kinematically equivalent to simple revolute joints. Another example is the universal joint shown in Figure 3.12b. This is a combination of revolute joints and has two rotational degrees of freedom.

Planar joints are much simpler to define mathematically than their counterparts in space. For example, a planar revolute joint allows relative rotation at a point P that is common to Bodies i and j, as shown in Figure 3.13a. If one body is held fixed, the other body has only a single rotational dof. Thus, a planar resolute joint eliminates two dof from the pair. This joint is defined by locating point P_i on Body i by $s_i'^P$ in the $x_i' - y_i'$ frame (fixed to body i) and locating P_j on Body j by $s_j'^P$ in the $x_j' - y_j'$ frame (fixed to body j), respectively. The constraint equation can be defined as

$$\boldsymbol{\Phi}^{r(i,j)} = \boldsymbol{r}_i + \boldsymbol{s}_i^P - \boldsymbol{r}_j - \boldsymbol{s}_j^P$$
$$= \boldsymbol{r}_i + \boldsymbol{A}_i \boldsymbol{s'}_i^P - \boldsymbol{r}_j - \boldsymbol{A}_j \boldsymbol{s'}_j^P \qquad (3.88)$$

where $\boldsymbol{A}_i$ and $\boldsymbol{A}_j$ are the transformation matrices that transform position vectors $\boldsymbol{s'}_i^P$ and $\boldsymbol{s'}_j^P$ from their respective frames $x_i' - y_i'$ and $x_j' - y_j'$ to the global frame X-Y, respectively. Note that

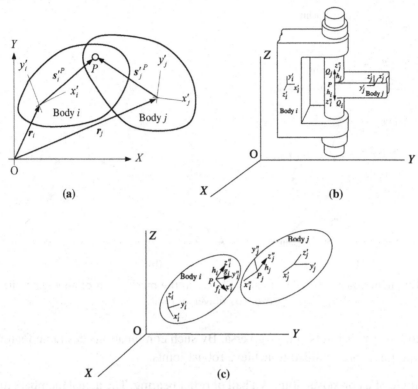

FIGURE 3.13: Mathematical formulation of a revolute joint: (a) planar joint, (b) spatial joint, and (c) dot-1 constraint.

there are two equations in Eq. 3.88; therefore, motion is constrained in both the X- and Y-directions. Also note that the constraint equations of Eq. 3.78 in Example 3.8 define a revolute joint at points P_1 and P_2 of the crank and rod, respectively.

A spatial revolute joint between Bodies i and j allows relative rotation about a common axis, but precludes relative translation along this axis, as shown in Figure 3.13b. To define the revolute joint, the joint center is located on Bodies i and j by points P_i and P_j. The axis of relative rotation is defined in Bodies i and j by points Q_i and Q_j and hence unit vectors h_i and h_j along the respective z''-axes of the joint reference frames. The remaining joint definition frame axes are defined at the convenience of the designer. The analytical formulation of the revolute joint is that points P_i and P_j coincide and that body-fixed vectors h_i and h_j are parallel, leading to the constraint equations

$$\Phi^s(P_i, P_j) = r_i + A_i s'^P_i - r_j - A_j s'^P_j = 0 \qquad (3.89a)$$

$$\Phi^p(h_i, h_j) = \begin{bmatrix} \Phi^d(f_i, h_j) \\ \Phi^d(g_i, h_j) \end{bmatrix} = 0 \qquad (3.89b)$$

Table 3.1: Common lower pair joint employed in motion simulations.

| Joint Type | dof Removed | | | Remarks |
	Translation	Rotation	Total	
Revolute	3	2	5	Rotates about axis
Translational	2	3	5	Translates along axis
Cylindrical	2	2	4	Translates along and rotates about axis
Spherical	3	0	3	Rotates in any direction
Planar	2	1	3	Bodies connected by planar joint move in plane w.r.t. each other. Rotation about axis perpendicular to plane
Universal	3	1	4	Rotates about two axes

There are three scalar equations in Eq. 3.89a, which eliminate three relative translational dof between Bodies i and j at points P_i and P_j. Eq. 3.89b defines a parallel constraint, consisting of two so-called dot-1 constraints. A dot-1 constraint is defined by a dot product of two perpendicular vectors. In this case, f_i is perpendicular to h_j and g_i is perpendicular to h_j. Also, as shown in Figure 3.13c, h_i is perpendicular to both f_i and g_i; therefore, h_i and h_j are in parallel. There are, overall, five constraint equations in Eqs. 3.89a and 3.89b; all are independent and allow only rotation along h_i or h_j—thus, a revolute joint in space. The dof that each spatial joint eliminates are summarized in Table 3.1.

3.3.3 Multibody Dynamic Analysis

Consider mechanical systems that are made up of a collection of rigid bodies in a plane, with kinematic constraints between them. The variational approach commonly found in dynamics textbooks is employed to formulate differential equations of motion. The key idea is to couple the differential equations of motion with the kinematic constraint equations by introducing Lagrange multipliers to account for the constraints.

The variational equation of motion for a rigid body in a plane can be formulated as

$$\delta r^T [m\ddot{r} - F] + \delta\theta^T [J'\ddot{\theta} - n] = 0 \tag{3.90}$$

where

r is the position vector to the mass center of the body.
F and n are external forces and torque, respectively.
J' is the polar moment of inertia at the centroid of the body referring to the body-fixed reference frame x'-y' (see Figure 3.13a).
δr and $\delta\theta$ are the virtual displacements and rotation of the rigid body, respectively.

The body-fixed reference frame x'-y' is defined at the centroid of the body.

Equation 3.90 can be written as

$$\delta q^T [M\ddot{q} - Q] = 0 \tag{3.91}$$

where

$q = [r, \theta]^T$ is the vector of generalized coordinates.

$Q = [F, n]^T$ is the vector of generalized forces.

M is the diagonal mass matrix consisting of mass m and moment of inertia J' for the rigid body.

The variational equations of motion for each Body i in a planar multibody system of nb bodies, given by Eq. 3.91, may be summed to obtain the system variational equations of motion:

$$\sum_{i=1}^{nb} \delta q_i^T [M_i \ddot{q}_i - Q_i] = 0 \tag{3.92}$$

Redefine q, M, and Q as the composite state variable vector, composite mass matrix, and composite vector of generalized forces, respectively:

$$q = \left[q_1^T, q_2^T, ..., q_{nb}^T \right]^T \tag{3.93a}$$

$$\mathbf{M} = \text{diag}(\mathbf{M}_1, \mathbf{M}_2, ..., \mathbf{M}_{nb}) \tag{3.93b}$$

$$Q = [Q_1^T, Q_2^T, ..., Q_{nb}^T]^T \tag{3.93c}$$

The variational equations of Eq. 3.93 can be written as

$$\delta q^T \left[M\ddot{q} - Q^A \right] - \delta q^T Q^C = 0 \tag{3.94}$$

where the generalized force vector Q has been separated into two parts, Q^A and Q^C, representing the externally applied forces and constraint forces due to the joints defined between bodies.

Note that by Newton's law of action and reaction, if there is no friction in kinematic joints, constraint forces act perpendicular to contacting surfaces and are equal in magnitude. Thus, if attention is restricted to virtual displacements that are consistent with the constraints that act on the system, then the virtual work of all constraint forces is zero:

$$\delta q^T Q^C = 0 \tag{3.95}$$

Hence, Eq. 3.94 becomes

$$\delta q^T \left[M\ddot{q} - Q^A \right] = 0 \tag{3.96}$$

Now we introduce Lagrange multipliers λ to couple the equation of motion of Eq. 3.96 with the constraint equation of Eq. 3.71:

$$\left[M\ddot{q} - Q^A \right]^T \delta q + \lambda^T \Phi_q \delta q = \left[M\ddot{q} + \lambda^T \Phi_q - Q^A \right]^T \delta q = 0 \tag{3.97}$$

for arbitrary δq. Therefore, the Lagrange multiplier form of the equations of motion is

$$M\ddot{q} + \lambda^T \Phi_q - Q^A = 0 \tag{3.98}$$

In addition to these equations of motion, recall that the velocity and acceleration equations of Eqs. 3.76 and 3.77, respectively, comprise the complete set of constrained equations of motion for the system.

Equations 3.98 and 3.77 may be written in matrix form as

$$\begin{bmatrix} M & \Phi_q^T \\ \Phi_q & 0 \end{bmatrix} \begin{bmatrix} \ddot{q} \\ \lambda \end{bmatrix} = \begin{bmatrix} Q^A \\ \gamma \end{bmatrix} \tag{3.99}$$

This is a mixed system of differential-algebraic equations (DAE) since no derivatives of the Lagrange multipliers λ appear. Note that the Lagrange multipliers are related to the reaction forces and torque at the joint. These reactions are critical for the structural integrity and durability of the product.

Example 3.9

Derive the differential-algebraic equation for the pendulum from Example 3.7.

Solution
Successively differentiate the following constraint equations:

$$\Phi^K(q) \equiv \begin{bmatrix} x_1 - \ell \sin \theta \\ y_1 + \ell \cos \theta \end{bmatrix} = 0 \tag{3.100}$$

The velocity and acceleration equations are, respectively,

$$\Phi_q \dot{q} = \begin{bmatrix} 1 & 0 & -\ell \cos \theta \\ 0 & 1 & -\ell \sin \theta \end{bmatrix} \begin{bmatrix} \dot{x}_1 \\ \dot{y}_1 \\ \dot{\theta} \end{bmatrix} = 0 \equiv \nu \tag{3.101}$$

$$\Phi_q \ddot{q} = \begin{bmatrix} -\ell\dot{\theta}^2 \sin\theta \\ \ell\dot{\theta}^2 \cos\theta \end{bmatrix} \equiv \gamma \tag{3.102}$$

From Eqs. 3.99 and 3.102, the differential-algebraic equations are

$$\begin{bmatrix} m & 0 & 0 & 1 & 0 \\ 0 & m & 0 & 0 & 1 \\ 0 & 0 & 0 & -\ell\cos\theta & -\ell\sin\theta \\ 1 & 0 & -\ell\cos\theta & 0 & 0 \\ 0 & 1 & -\ell\sin\theta & 0 & 0 \end{bmatrix} \begin{bmatrix} \ddot{x}_1 \\ \ddot{y}_1 \\ \ddot{\theta} \\ \lambda_1 \\ \lambda_2 \end{bmatrix} = \begin{bmatrix} 0 \\ -mg \\ 0 \\ -\ell\dot{\theta}^2 \sin\theta \\ \ell\dot{\theta}^2 \cos\theta \end{bmatrix} \tag{3.103}$$

Solve Eq. 3.103 to obtain

$$\begin{bmatrix} \ddot{x}_1 \\ \ddot{y}_1 \\ \ddot{\theta} \\ \lambda_1 \\ \lambda_2 \end{bmatrix} = \begin{bmatrix} -\sin\theta\left(\ell\dot{\theta}^2 + g\cos\theta\right) \\ -g\sin^2\theta + \ell\cos\theta\dot{\theta}^2 \\ -\dfrac{g}{\ell}\sin\theta \\ -m\ddot{x}_1 \\ -m\left(g + \ddot{y}_1\right) \end{bmatrix} \tag{3.104}$$

Note that the third equation in Eq. 3.104 is identical to that in Example 3.1, which was derived using Newton's method. Also, in this example, λ_1 and λ_2 are the reaction forces at the pin joint.

The approach just shown is much more general than those found in classical dynamics (some of the methods were discussed in Section 3.2). Any mechanical system that is characterized by bodies and kinematic joints can be formulated as a set of differential equations (Eqs. 3.86) and as a mixed system of differential-algebraic equations, as shown in Eq. 3.99 for kinematic and dynamic analysis, respectively. Equations of motion can be solved using numerous numerical algorithms. Interested readers are referred to, for example, Haug (1989) and Kane and Levinson (1985) for a comprehensive discussion of this subject.

3.4 Motion Simulation

The overall process of using computer tools for motion analysis consists of three main steps: model generation (or preprocessing), analysis (or simulation), and result visualization (or postprocessing), as illustrated in Figure 3.14. Key entities that constitute a motion model include servomotors that drive the mechanism for kinematic analysis, external loads (force and torque), force entities such as spring and damper, and the mechanism's initial conditions. Most important, assembly mates must be properly defined in CAD for the mechanism so that the motion model captures essential characteristics and closely resembles the physical behavior of the mechanical system. Note that most motion analysis software accepts

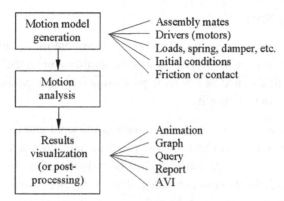

FIGURE 3.14: General process for use of computer tools for motion analysis.

assembly mates defined in CAD and converts them into kinematic joints for support of motion analysis.

The analysis or simulation is carried out by a simulation engine, which is a numerical solver that solves the equations of motion for the mechanism. The solver calculates the position, velocity, acceleration, and reaction forces acting on each of the mechanism's moving parts. Typical problems, such as static (equilibrium configuration) and motion (kinematic and dynamic) are supported.

The analysis results can be visualized in various forms. The motion of the mechanism may be animated, or graphs for more specific information, such as the reaction force of a joint in the time domain, may be generated. The user may also query results at specific locations for a given time and may ask for a report on specified results, such as the acceleration of a moving part in the time domain. The motion animation may be saved to an AVI for faster viewing and file portability.

3.4.1 Creating Motion Models

The basic entities of a valid motion simulation model include ground parts (or ground body), moving parts (or moving bodies), constraints (imposed by assembly mates in CAD), initial conditions (usually the position and velocity of a moving body), and forces and/or drivers. Each of the basic entities is briefly discussed in the following subsections.

Ground Parts (or Ground Bodies)

A ground part, or a ground body, represents a fixed reference in space. The first component brought into the CAD assembly is usually stationary, therefore becoming a ground body. Parts (or subassemblies) assembled to the stationary components with no possibility of moving become part of the ground body. Moving and nonmoving parts in the assembly must be identified and assembly mates must be defined to completely constrain the movement of the nonmoving parts.

Moving Parts (or Moving Bodies)

A moving part or body represents a single rigid component that moves relative to other parts (or bodies). It may consist of a single part or a subassembly composed of multiple parts. When a subassembly is designated as a moving part, none of its composing parts is allowed to move relative to one another within it.

A moving body has six degrees of freedom—three translational and three rotational—while a ground body has none. That is, a moving body can translate and rotate along the X-, Y-, and Z-axes of a coordinate system. Rotation of a rigid body is measured by referring the orientation of its local coordinate system to the global coordinate system, which is usually the default system of the assembly in CAD. In motion simulation software, the local coordinate system is assigned automatically, usually at the mass center of the part or subassembly. Mass properties, including total mass, moment of inertia, and so forth, are calculated using part geometry and material properties referring to the local coordinate system.

Constraints

As mentioned earlier, an unconstrained rigid body in space has six degrees of freedom: three translational and three rotational. When a joint (or a constraint) is added between two rigid bodies, degrees of freedom between them are removed. In CAD, more commonly employed joints (e.g., revolute, translation, cylindrical) have been replaced by assembly mates. Like joints, assembly mates remove degrees of freedom between parts.

Each independent movement permitted by a constraint (either a joint or a mate) is a free degree of freedom. The free degrees of freedom that a constraint allows can be translational or rotational along the three perpendicular axes. For example, a concentric mate between the propeller assembly and the case of a single-piston engine shown in Figure 3.15 allows one translational dof (movement along the center axis—in this case the X-axis) and one rotational dof (rotating along X-axis). Since the case assembly is stationary, serving as the ground body, the propeller assembly has two free dof. Adding a coincident mate between the two respective faces of the engine case and the propeller shown in Figure 3.15 removes the remaining translational dof, yielding a desired assembly that resembles the physical situation—that is, with only the rotational dof (along the X-axis).

In creating a motion model, instead of all movements being completely fixed, certain dof (translational and/or rotational) are left to allow desired movement. Such a movement is either driven by a motor, resulting in a kinematic analysis, or determined by a force, leading to a dynamic analysis. For example, a rotary motor is created to drive the rotational dof of the propeller in the engine example. This motor rotates the propeller at a prescribed angular velocity. In addition to a prescribed velocity, the motor may be used to drive a dof at a prescribed displacement or velocity, either translational (using a linear motor) or rotational (using a rotary motor).

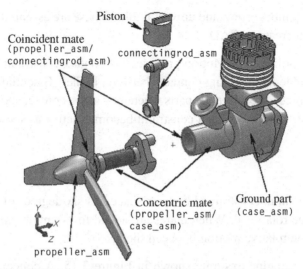

FIGURE 3.15: Assembly constraints defined for the engine model (exploded view).

It is extremely important to understand assembly mates in order to create successful motion models. In addition to standard mates such as concentric and coincident, CAD (SolidWorks, for example) provides advanced and mechanical mates. Advanced mates provide additional ways to constrain or couple movements between bodies. For example, a linear coupler allows the motion of a translational or a rotational dof of a given mate to be coupled with that of another mate. A coupler removes one additional degree of freedom from the motion model. Also, a path mate allows a part to move along a curve slot, a groove, or a fluting, varying its moving direction specified by the path curve, as shown in Figure 3.16. Such a capability supports animation and motion analysis of a whole new set of applications involving curvilinear motion. In addition to the path mate, the mechanical mates include cam-follower,

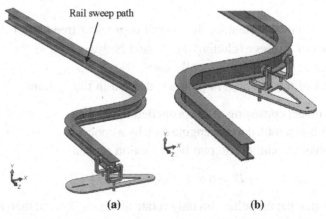

FIGURE 3.16: Rail with path mates: (a) sweep path and (b) rail moving along the sweep path.

gear, hinge, rack and pinion, screw, and universal joint. These are essential for motion models yet extremely easy to create in CAD.

In addition to mates, SolidWorks Motion provides 3D contact constraint, which helps to simulate physical problems involving contacts between bodies. Essentially, 3D contact constraint applies a force to separate the parts when they are in contact and prevent them from penetrating each other. The 3D contact constraint becomes active as soon as the parts are in contact.

Degrees of Freedom

As mentioned before, an unconstrained body in space has six degrees of freedom: three translational and three rotational. When mates are added to assemble parts, constraints are imposed to restrict the relative motion between them.

Let us go back to the engine example shown in Figure 3.15. A concentric mate between the propeller and the engine case restricts movement on four dof (Ty, Tz, Ry, and Rz) so that only two movements are allowed, one translational (Tx) and one rotational (Rx). To restrict the translational movement, a coincident mate is added. A coincident mate between two respective faces of the propeller and the engine case restricts movement of three dof: Tx, Ry, and Rz. Even though combining these two mates achieves the desired rotational motion between the propeller and the case, redundant dof are imposed: Ry and Rz in this case.

It is important to understand how to count the overall degrees of freedom for a motion model. For a given motion model, the number of degrees of freedom can be determined using Gruebler's count, defined as

$$D = 6M - N - O \tag{3.105}$$

where

> D is Gruebler's count representing the overall degrees of freedom of the mechanism.
> M is the number of bodies excluding the ground body.
> N is the number of dof restricted by all mates.
> O is the number of motion drivers (motors) defined in the system.

Consider a motion model consisting of the propeller, the engine case, and the rotary motor, in which the propeller is assembled to the engine case by a concentric and a coincident mate (see Figure 3.15). Gruebler's count of the two-body motion model is

$$D = 6 \times 1 - (4 + 3) - 1 = -2$$

However, we know that the propeller can only rotate along the X-axis; therefore, there is only 1 dof for the system (Rx), so the count should be 1. After adding the rotary motor, the count becomes 0. The calculation gives us -2 because there are two redundant dof, Ry and Rz, which are restrained by both concentric and coincident mates. If we remove the redundant dof, the count becomes

$$D = 6 \times 1 - (4 + 3 - 2) - 1 = 0$$

Another example is a door assembled to a door frame by two hinge joints. Each hinge joint allows only one rotational movement along the axis of the hinge. The second hinge adds five redundant dof. Gruebler's count becomes

$$D = 6 \times 1 - 2 \times 5 = -4$$

Again, if we remove the redundant dof (by removing the second hinge), the count becomes, as it should,

$$D = 6 \times 1 - (2 \times 5 - 5) = 1$$

This is before any motor is added. Usually, redundant constraints are present in a CAD assembly. When a CAD assembly is properly assembled with the desired kinematics, Gruebler's count is often less than 0 because of the redundant dof constrained.

For kinematic analysis, Gruebler's count must be equal to 0 after adding motors. The solver recognizes and deactivates redundant constraints during analysis. For a kinematic analysis, if you create a model and try to animate it with a Gruebler's count greater than 0, the motion analysis does not run and an error message appears. For the door example, the vertical movement constrained by the second hinge is identified as redundant and removed from the solution. As a result, in a dynamic simulation the entire vertical force is carried by the first hinge. No reaction force is calculated at the second hinge.

To get Gruebler's count to 0, it is often necessary to replace mates that remove a large number of dof with mates that remove fewer dof but still restrict the mechanism motion in the same way. Most motion solvers detect redundancies and ignore redundant dof in all but dynamic analyses. In dynamic analysis, the redundancies can lead to possibly incorrect reaction results, yet the motion is correct. For complete and accurate reaction forces, it is critical to eliminate redundancies from the mechanism. The challenge is to find the mates that impose nonredundant constraints and still allow the intended motion. A combination of a concentric and a coincident mate is kinematically equivalent to a revolute joint, as illustrated in Figure 3.15, between the propeller and the engine case. A revolute joint removes five dof (with no redundancy); however, combining a concentric and a coincident mate removes seven dof, among which two are redundant. Using assembly mates to create motion models almost guarantees redundant dof.

The best strategy is to create an assembly that closely resembles the physical mechanism by using mates that capture the characteristics of the motion revealed in the physical model. That is, an assembly should first be created that correctly captures the mechanism's kinematic behavior. If the purpose of a dynamic analysis is to capture reaction forces at critical components, the assembly mates must be examined to identify redundant dof. Then, reaction forces at all mates should be checked for the component of interest and only those that make sense should be taken—that is, mostly nonzero forces. Note that zero reaction force is reported at the redundant dof in most motion simulation software.

Forces

Forces are used to operate a mechanism. Physically, forces are produced by motors, springs, dampers, gravity, and so forth. A force entity in motion software can be a force or a torque. Usually motion software provides three types of force: applied forces, flexible connectors, and gravity. Applied forces are those that cause the mechanism to move in certain ways. They are very general, but the force magnitude must be defined by specifying a constant force value or expression function, such as a harmonic function.

Flexible connectors resist motion and are simpler and easier to use than applied forces because only constant coefficients for the forces—a spring constant for example—are supplied. The flexible connectors include in general translational springs, torsional springs, translational dampers, torsional dampers, and bushings.

A magnitude and a direction must be included for a force definition. A predefined function, such as a harmonic function, may be selected to define the magnitude of the force or moment. For springs and dampers, motion software makes the force magnitude proportional to the distance or velocity between two points, based, respectively, on the spring constant and damping coefficient entered. The direction of a force (or moment) can be defined either along an axis defined by an edge or along the line between two points, where a spring or a damper is defined.

Initial Conditions

In motion analysis, initial conditions consist of the initial configuration of the mechanism and the initial velocity of one or more of the mechanism's components. Motion analysis must start with a properly assembled solid model that determines an initial mechanism configuration, composed of position and orientation of individual components. The initial configuration can be completely defined by assembly mates. However, one or more assembly mates must be suppressed, if the assembly is fully constrained, to provide adequate movement.

Motion Drivers

Motion drivers (or motors) impose a particular movement on a free dof over time. A motion driver specifies position, velocity, or acceleration as a function of time, and can control either translational or rotational motion. When properly defined, a motion driver accounts for the remaining dof of the mechanism that brings Gruebler's count to zero (exactly zero after removing all redundant dof) or fewer for a kinematic analysis.

3.4.2 Motion Analysis

The motion solver is capable of solving typical engineering problems, such as static (equilibrium configuration), kinematic, and dynamic. Three common numerical solvers are provided in motion software (e.g., SolidWorks Motion). They are GSTIFF, SI2_GSTIFF, and WSTIFF. GSTIFF is the default integrator and is fast and accurate for displacements. It is used for wide range of motion simulations. SI2_GSTIFF provides better accuracy of velocities and accelerations, but can be significantly slower. WSTIFF provides better accuracy for special problems, such as discontinuous forces.

Static analysis is used to find the rest position (equilibrium condition) of a mechanism in which none of the bodies are moving. A simple example of static analysis is illustrated in Figure 3.17a, in which an equilibrium position of a block is to be determined according to its own mass m, the two spring constants k_1 and k_2, and the gravity g. Very often, a static analysis is carried out to find the initial equilibrium configuration of the system before a kinematic or dynamic analysis is conducted.

As discussed earlier, kinematics is the study of motion without regard to the forces that cause it. A mechanism can be driven by a motion driver for a kinematic analysis, where the position, velocity, and acceleration of each link of the mechanism can be analyzed for a given period. Figure 3.17b shows a servomotor driving a mechanism at a constant angular velocity. Dynamic analysis is employed for studying the mechanism motion in response to loads, as illustrated in Figure 3.17c. This is the most complicated and common, and usually more time-consuming, analysis.

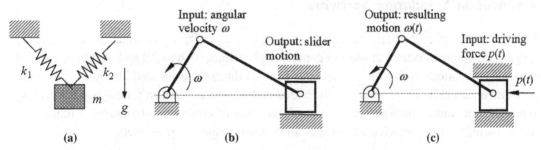

FIGURE 3.17: Common motion analyses: (a) static, (b) kinematic, and (c) dynamic.

3.4.3 Results Visualization

In motion analysis software, the results of the analysis can be realized using animations, graphs, reports, and queries. Animations show the configuration of the mechanism in consecutive time

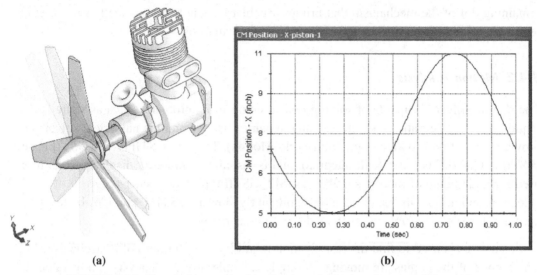

(a)   (b)

FIGURE 3.18: Simulation result visualization: (a) animation and (b) graph.

frames. They give a global view of the mechanism's behavior, for example, a single-piston engine shown in Figure 3.18a. The animation may be exported to AVI for other needs.

A joint or a part may be chosen to generate result graphs. An example is the position versus time of the piston in the engine example, as shown in Figure 3.18b. Graphs provide a quantitative understanding on the characteristics of the mechanism. In addition, most motion simulation software allows checking of interference between bodies during motion. Furthermore, the reaction forces calculated can be used to support finite element analysis of a component using, for example, ANSYS®.

3.5 Motion Simulation Software

A large number of commercial motion software tools are currently obtainable. Among them are general-purpose codes that support general applications, such as 2D and 3D kinematic and dynamic simulations. Some have strong ties to CAD through designated interfaces; some are even embedded in it. There are also specialized codes, such as CarSim®, designed for vehicle dynamic simulations. In this section, a brief overview of commercially available motion codes, including their relative advantages and disadvantages, is presented.

3.5.1 General-Purpose Codes

Adams® and dynamic analysis and design systems (DADS) are the earliest general-purpose codes, becoming available commercially in late 1970s. They were text based and available only on UNIX systems. Users had to create input data files that defined the mechanical system, including bodies, mass properties, joints, initial conditions, and forces. Analysis results were presented in text files and X-Y graphs. Since early 1990s, both codes migrated to PCs and incorporated graphics-based user interfaces for pre- and post-processing, which significantly simplified model creation and improved result visualizations. In the early 2000s Adams and DADS were integrated with CAD. For example, LMS released CAT/DADS (integrated with CATIA®) in 2000, which is now part of the LMS® Virtual.Lab.

Today, these two general-purpose codes remain the most widely used in academia and industry. Both offer capabilities for vehicle dynamic simulations with excellent tire models as well as flexible-body dynamic simulations. These are basically analysis tools for engineers working in dynamic analysis. They are used mainly for support of detail design.

In the 1990s CAD began offering motion simulation capabilities embedded in respective CAD systems, including Pro/ENGINEER Mechanism Design, SolidWorks COSMOSMotion™ (renamed SolidWorks Motion after 2008), CATIA Motion, NX™ Motion Simulation-RecurDyn, and Solid Edge® Motion. Very recently, IN-Motion became a new add-in module for AutoCAD Inventor. All of these codes provide a seamless connection from and to their respective CAD systems without the need for any geometric translators. All operations are performed within the CAD environment, including pre-processing, analysis, and post-processing. The learning curve on these software tools is usually relatively flat if the user has CAD experience. However, all of them provide very basic kinematic and dynamic simulation capabilities. CAD-embedded tools are in general less powerful and limited to rigid bodies. They are usually more error prone but much easier to learn because of their simplicity and CAD connections. They are more or less designers' tools, mainly for support of mostly conceptual design.

3.5.2 Specialized Codes

Specialized codes usually offer capabilities in specific engineering fields in addition to standard capabilities. Two popular specialized codes are commercially available: CarSim (Mechanical Simulation Corp.) and Adams/Car (MSC Software Corp.). Both support vehicle dynamics simulations.

The Windows-based CarSim version was first released in 1997. CarSim simulates vehicle dynamics, where the vehicle model consists of braking, acceleration, handling, and riding. CarSim's vehicle model can respond to driver control, ground, and aerodynamics. This software supports scenarios such as rollover of double lane changes and stability of trailer towing.

Adams/Car is a template-based modeling and simulation tool that helps journeyman engineers (especially students) speed up and simplify the vehicle modeling process. With Adams/Car, users can simply enter vehicle model data into the templates and the program automatically constructs subsystem models (e.g., engine, shock absorbers, tires) as well as

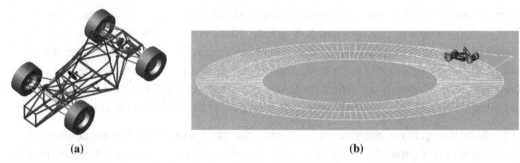

(a)   (b)

FIGURE 3.19: Vehicle dynamic simulation of a Formula SAE racecar: (a) 15-dof Adams/Car model, and (b) skid pad racing (constant-radius cornering simulation).

full vehicle assemblies. Once these templates are created, they can be made available to novice users of the software, enabling them to perform standardized vehicle maneuvers.

A Formula SAE racecar model developed by engineering students was converted into an Adams/Car model using the provided templates, as shown in Figure 3.19a. The vehicle model was then simulated for skid pad racing, a constant-radius cornering simulation shown in Figure 3.19b.

3.6 Case Studies

In this section, four case studies and two tutorial examples are presented. The case studies involve a broad range of applications, including a kinematic study of a racecar suspension, the design of a HMMWV suspension, driving simulators, and recreational waterslides. The purpose of these case studies is to demonstrate the engineering capabilities of motion analysis software and some of its common industry applications. Tutorial examples, including a sliding block mechanism and a single-piston engine, are also presented. Step-by-step instructions for creating these tutorial examples are given in Project S2 and P2. Model files are available for download on this book's companion website (http://booksite.elsevier.com/9780123984609).

3.6.1 Formula SAE Racecar

The Formula SAE (Society of Automotive Engineers) racecar study involves kinematic and dynamic analyses of a racecar-style suspension, as shown in Figure 3.20. Each year engineering students throughout the world design and build formula-style racecars and participate in annual Formula SAE competitions (students.sae.org). The competition is a meaningful engineering experience that provides an opportunity for students to work in a dedicated team environment.

The suspension of the entire racecar was modeled for both kinematic and dynamic analyses. The analysis results were validated using experimental data, acquired by mounting a data acquisition system on the racecar and driving the car on the test track following specific

(a)   (b)

FIGURE 3.20: Formula SAE racecar designed and built by engineering students: (a) physical racecar and (b) virtual racecar designed in Pro/ENGINEER.

driving scenarios that were consistent with those of the simulations. The results were used to aid the suspension design for handling and cornering (Wheeler 2006). Assembling an entire vehicle suspension for motion analysis is nontrivial and beyond the scope of this book. Therefore, only the right front quarter of the suspension, as shown in Figure 3.21, is included in this section.

The purpose of this case study is mainly to highlight capabilities in Pro/ENGINEER Mechanism Design and SolidWorks Motion for supporting design of the kinematic characteristics of vehicle suspension. The motion model was first created in Pro/ENGINEER for kinematic analysis, and then imported into SolidWorks for dynamic analysis and design studies. The road profile is characterized by the geometric shape of a profile cam, which is assembled to the tire using a cam-follower connection.

The quarter suspension consists of major components that essentially define the kinematic and dynamic characteristics of the racecar. These components include upper and lower control arms, upright, rocker, shock, push rod, tie rod, and wheel and tire, as shown in

FIGURE 3.21: Right front quarter of the racecar suspension.

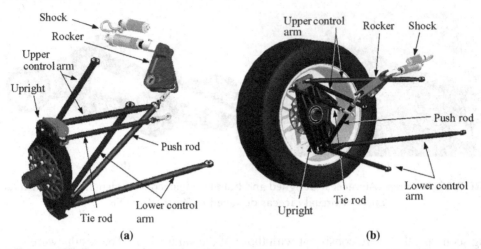

FIGURE 3.22: Major components of the quarter suspension: (a) saved view (View A) and (b) saved view (View B).

Figure 3.22. The dangling end of the shock, both control arms, rocker, and tie rod are connected to the chassis frame using numerous joints. The chassis frame was assumed fixed and the tire is pushed and pulled by the profile cam (not shown), mimicking the road profile. Two views, shown in Figure 3.22, were created to aid the visualization of the assembled model.

The tire of the quarter suspension is in contact with the profile cam that characterizes the road profile. As shown in Figure 3.23a, the geometry of the cam consists of two circular arcs of radius 7.65 in. (AB and FG), which are concentric with the cam center. Therefore, when the cam rotates, these two circular arcs do not push or pull the tire; the result is two flat segments of the road profile, as shown in Figure 3.23b. In addition, the circular arc CDE is centered 4 in. above the cam center with a radius of 4 in. When the cam rotates, arc CDE pushes the tire up, mimicking a hump of 1.35 in. (that is $8 - 7.65 = 1.35$ in., peak at point D). A ditch is characterized by an 8 in. arc (HIJ) centered at 3 in. above the cam center. As the cam rotates, arc HIJ creates a ditch 1.65 in. deep (that is $7.65 - (8 - 3) = 1.65$ in.). The remaining straight lines and arcs provide smooth transitions between flats, humps, and ditches in the road profile.

Based on the geometry of the profile cam, this quarter suspension goes over a 1.35 in. hump and a 1.65 in. ditch in one complete rotation of the cam. Note that since the radius of arc AB is 7.65 in. the cam causes the quarter suspension to travel roughly 41.8 in. (3.48 ft) in one complete rotation. Since the profile cam rotates two complete cycles in one second, the suspension travels about 6.96 ft/sec (i.e., 4.74 MPH), which is very slow.

There are nine bodies in this motion model, including the ground body. There are two rigid (no symbol) joints, three pin joints, eight ball joints, and one cylinder joint, as shown in Figure 3.24. A servomotor that rotates the profile cam for 1 sec was added to conduct

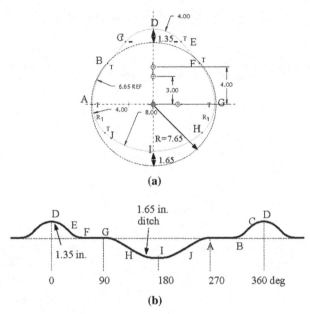

(a)

(b)

FIGURE 3.23: Road profile: (a) geometry of the profile cam and (b) road profile generated by the profile cam.

a kinematic analysis. Several measures are critical in determining the pros and cons of the suspension design. Among them, the most important one is probably the camber angle. We will show the camber angle results momentarily. Note that the camber angle is defined as the rotation of the upright along the X-axis of *WCS* (World Coordinate System, which is the reference frame of the motion model). First, we look at the shock travel distance.

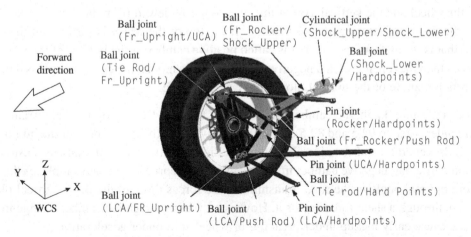

FIGURE 3.24: Joints defined for the quarter suspension assembly. UCA = upper control arm; LCA = lower control arm.

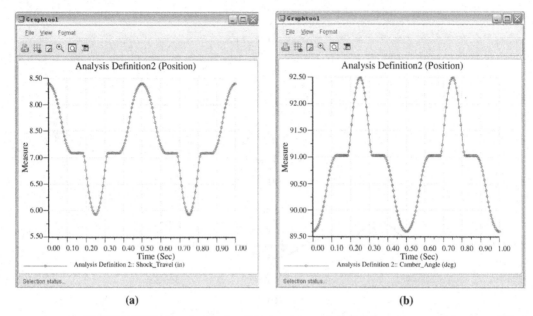

FIGURE 3.25: Result graphs: (a) shock travel and (b) camber angle.

The graph in Figure 3.25a shows that the shock travels between about 6 in. and 8.5 in. The overall travel distance is about 2.5 in., which is probably too large for such a small hump or ditch. In fact, in the simulation it appeared that the shock was compressed too much, allowing the piston to penetrate its reserve cylinder. In reality, this would not happen; however, the simulation raised a flag indicating that there could be severe contact within the shock, leading to potential part failure.

The camber angle is made by the wheel of an automobile—specifically, between the vertical axis of the wheel and the vertical axis of the vehicle when viewed from the front or rear. It is used in the design of steering and suspension. If the top of the wheel is further out than the bottom (that is, away from the axle), it is called positive camber; if the bottom of the wheel is further out than the top, it is called negative camber. In this model, the camber angle is defined as the rotation angle of the upright along the *X*-axis of *WCS*.

As shown in Figure 3.25b, the camber angle was set to about 91 deg. on the flat terrain. The camber angle varied to 92.5 and 89.5 deg. respectively, when the tire went over the hump and the ditch. In general, camber angle alters the handling quality of a particular suspension design; in particular, negative camber improves grip when cornering because it places the tire at a more optimal angle to the road, transmitting the forces through the tire's vertical plane rather than through a shear force across it. However, excessive negative camber change in the hump can cause early lockup under breaking or wheel spin under acceleration.

The Pro/ENGINEER model of the quarter suspension was imported into SolidWorks and a motion model was constructed with all assembly mates defined according to the Pro/

ENGINEER motion joints. A guide cylinder was used to define the cam mate in SolidWorks. The motion study was carried out in SolidWorks Motion, where three measures, consistent with those defined in the Pro/ENGINEER kinematic analysis, were recorded, including vertical wheel travel, shock travel, and camber angle (Figure 3.26). The result graphs show

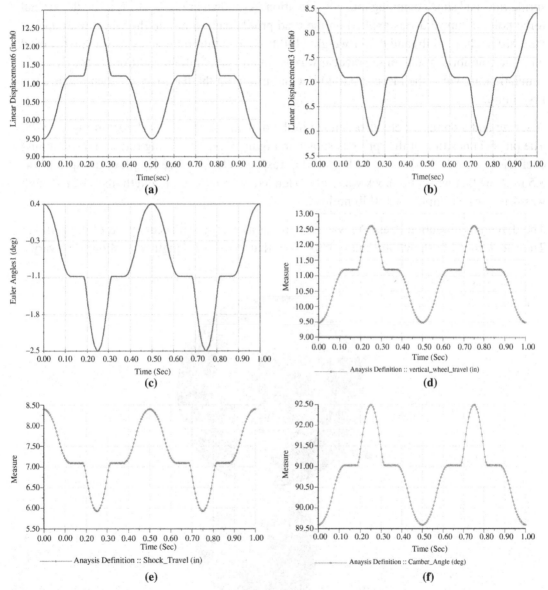

FIGURE 3.26: Verification of kinematic analysis results: (a) SolidWorks vertical wheel travel; (b) SolidWorks shock travel; (c) SolidWorks camber angle; (d) Pro/ENGINEER vertical wheel travel; (e) Pro/ENGINEER shock travel; (f) Pro/ENGINEER camber angle.

that the kinematic analysis using SolidWorks Motion yielded identical results compared with the Pro/ENGINEER analysis. These results indicate the accuracy of the model translation and reassembly and imply that the quarter suspension kinematic analysis can be duplicated in SolidWorks.

The dynamic analysis of the quarter suspension was performed by taking racecar weight, spring rate, and shock damping into consideration. As shown in Figure 3.27, a 150 lb external force pointing upward was applied on the road profile cam to mimic the wheel load due to racecar weight (445 lb) and driver weight (150 lb). An equilibrium analysis was first carried out. The equilibrium state of the racecar was assumed as the initial condition for the dynamic simulation, in which the racecar started in equilibrium on the flat road and then reached the first hump.

A spring and a damper were also defined in the dynamic analysis, as shown in Figure 3.28. The physical position of the spring is shown in Figure 3.29. The spring rate and the damping coefficient were 100 lb/in. and 10 lb/(in./sec), respectively. The free length of the spring was 5.5 in. Note that when the shock was fully extended to its maximum length, the spring length was 4 in., which implies a 150 lb preload.

The dynamic simulation (Case A) was carried out assuming a racecar speed of 4.74 MPH. The shock travel is shown in Figure 3.30. Note that the shock length was allowed to vary

FIGURE 3.27: External force of 150 lb applied to the tire.

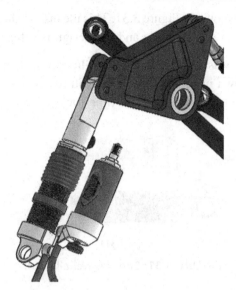

FIGURE 3.28: Spring and damper in the suspension.

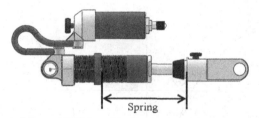

Spring

FIGURE 3.29: Physical position of the spring.

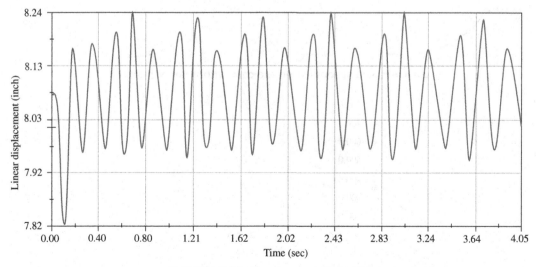

FIGURE 3.30: Shock travel (Case A).

between 7.3 in. and 9 in., as shown in Figure 3.31. This means the shock travel obtained in this dynamic simulation (Case A) is acceptable and the design is safe.

Another scenario (Case B) was created where a modified profile cam with a larger hump (4.35 in.) was used, as shown in Figure 3.32. To avoid resonance, a segment velocity (Figure 3.33) was assigned to ensure adequate time for the suspension to return to the

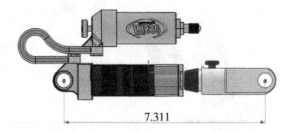

FIGURE 3.31: Shock travel allowed.

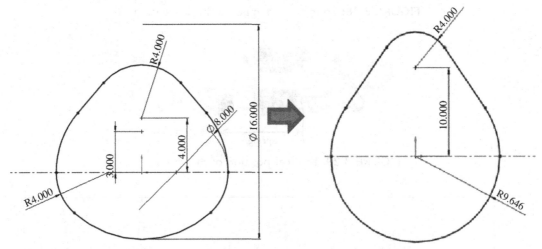

FIGURE 3.32: Modified profile cam with a larger hump.

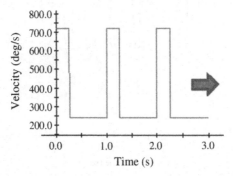

FIGURE 3.33: Segment velocity added for Case B.

equilibrium state between each hump. As can be seen from the resultant shock travel graphs (Figures 3.34a and 3.34b), in Case B the shock was compressed too much and the shock travel exceeded the permitted range, which might have led to a part failure. Making design changes therefore became necessary to bring the shock travel back to the safe range.

In this study, the three design variables investigated were the spring preload, the shape of the rocker, and the length of the push rod. The preload of the spring was 150 lb at the current design, which means that the spring was compressed by 1.5 in. when the shock was fully extended. The spring preload was increased from 150 lb (current design) to 200 lb and 250 lb; the resultant shock travels are shown in Figure 3.35. As can be seen, as the preload increased, the overall shock length slightly decreased but the piston moved further away from the cylinder. This is because as long as the compressive force on the shock is larger than the preload, the shock with a relatively large preload will be less compressed than one with a small preload under the action of the same force.

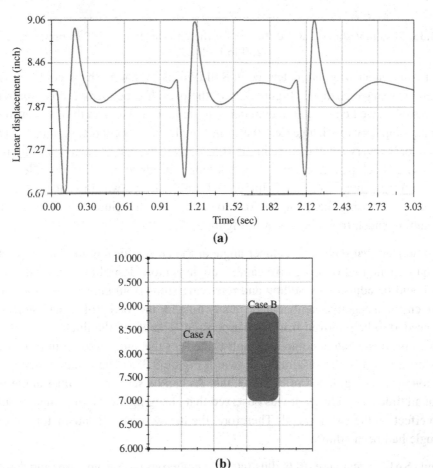

(a)

(b)

FIGURE 3.34: Shock travel: (a) Case B in the time domain and (b) comparison of Cases A and B.

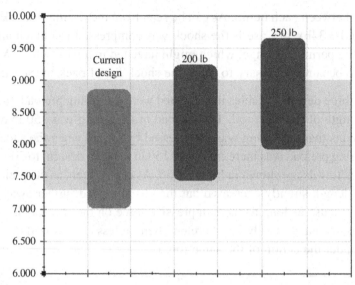

FIGURE 3.35: Resultant shock travel distances for the current design (150 lb preload), 200 lb and 250 lb preload.

The rocker component was parameterized in SolidWorks so that its shape could be adjusted. When the rocker shape changed (Figure 3.36a), the overall shock length could be reduced while the piston moved closer to the cylinder (Figure 3.36b). The length of the push rod did not have much capacity for change (less than 1 in.). It also turned out that varying the push rod length led to only a very small change in the shock travel. As can be seen from Figure 3.34b, to ensure a safe shock travel, the overall shock length needs to be reduced while the piston should move further away from the cylinder. A feasible design was then achieved by simultaneously increasing the spring preload to 250 lb and changing the rocker height to 2.7 in. The resulting shock travel is shown in Figure 3.37.

It was also noticed that the wheel camber angle at the current design might not meet the racecar requirement; that is, a negative camber angle is more desirable. The camber angle of the wheel could be adjusted by adding and removing slims from between the toe link and upright. To ensure a negative camber angle at any time for the road profile in Case B, some of the slims needed to be removed from the suspension. Therefore, the thickness of the slim block, which is an equivalent component substituted for the slims, was reduced so that only one thin slim was left in the system, as shown in Figure 3.38a. The resultant camber angle change through time is graphed in Figure 3.38b. As can be seen, the camber angle became negative at all times. In addition, it was observed that changing the camber angle would have almost no effect on the shock travel. Therefore, the design was still satisfactory after the camber angle had been adjusted.

The Formula SAE racecar case study illustrates the numerous motion analyses and designs that can be carried out using Pro/ENGINEER and SolidWorks. However, although much useful

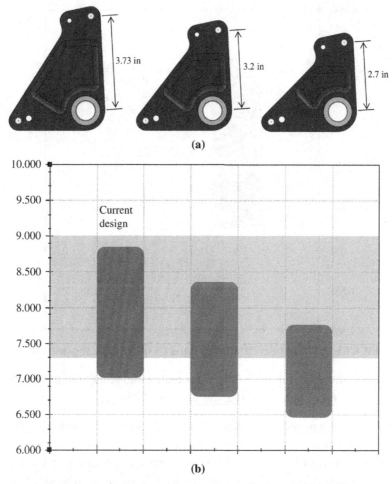

(a)

(b)

FIGURE 3.36: (a) Rocker shape change and (b) impact of the rocker shape change on the overall shock travel.

information was obtained by the kinematic and dynamic analyses using the quarter suspension model, ultimately a full-vehicle dynamic simulation must be carried out to fully understand the suspension design and, hopefully, to develop a strategy for design improvement.

3.6.2 High-Mobility Multipurpose Wheeled Vehicle

The high-mobility multipurpose wheeled vehicle (HMMWV) example discussed in Chapter 1 is presented here, in more detail, to illustrate dynamic simulation and design. More than 200 parts and assemblies were created in the CAD model, as shown in Figure 3.39a. The suspension was modeled in detail (Figure 3.39b) such that the main aspects of the vehicle's performance could be captured accurately in motion simulation. A more detailed view of the front right suspension quarter is shown in Figure 3.40a. A dynamic simulation model of 18

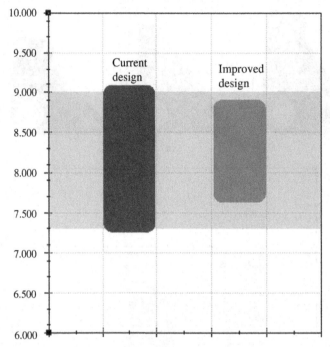

FIGURE 3.37: Impact of adding slims to the overall shock travel.

bodies and 21 joints, shown in Figure 3.40b, was created and simulated in DADS with a total of 17 sec and a time step of 0.001 sec (Chang and Joo 2006).

A 100 ft × 100 ft terrain was used for simulation (see Figure 3.41). Note that the terrain was fairly bumpy. The maximum height of the bumps on the terrain was 7.68 in. The vehicle vibrated significantly toward the later stage of the simulation because of bumpy road conditions. In this model, the vehicle was "driven" by a constant angular velocity of 1.53 rev/ sec applied at the four wheels, which produced a path that went through both bumpy areas, as shown in Figure 3.41. In this case, the vehicle was moving at a constant speed of 9.54 MPH.

The design problem was defined as follows:

Minimize

$$\phi(b) = \frac{1}{T} \int_0^T [F(\mathbf{p}(t))]^2 dt$$

Subject to

$$\psi_\alpha(b) = (z_{w_i}(t) - 18) - h_i(t) \le \psi_\alpha^u, \quad \alpha = 1,4$$

$$\psi_\beta(b) = -[(z_{w_i}(t) - 18) - h_i(t)] \le \psi_\beta^u, \quad \beta = 1,4 \qquad (3.106)$$

$$\psi_9(b) = |\ddot{z}_{ds}(t)| \le \psi_9^u$$

$$\psi_{10}(b) = |z_{ds}(t) - z_{ch}(t)| \le \psi_{10}^u$$

$$\psi_\gamma(b) = |z_{w_i}(t) - z_{ch}(t)| \le \psi_\gamma^u, \quad \gamma = 1,4$$

$$b_j^\ell \le b_j \le b_j^u, \quad j = 1,3$$

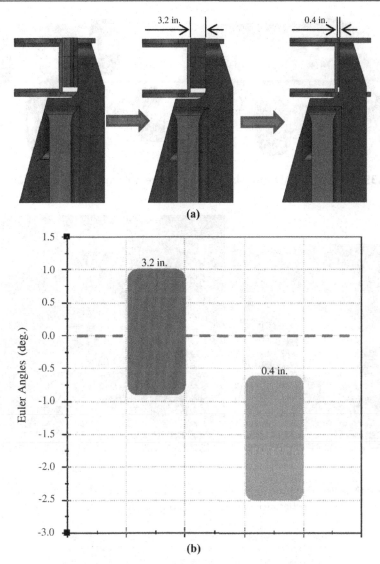

FIGURE 3.38: Impact of adding slims to the camber angle.

where

$\phi(\boldsymbol{b})$ represents energy absorption ability of the vehicle suspension at the driver's seat.

$z_{wi}(t)$ is the z-coordinate of the ith wheel center.

$h_i(t)$ is the height of the road profile corresponding to the ith wheel at the given time t.

$z_{ds}(t)$ and $\ddot{z}_{ds}(t)$ are the driver seat position and acceleration, respectively, in the z-direction of the global coordinate system (vertical).

$z_{ch}(t)$ is the z-displacement of the chassis with respect to the global coordinate system (shown in Figure 3.40b).

FIGURE 3.39: HMMWV CAD model: (a) vehicle assembly and (b) suspension assembly.

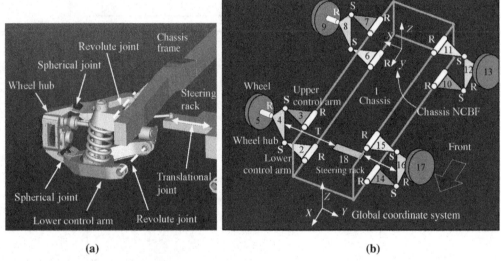

FIGURE 3.40: HMMWV dynamic simulation: (a) CAD model of the front suspension and (b) schematic view of the simulation model.

FIGURE 3.41: HMMWV dynamic simulation performed on a bumpy terrain.

Note that in Eq. 3.106, $\psi_\alpha(\boldsymbol{b})$ and $\psi_\beta(\boldsymbol{b})$ essentially characterize the deformation of the tires. ψ_β^u simply represents the lower bound of the constraints $\psi_\alpha(\boldsymbol{b})$. Note also that the tire radius is 18 in. and $z_{wi}(t) - 18$ is 0 if no deformation occurs in the tire. $\psi_\gamma(\boldsymbol{b})$ specifies the wheel center position with respect to the chassis in the vertical direction.

The function in the integrand $F(\boldsymbol{p}(t))$ of Eq. 3.106 is defined as follows (U.S. Tank-Automotive Research and Development Command 1979),

$$F(\boldsymbol{p}(t)) = p_1(t) - 0.108\, p_4(t) + 0.25\, p_6(t) - p_7(t) \tag{3.107}$$

where $p_i(t)$ can be computed from the absorbed power equations,

$$\begin{bmatrix} \dot{p}_1(t) \\ \dot{p}_2(t) \\ \dot{p}_3(t) \\ \dot{p}_4(t) \\ \dot{p}_5(t) \\ \dot{p}_6(t) \\ \dot{p}_7(t) \end{bmatrix} = \begin{bmatrix} -29.8p_1(t) - 497.49\ddot{z}_s(t) - 100.0\, p_2(t) \\ 10.0p_1(t) \\ 1736.9p_1(t) - 108.0p_4(t) \\ 100.0p_1(t) - 35.19p_3(t) - 39.1p_4(t) \\ -315.7p_1(t) + 34.0956p_4(t) + 171.075p_6(t) \\ -80.0p_1(t) - 91.36p_4(t) - 30.28p_5(t) \\ p_1(t) - 0.108p_4(t) + 0.25p_6(t) - 6.0p_7(t) \end{bmatrix} \tag{3.108}$$

for $0 \leq t \leq T$, with initial conditions

$$p_i(0) = 0, \quad i = 1, 7 \tag{3.109}$$

The initial conditions were reset at each time step during numerical calculation.

Three design variables were defined for the HMMWV: vehicle track, wheelbase, and percentage change in thickness of the lower control arm. Constraint function bounds are shown in Table 3.2. The constraint functions were evaluated at every 0.01 seconds during the total 17-second simulation period, except for the z-accelerations at the driver seat, where the step size was refined to be 0.001 seconds since the z-accelerations directly contributed to the objective function. Therefore, there were 39,100 ($13 \times 100 \times 17 + 1 \times 1000 \times 17$) constraint functions to process at each design iteration. Note that most of the constraint

Table 3.2: Upper bounds of constraint functions.

Performance Function	Description	Upper Bound
ψ_a^u	Jounce of each wheels	1.25 in.
ψ_b^u	Rebound of each wheel	3.5039 in.
ψ_9^u	Driver's seat acceleration	0.75 G
ψ_{10}^u	Driver's seat position w.r.t. chassis	3.5 in.
ψ_g^u	Wheel center position w.r.t. chassis	12.0 in.

function values at the initial design were less than their respective upper bounds, except for a few time steps of driver seat acceleration $\psi_9(\boldsymbol{b})$ and driver seat position $\psi_{10}(\boldsymbol{b})$ (see Figures 3.42a and 3.42b, respectively). This meant that the initial design was infeasible.

The optimization took 18 iterations to converge using the modified feasible direction (MFD) method. At the optimal design the objective function was reduced by 31.3%, and all performance constraints were satisfied. The track (b1) and wheelbase (b2) design variables increased by 17.6% and 14.6%, respectively. The percentage thickness of the lower control arm (design variable b3) decreased by 17.9%, as summarized in Table 3.3.

Note that the reduction in objective function value was due to the significant decreasing z-acceleration values at the driver seat (Figure 3.42a). Also, the distance between the driver

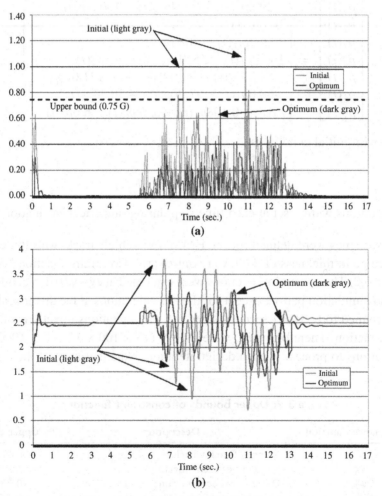

FIGURE 3.42: Change of vehicle performance: (a) driver's seat accelerations $\psi_9(\boldsymbol{b})$ (G), and (b) driver seat position $\psi_{10}(\boldsymbol{b})$ (in.).

Table 3.3: Design optimization of the HMMWV dynamic model.

Measure	Initial Design	Optimal Design	% Change
$\phi(\boldsymbol{b})$	0.00595 W	0.00405 W	−31.3
b1	21.68 in.	25.52 in.	+17.6
b2	50.69 in.	58.12 in.	+14.6
b3	100% original thickness	82.0% original thickness	−17.9

seat and the chassis in the z-direction was significantly reduced (Figure 3.42b). The rest constraint functions also indicated that the vehicle became smoother while moving along the same paths. This was due to the fact that both vehicle track and wheelbase were increased, which contributed to a wider and longer chassis and therefore more stability and less vibration. The change in HMMWV suspension geometry is shown in Figure 3.43.

3.6.3 Driving Simulators

Driving simulators are probably the most sophisticated application of computer-aided kinematic and dynamic simulations, as well as one of the biggest triumphs in their development. Similar to flight simulators, driving simulators place the driver in an artificial environment believed to be a valid substitute for one or more aspects of the actual driving experience. However, unlike flight simulators developed mainly for pilot training, driving simulators support much more than driver training. Advanced driving simulators today are used by engineers and researchers in vehicle design, intelligent highway design, and human factors studies such as driver behaviors under the influence of drugs, alcohol, and severe weather conditions. They provide a safe environment for testing in which controlled, repeated measurements can be undertaken cost-effectively. Researchers and engineers believe that the

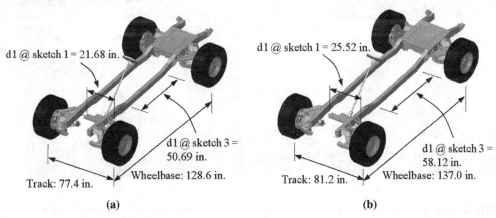

FIGURE 3.43: HMMWV suspension: (a) initial design and (b) optimal design.

measurements obtained can help them predict equivalent measurements in the real world that lead to a better understanding of the complex driver-vehicle-roadway interaction in critical driving situations. The results of such studies will ultimately lead to reductions in the number of traffic-related deaths and injuries on the nation's highways.

From its appearance, a driving simulator—for example, the University of Iowa's National Advanced Driving Simulator (NADS) in Iowa City—consists of a dome on top of a Stewart platform mounted on longitudinal and lateral rails on the ground, as shown in Figure 3.44a (Center for Computer-Aided Design, National Advanced Driving Simulator 1994). The motion system, on which the dome is mounted, provides horizontal and longitudinal travel and rotation in either direction, so the driver feels acceleration, braking, and steering cues as if he or she were actually driving a real car, truck, or bus. Inside the dome is a vehicle cab (or a full-sized vehicle body) equipped electronically and mechanically with instrumentation specific to its make and model. The 360-degree visual displays offer traffic, road, and weather conditions; a high-fidelity audio subsystem completes the driving experience. The driver is immersed in sight, sound, and movement so real that impending crash scenarios can be convincingly presented without the driver being endangered.

The key technology in a driving simulator is the real-time vehicle dynamic simulation (Bae and Haug 1987). The driver's interaction with the system through the steering wheel and gas and brake pedals is captured by sensors and electronics. The signals are converted into inputs to the underlying vehicle dynamic model. The equations of motion of the vehicle dynamic model must be solved faster than in real time, so that the actuators underneath the Stewart platform can perform the pushes or pulls that mimic various driving conditions. In the meantime, the simulation results provide large excursions in longitudinal and lateral

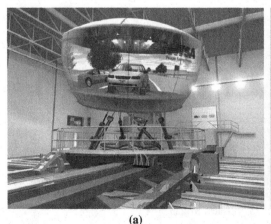
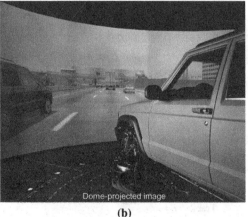

(a)   (b)

FIGURE 3.44: National Advanced Driving Simulator: (a) dome on top of motion platform mounted on rails and (b) vehicle cab and image projection inside the dome.

directions that are used to give accelerations and motions cues to the driver inside. Without the real-time dynamic simulation, the driving experience cannot claim to be truly physics based.

There are several notable driving simulators and associated research groups around the world, including those in the United Kingdom, France, Sweden, and the United States, in addition to the NADS in Iowa City (Figures 3.45a through 3.45d). Car manufacturers have built their own

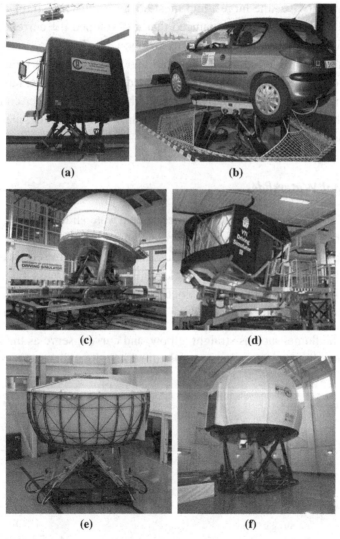

(a)   (b)

(c)   (d)

(e)   (f)

FIGURE 3.45: Driving simulators: (a) Center for Advanced Transportation Systems (University of Central Florida); (b) Valenciennes University (France); (c) Leeds University (United Kingdom); (d) Swedish National Road and Transport Research Institute; (e) Daimler-Benz (Berlin); (f) VIRTTEX (Ford Motor Co.).

simulators in the past 20 years, among them the pioneering Daimler-Benz Driving Simulator that was put into operation in 1985 and which has been enhanced since that time in many details and has been used in many different applications. Located in Berlin, this simulator went through a major modernization in 1994 (Figure 3.45e), including extension of the platform motion in a lateral direction to improve motion simulation quality. Ford Motor Company. developed a motion-based driving simulator, called Virtual Test Track Experience (VIRTTEX) to test the reactions of sleepy drivers (Figure 3.45f).

The NADS in Iowa City was the largest and most advanced driving simulator in the world until December 2007, when Toyota announced that it developed the current world's largest simulator at its Higashifuji Technical Center in Susono City. Toyota's simulator houses an actual car on a platform inside a dome 15 feet tall and 23 feet wide. It includes 360-deg. concave video screen that projects computer-generated images of roads, landscapes, street signs, and pedestrians. Also, it allows driving tests to be replicated under conditions that are too dangerous in the real world, such as the effects of drowsiness, fatigue, inebriation, illness, and inattentiveness.

3.6.4 Recreational Waterslides

This case study involves verifying the safety of recreational waterslides using motion simulation. Safety is the top priority in the construction of recreational waterslides. Safety problems discovered after the slide is built and installed are usually too late and too costly to correct. In this study the riding object was assumed to be a particle with concentrated mass. Flume sections were represented in a CAD environment using geometric dimensions such as height and width. Friction forces between the riding object and the flume surface were also incorporated.

Basic sections of the flume, such as straight, elbow, and curved, serve as the building blocks for composing waterslide configurations (see Figure 3.46). In addition, guard sections

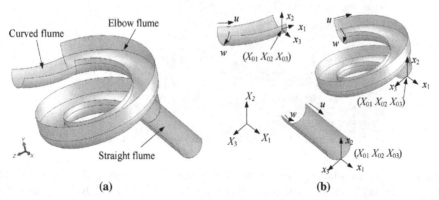

FIGURE 3.46: Geometric representation of a waterslide in flume sections: (a) assembled configuration and (b) individual flume sections.

(essentially vertical walls) are added to reinforce safety requirements, especially for elbow sections. The geometry of all sections is expressed in parametric surface forms in terms of the parametric coordinates u and w, using CAD geometric dimensions.

The overall waterslide configuration can be expressed mathematically as

$$\overline{X}(u,w) = \sum_{i}^{N} X^i\left(u^i, w^i\right) \tag{3.110}$$

where

$X^i(u^i, w^i)$ is the parametric equation of the ith flume section.
N is the total number of sections.

For each flume section (with superscript i removed for simplicity),

$$X(u,w) = [X_1(u,w), X_2(u,w), X_3(u,w)]^T \tag{3.111}$$

where $X_j(u, w)$ is the jth coordinate of any given point on the surface with prescribed parameters (u, w). Note that these sections are translated and properly oriented to compose an overall waterslide configuration by the following translation and rotation operations:

$$X(u,w) = X_0 + T(\theta)x(u,w) = \begin{bmatrix} X_{0_1} \\ X_{0_2} \\ X_{0_3} \end{bmatrix} + \begin{bmatrix} \cos\theta & 0 & \sin\theta \\ 0 & 1 & 0 \\ -\sin\theta & 0 & \cos\theta \end{bmatrix} \begin{bmatrix} x_1(u,w) \\ x_2(u,w) \\ x_3(u,w) \end{bmatrix} \tag{3.112}$$

where

$T(\theta)$ is the rotational matrix that orients the section by rotating through an angle θ about the X_2 (or Y) axis.
X_{0i} is the location of the local coordinate system of the section in the waterslide configuration.
$x(u,w)$ is the surface function of the flume section referring to its local coordinate system.

As discussed in Section 3.2.1, the Lagrange equation of motion based on Hamilton's principle (Kane 1985) for this particle dynamic problem, shown in Figure 3.47, can be stated as

$$\frac{d}{dt}\left(\frac{\partial L}{\partial \dot{q}}\right) - \frac{\partial L}{\partial q} = Q \tag{3.113}$$

where

The Lagrangian function L is defined as $L \equiv T - V$, $\dot{q} = \partial q / \partial t$.
The generalized coordinates q in this waterslide application are the parametric coordinates of the surface (i.e., $q = [u, w]^T$).

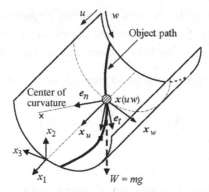

FIGURE 3.47: Object path and unit vectors for friction forces on a straight flume section.

When the system is conservative, $Q = 0$. For a nonconservative system, $Q = F$, where F is the vector of generalized friction forces. For this motion analysis problem, the kinetic energy T and the potential energy U are, respectively,

$$T = \frac{m}{2}\left\|\dot{X}(u, w)\right\|^2 \quad \text{and} \quad U = mgX_2\left(u, w\right) \tag{3.114}$$

where m is the particle mass and g is the gravitational acceleration.

For friction cases, the generalized friction forces $Q = [f_u, f_w]^T$ can be derived as (Chang 2007)

$$f_u = -\mu(g + a_n) \cdot n(e_t \cdot X_{,u}), \quad \text{and} \quad f_w = -\mu(g + a_n) \cdot n(e_t \cdot X_{,w}) \tag{3.115}$$

where

μ is the friction coefficient.
n is the unit normal surface vector (shown in Figure 3.47).
a_n is the normal acceleration of the riding object.
e_t is the unit vector along the tangential direction of the object's path, which is also the direction of the object's velocity $\dot{X}$ and the tangential acceleration a_t.

Following Eq. 3.113, two coupled second-order ordinary differential equations that govern the particle motion can be obtained as

$$k_0\ddot{u} = k_1\dot{u}^2 + k_2\dot{w}^2 + k_3\dot{u}\dot{w} + k_4 \tag{3.116a}$$

$$k_0\ddot{w} = k_5\dot{u}^2 + k_6\dot{w}^2 + k_7\dot{u}\dot{w} + k_8 \tag{3.116b}$$

where k_0 through k_8 consist of polynomials of u and w and their products. Note that X must be at least second-order differentiable with respect to u and w. These requirements are satisfied within individual flume sections but not necessarily across sections.

The initial conditions, including initial position and velocity of the riding object, must be provided to solve the equations of motion:

$$u(0) = u^0, w(0) = w^0, \dot{u}(0) = \dot{u}^0, \text{ and } \dot{w}(0) = \dot{w}^0 \qquad (3.117)$$

The system of ordinary differential equations can be solved numerically for positions $u(t)$ and $w(t)$, velocities $\dot{u}(t)$ and $\dot{w}(t)$, and accelerations $\ddot{u}(t)$ and $\ddot{w}(t)$ of the riding object using, for example, Wolfram's *Mathematica* (1998).

A waterslide configuration consisting of 20 flume sections, shown in Figure 3.48, was modeled and analyzed (Chang 2008). The overall size of the waterslide was about 300 in. × 1150 in. × 378 in. Note that the riding object started at the center of the cross section ($w = 0.5$) of the top section. The friction coefficient was assumed to be $\mu = 0.08$.

The path of the riding object can be seen in Figure 3.48, which shows the object running over edge of the flume section at three critical areas A, B, and C which posed a safety hazard to the rider. The design had to be revisited by either changing the composition of the configuration or using closed-flume (360 deg.) instead of open-flume sections (180 deg.) currently employed. The overall riding time was 21.2 seconds which was very close

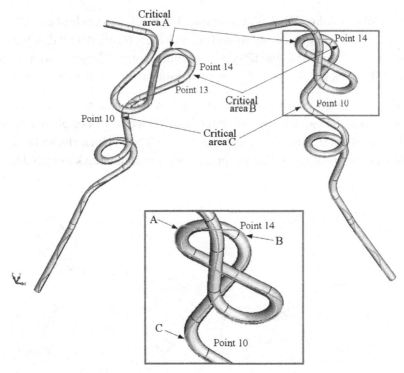

FIGURE 3.48: Object path on a large-scale waterslide showing critical areas of safety concern.

to what was reported by the company that designed the waterslide (20 seconds). The maximum acceleration and velocity were 2.7 g and 12.8 MPH, respectively; these are useful in characterizing the riding experiences (e.g., rider excitement level).

3.7 Tutorial Examples

A sliding block and a single-piston engine, mentioned in Section 3.1, are included in the tutorial lessons. The sliding block example is simply prepared for an easy start with both SolidWorks Motion and Pro/ENGINEER Mechanism Design. Default options and values are mostly used. Once readers are more familiar with either of these two software tools, they may move to the second tutorial example, the single-piston engine.

The first example simulates a block sliding down a 30-deg. slope with no friction. Because of gravity, the block slides and hits the ground, as depicted in Figure 3.49. Simulation results obtained from motion software can be verified using particle dynamics theory learned in a physics class.

3.7.1 Sliding Block

The physical model of a sliding block is very simple. The block was made of cast alloy steel with a size of 10 in. × 10 in. × 10 in. As shown in Figure 3.49, it traveled a total of 9 in. The units system employed for this example was IPS (inch, pound, second). The gravitational acceleration was 386 in./sec^2. The block was released from a rest position (that is, the initial velocity is zero).

The block and slope (or ground) were assumed rigid. A *limit distance mate* (in SolidWorks) was defined to prevent the block from sliding out of the slope face. The block reached the end of the slope face in about 0.3 sec, as indicated in Figure 3.50a, which shows the *Y*-position of the mass center of the block. Note that the graph shows that the block bounced back when it

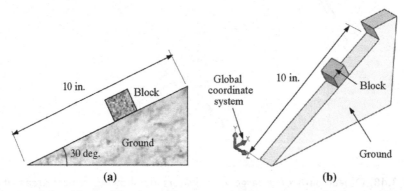

(a)   (b)

FIGURE 3.49: Sliding block: (a) schematic view and (b) motion model in CAD (SolidWorks).

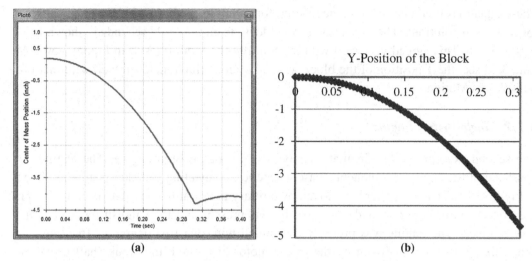

FIGURE 3.50: Graph of the *Y*-position of the mass center of the block, (a) obtained from SolidWorks Motion and (b) from spreadsheet calculations.

reached the end, which was because of the limit distance mate and was artificial. The *Y*-position of the mass center of the block was about 0.18 in. and traveled down to about −4.31 in. at 0.306 sec. The graph may be exported to an Excel file to check these numbers in SolidWorks Motion. The total vertical travel distance was 4.49 in.

The motion simulation result could be verified. Two assumptions had to be made to apply the particle dynamics theory to this sliding problem:

- The block was of a concentrated mass.
- No friction was present.

It is well known that the equations of motion for the sliding block can be derived from Newton's second law. By sketching a free-body diagram, we have the following acceleration equation:

$$F = ma = mg \sin 30 = 0.5mg; \text{ hence } a = 0.5g \qquad (3.118a)$$

The block's velocity and distance could be obtained by integrating Eq. 3.118a:

$$v = at = 0.5gt \qquad (3.118b)$$

$$s = 0.5at^2 = 0.25gt^2 \qquad (3.118c)$$

The *Y*-position of the block could be obtained as

$$P_y = -\frac{1}{2}at^2 \sin 30° = -\frac{1}{8}gt^2 \qquad (3.118d)$$

These equations could be implemented using, for example, a Microsoft® Excel spreadsheet for numerical solutions. The *Y*-position of the block, from 0 to 3.05 seconds is shown in Figure 3.50b. This agreed very well with the SolidWorks Motion result in Figure 3.50a. At time 3.05 sec, the *Y*-position of the block was −4.49 in., which matched well with that of SolidWorks Motion.

3.7.2 Single-Piston Engine

The second example is the kinematic analysis of a single-piston engine. The engine consisted of four major components: case, propeller, connecting rod, and piston, as shown in Figure 3.51a. The propeller was driven by a rotary motor at the angular speed of 60 rpm (i.e., one revolution per second). No gravity was present and the English units system was assumed. The engine was properly assembled with one free degree of freedom. When the propeller was driven by the rotary motor, it rotated, the crank shaft drove the connecting rod, and the connecting rod pushed the piston up and down within the piston sleeve.

The engine assembly consisted of three subassemblies (engine case, propeller, and connecting rod) and one part (piston). The engine case was fixed (ground body). The propeller was assembled to the engine case using concentric and coincident mates, as shown in Figure 3.51b. It was free to rotate along the *X*-direction. The connecting rod was assembled to the propeller (at the crankshaft) using concentric and coincident mates. It was free to rotate relative to the

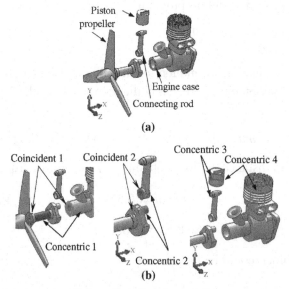

FIGURE 3.51: Single-piston engine: (a) exploded view and (b) constraints defined between bodies (or subassemblies).

propeller (at the crankshaft) along the X-direction. Finally, the piston was assembled to the connecting rod (through the piston pin) using a concentric mate. The piston was also assembled to the engine case using another concentric mate. This mate restricted the piston's movement along the Y-direction, which in turn restricted the top end of the connecting rod to move vertically.

The position and velocity of the piston obtained from motion analysis are shown in Figures 3.52a and 3.52b, respectively. From the graph, we see that the piston moved between about 1.0 and 2.1 in. vertically. The total travel distance was about 1.1 in., which could be easily verified by the radius of the crankshaft, which was 0.58 in. The piston travel distance was two times the radius of the crankshaft, which was 1.16 in.

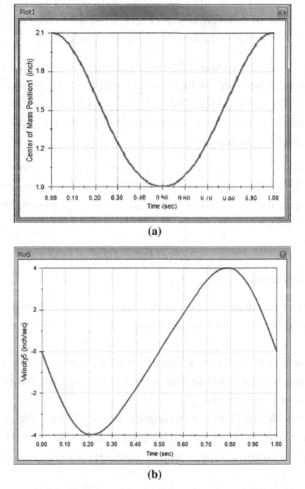

(a)

(b)

FIGURE 3.52: Result graphs: (a) Y-position of the piston and (b) Y-velocity of the piston.

3.8 Summary

In this chapter, we discussed methods for motion analysis, both analytical and computer aided. Analytical methods employ numerous concepts and physics laws to support dynamic analysis for relatively simple problems. They are not adequate to support general problems encountered in engineering design. Even for a two-rigid-body problem, as discussed in Example 3.5, in which it is extremely difficult, if not entirely impossible, to formulate the equations of motion. However, analytical methods are important for mechanical engineers to understand the fundamentals of motion analysis and to be able to develop references for verifying motion analysis results obtained from computer software.

On the other hand, computer-aided methods are formulated in a general and systematic setting. Any mechanical systems that are characterized by bodies and kinematic joints can be formulated in the same type of differential (and algebraic) equations. No matter how complex the motion problem, it can be solved numerically using computers.

A number of software tools are available for support of motion simulation, including codes imbedded in CAD, standalone codes with CAD connections, and specialized codes. It is hoped that this chapter has provided the reader with a good understanding of how the motion analysis method works, how to create motion analysis models, and how to choose the right software tool for the problem at hand.

The chapter also presented case studies involving design aspects of motion analysis, including a Formula SAE racecar and an HMMWV suspension. These two cases should provide a general idea of the applications that lend themselves to simulation for motion analysis and design in general.

Questions and Exercises

3.1. Derive equations of motion for the particle sliding along a straight line shown in Figure 3.4b using Newton's method. Compare your results with those obtained in Example 3.2.

3.2. Repeat Exercise 3.1; that is, derive equations of motion for the particle shown in Figure 3.4b using Newton's method as well as Lagrange's equations, assuming that friction coefficient μ is nonzero.

3.3. Equation 3.10 was derived with an assumption of no friction. If friction is present, derive the generalized forces Q for the particle sliding along the curve $x(u)$.

3.4. Repeat Example 3.3 using an initial angular velocity that is greater than $\omega_{min} = \sqrt{\dfrac{g}{r_0}}$, for example, $\omega_0 = \sqrt{\dfrac{8g}{3r_0}}$.

3.5. Derive the acceleration equations for the slider-crank mechanism by taking derivatives of Eqs. 3.50a and 3.50b with respect to time. Solve these equations for the linear acceleration of the piston and the angular acceleration of the mate Concentric 2, using a spreadsheet.

3.6. Show that Φ_q of Eq. 3.81 in Example 3.8 is nonsingular.

3.7. Solve for $\ddot{q}$ of Eq. 3.83.

3.8. Several driving simulators were mentioned in Section 3.6.3. Please review three or more, identify their main uses, and compare their strengths and weaknesses.

References

Bae, D.S., Haug, E.J., 1987. A recursive formulation for constrained mechanical systems. Part I: Open loop. Mechanics Based Design of Structures and Machines. An International Journal 1987 15 (3), 359–382.

Chang, K.H., 2007. Computer-Aided Modeling and Simulation for Recreational Waterslides. Mechanics Based Design of Structures and Machines. An International Journal 35 (3), 229–243.

Chang, K.H., 2008. Modeling and Simulation for Waterslides in CAD Flume Sections. International Journal of Pure and Applied Mathematics vol. 42 (No. 3), 345–352.

Chang, K.H., Joo, S.-H., 2006. Design Parameterization and Tool Integration for CAD-Based Mechanism Optimization. Advances in Engineering Software 37, 779–796.

Greenwood, D.T., 1987. Principles of Dynamics. Prentice Hall.

Haug, E.J., 1989. Computer Aided Kinematics and Dynamics of Mechanical Systems, vol. 1: Basic Methods. Allyn & Bacon.

Kane, T.R., Levinson, D.A., 1985. Dynamics: Theory and Applications. McGraw-Hill.

NADS Vehicle Dynamics Simulation, Release 4.0, 1994. Center for Computer-Aided Design. National Advanced Driving Simulator. University of Iowa.

Stewart, D.A., 1965/1966. Platform with six degrees of freedom. Proceedings of the Institute of Mechanical Engineering 180 (15, part I), 371–387.

U.S. Tank-Automotive Research and Development Command, 1979. NATO Reference Mobility Model, Edition I. Technical Report 12503 prepared for the North Atlantic Treaty Organization.

Wheeler III, R.M., 2006. Vehicle Dynamic Simulation and Validation of a Formula SAE Car. M.S. Thesis. The University of Oklahoma, Norman, OK.

Wolfram, S., 2003. The Mathematica Book, fifth ed. ISBN 1-57955-022-3 (reference.wolfram.com/legacy/v5_2/).

Sources

Adams/Car: www.mscsoftware.com

ANSYS: www.ansys.com

AutoCAD Inventor: www.usaautodesk.com

CarSim: www.carsim.com

CAT/DADS: www.lmsintl.com

CATIA Motion: www.3ds.com

COSMOSMotion: www.cosmosm.coms

IN-Motion: www.smallguru.com/2010/02/in-motion-released-for-autodesk-inventor-2010

LMS Virtual.Lab: www.lmsintl.com

NX Motion Simulation-RecurDyn: www.plm.automation.siemens.com
Pro/ENGINEER Mechanism Design: www.ptc.com
Solid Edge Motion: www.plm.automation.siemens.com
SolidWorks Motion: www.solidworks.com
VIRTTEX (Google Images): www.google.com/search?q=VIRTTEX&hl=en&tbo=u&tbm=isch
&source=univ&sa=X&ei=ZpW-UKWVOMry2gWLw4CwCg&ved=0CFUQsAQ&biw=1206
&bih=880

Fatigue and Fracture Analysis

One of the most technically challenging issues facing aerospace and mechanical engineers is structural failure due to fatigue and fracture, which causes mechanical failures and safety hazards. Because of the lack of adequate simulation tools for crack growth analysis, especially for complex three-dimensional structural components, heavy emphasis has been placed on physical testing, which is costly and time consuming. Very recently, computational methods were expanded to support fatigue and fracture simulations, especially the finite element method (FEM) and other advanced methods, such as extended FEM (XFEM). It is important to understand and incorporate crack initiation and crack growth simulation techniques into the design of structural components.

In general, fatigue and fracture occur in three stages: crack initiation, crack propagation, and fracture. As design engineers, the common questions we should ask are:

- When and where cracks initiate and after how many load cycles.
- How a crack grows—in which direction and at what rate.
- When the crack size reaches a critical point where the structure is too much damaged to withstand the operating load, leading to fracture.

Different methods have been developed to predict fatigue and fracture for all three stages, with the aim of answering the questions just posed. Some are empirical and largely rely on experimental results in the form of graphs and data. These methods are usually simple and can be carried out by hand or by using lookup tables when nominal stress is calculated. However, they are usually limited to simple cases in terms of loads, structural geometry, and crack size and type. More theoretically sound approaches involve more computations, usually providing more accurate predictions. They also are able to support more general applications. It is important to note that there is no single theory or method that is universally superior to the rest. All involve certain limitations under prescribed assumptions.

This chapter provides a brief overview on the computational methods that help us answer questions about structural fatigue and fracture in various stages. The focus here is not a thorough review of fatigue theory but a discussion of the essential elements of fatigue and fracture simulations from a practical perspective. The goal of this chapter is to provide confidence and competency in the use of software tools for creating adequate models and obtaining reasonable results to support design. Readers who are interested in fatigue and fracture theory can refer to excellent treatments such Collins (1993). Simple examples are introduced to illustrate the theory and computational methods. In addition, practical examples are introduced, including a tracked vehicle roadarm and an engine connecting rod, to illustrate and demonstrate computational methods for structural fatigue and fracture analyses. Overall the objectives of this chapter are as follows:

- To provide basic fatigue and fracture theory using simple examples to promote understanding of how the theory works.

- To promote familiarity with fatigue and fracture modeling and computations for effective use of existing commercial software tools.
- To promote use of either Pro/MECHANICA® Structure or SolidWorks® Simulation for basic applications (after the tutorial lessons have been completed).

4.1 Introduction

Failure of structural components due to fatigue and fracture is a major issue that spans several engineering disciplines and costs hundreds of billions of dollars (National Bureau of Standards 1983). Structural components commonly observed in the aerospace and mechanical engineering fields are obvious examples where crack growth can lead to downtime or failure and may even result in substantial damage and loss of life.

It is widely recognized that about 80% of the failures of mechanical/structural components and systems are related to fatigue (Bannantine et al. 1990). Structural fatigue has produced many losses of aging aircraft. One of the most well known is Aloha Airlines Flight 243, a scheduled flight between Hilo and Honolulu in Hawaii on a Boeing 737-297. On April 28, 1988 the aircraft suffered extensive damage after an explosive decompression in flight (National Transportation Safety Board 1989). Nearly 6 meters of cabin skin and structure aft of the cabin entrance door and above the passenger floor line separated from the aircraft (Figure 4.1a) while cruising at 24,000 feet. One flight attendant was sucked from the airplane and another 65 passengers and crew members were injured. Amazingly, the plane made a safe emergency landing at Kahului Airport on Maui. The subsequent investigation found that debonding and fatigue damage had led to the failure. The aircraft involved had completed 89,680 flight cycles with an average flight time of only 25 minutes, almost all of them in the marine environment of the Hawaiian Islands—a somewhat atypical service life that was considered to have allowed corrosion to increase the likelihood of fatigue.

Another incident involved United Airlines Flight 232, scheduled from Stapleton International Airport in Denver to O'Hare International Airport in Chicago. On July 19, 1989, the Douglas DC-10 suffered an uncontained failure of its number 2 engine, resulting in the crew's inability to move the flight controls (National Transportation Safety Board 1990). Only the thrust levers for the two remaining engines worked. The aircraft eventually broke up during an emergency landing on the runway at Sioux City, Iowa (Figure 4.1b), killing 111 of its 285 passengers and one of 11 crew members. The investigation attributed the cause of the accident to a failure of United Airlines maintenance processes to detect an existing fatigue crack at an inspection prior to flight.

Fatigue and fracture cause problems not only in aircrafts. Liberty ships (Bannerman and Young 1946) were lightly armed cargo vessels built in the United States for transporting desperately needed supplies across the U-boat—infested Atlantic to a beleaguered Europe during World War II. Some 2,700 vessels were built from 1942 until the end of the war. They suffered hull and deck cracks, and a few were lost to such structural defects. During the war, there were nearly 1,500 instances of significant brittle fractures. Twelve ships broke in half without warning, including the *John P. Gaines* (Figure 4.1c), which sank on November 24, 1943, with the loss of 10 lives.

Fatigue and cracks are critical issues to be addressed in product design, especially for products that are to endure heavy operating loads in a highly repetitive manner. Even though it is impractical to design a structure that is fatigue- and fracture-free, the design should be durable to prolong the service life of the structural components.

Figure 4.1: Major structural failures due to fatigue and fracture: (a) Aloha Airlines Flight 243 lost a third of its roof due to a stress fracture (*Source: Chapter 7; Bernard J. Hamrock, Steven R. Schmid, Bo O. Jacobson, Fundamentals of Machine Elements, 2nd Ed., McGraw-Hill, 2005*); (b) United Airlines Flight 232 broke up during an emergency landing on the runway at Sioux City, Iowa; (c) The *John P. Gaines* split in two off the Aleutians in 1943 (*Source: http://www.ntsb.gov/policy/policies.html*).

The two terms—*fatigue* and *fracture*—have been used interchangeably. In general, fatigue and fracture occur in three stages: crack initiation, crack propagation, and fracture failure. In this chapter, we specifically refer to *fatigue* at the crack initiation stage and keep *fracture* for crack propagation and fracture failure. Fatigue begins with microcracks usually due primarily to local stress concentration and the pile-up of dislocations. Local stress concentrations occur around pores, inclusions or impurities, unsmooth surfaces, and the like. Cracks can also be caused by a local decrease in fatigue strength due to a pile-up of dislocations, which form slip bands that grow and lead to cracks. Which of these two mechanisms dominates depends on the purity of the material, the nature of the loading, and so forth.

Crack initiation refers crack growth up to 0.08 in. (2 mm). Usually such small cracks do not affect the functionality and performance of structural components. However, cracks propagate after they are initiated since the geometry of a crack produces a very high concentration of stress at the crack tip. The crack continues to grow in a stable and predictable manner, which is referred to as crack propagation. When the crack reaches a critical size that is over the material strength, in this case the fracture toughness, it grows very fast and unstably, leading to sudden and unexpected catastrophic fracture failures. Note that for brittle materials there is no clear crack initiation. Once the crack initiates, it grows steadily; that is, it directly enters the crack propagation stage.

The questions are, first, when and where in the structure a crack initiates, and how many load cycles it takes for the component to reveal initial cracks; second, how the crack propagates, in which direction, and at what rate. The answers are critical since they offer the basis for inspection and maintenance schedules. The third question is when does the crack grow to a point where the structure is no longer able to withstand the loads, leading to fracture that causes catastrophic failure.

We discuss theory and computation methods that help us answer these critical questions for designing durable structural components. The discussion is brief, just enough to provide the basics of effective software tool use. Software tools that employ some of this theory and associated methods are introduced. In most of the discussion, we assume that finite element analysis (FEA) is employed for stress and strain calculation, where geometric stress concentration factors have been incorporated. We review first stress-based methods for high-cycle fatigue and then strain-based methods for low-cycle fatigue. Both predict fatigue life of the structure up to crack initiation. Theory and methods for crack propagation, such as linear elastic fracture mechanics (LEFM), which provides crack propagation rate and direction, are discussed. We also discuss commercial software tools. Finally, example problems modeled and solved using some of the commercial codes are presented.

4.2 The Physics of Fatigue

In materials science, fatigue is the progressive and localized structural damage that occurs when a material is subjected to cyclic loading. Usually, the maximum stress values are much less than the ultimate tensile strength and much below the yield strength of the material. This phenomenon was recognized in the mid-1800s; that is, that metal under a repetitive or fluctuating load fails at a stress level lower than required to cause failure under a single application of a similar load.

Why does metal fatigue under such low stress? When a component such as that shown in Figure 4.2 is subjected to a uniform sinusoidally varying force for a period of time, a crack can be seen to initiate on the circumference of the component. This crack propagates through the component, as illustrated in the figure, until the remaining intact section is incapable of sustaining the imposed stresses, in particular at the crack tip. At that point the component fails.

The physical development of a crack is generally divided into three stages: crack initiation (stage 1), crack growth or propagation (stage 2), and fracture (stage 3). Fatigue cracks initiate through the release of shear strain energy. Shear stresses result in local plastic deformation along slip planes or slip bands, as shown in Figure 4.3. As the loading is cycled sinusoidally, the slip planes move back and forth like a pack of cards, resulting in small extrusions and intrusions on the crystal surface (Halfpenny). These surface disturbances are approximately 1 to 10 microns in height and constitute embryonic cracks.

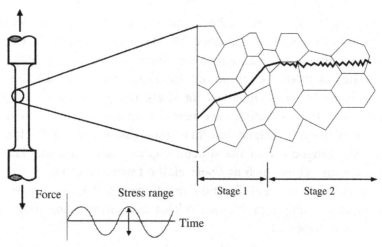

Figure 4.2: Component under a uniform sinusoidally varying force.

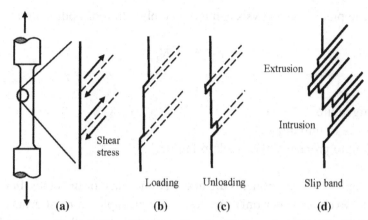

Figure 4.3: Physics of crack initiation: (a) component under a cyclic load, (b) slip under loading, (c) unloading, and (d) slip band form along planes of maximum shear giving rise to surface extrusion and intrusions.

A crack initiates in this way until it reaches the grain boundary. At this point the mechanism is gradually transferred to the adjacent grain. When the crack has grown through approximately 3 grains, it changes its direction of propagation. Stage1 growth follows the direction of the maximum shear plane, or 45 deg. to the direction of loading. During stage 2 the physical mechanism of fatigue changes. The crack is now sufficiently large to form a geometrical stress concentration. At this point, a tensile plastic zone is created at the crack tip and the crack propagates perpendicular to the direction of the applied load. When a crack grows to a critical size, at which the structure is not able to withstand the next cyclic load, fracture (stage 3) occurs.

Just as the physical mechanism of fatigue is divided into stages, the methods of analysis are conventionally divided into stages. Stage 1 is typically analyzed using the stress-strain (S-N) or local strain (ε-N) approach; stages 2 and 3 are analyzed using a fracture mechanics—based approach.

Even though S-N analysis is widely used in test-based fatigue analysis, it has one major drawback for fatigue life computation. Fatigue initiation is driven by local plastic strains, but S-N analysis uses elastic stress as the input; therefore, it is often limited to components with minimum and yet limited local plasticity areas. Such components often reveal high-cycle fatigue; that is, the load cycle is greater than 10^3 before cracks are initiated. On the other hand, when local strains are significant, the strain-life (or ε-N) approach is more suitable. In these cases, fatigue life is often less than 10^3, which is so-called low-cycle fatigue. Both approaches are discussed in this chapter (Sections 4.3 and 4.4, respectively).

A complete fatigue prediction analysis can use a combination of both methods:

$$N = N_i + N_g \tag{4.1}$$

where

> N is total fatigue life.
> N_i is fatigue life to initiation.
> N_g is life taken to propagate the crack to failure.

However, most engineering components are at either one stage or the other. In this case, it is normal to conservatively consider only one stage. For example, in most ground vehicle or heavy equipment designs, life of suspension components is typically governed by time to crack initiation. The components are relatively stiff and the load that the components bear is relatively heavy. Once the crack has initiated, it takes a relatively short time to propagate to failure.

By contrast, many aerospace applications such as skin panels of cowling of aircrafts use flexible components made of ductile materials. In this case, cracks propagate relatively slowly and so fracture mechanics approaches are usually more appropriate. Fatigue and fracture calculations normally assume that the structural component is under cyclic loading—that is, repetitive loads with the same magnitudes—as shown in Figure 4.4.

The most basic cyclic load is fully reversed (Figure 4.4a), which is often seen in fatigue experiments in a laboratory setup, where the material specimen is loaded with pure bending moment that reveals the identical stress magnitudes in tension (σ_{max}) and compression (σ_{min}). The mean (average) σ_m stress is therefore 0 and the alternating stress is $\sigma_a = 1/2(\sigma_{max} - \sigma_{min}) = \sigma_{max}$. This fully reversed cyclic load is also seen in machinery such as shafts rotating at constant speed with load exerted by mechanical components—a paired gear or driving belt, for example.

The second kind of cyclic load is nonfully reversed, where the magnitude of the maximum and minimum stresses are different. This can be like the repeated load shown in Figure 4.4b, where

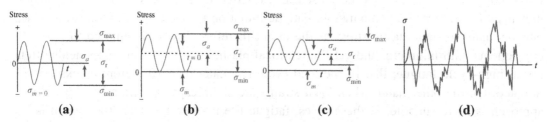

Figure 4.4: Cyclic loads: (a) fully reversed; (b) nonfully reversed, repeated; (c) nonfully reversed, fluctuated; (d) random load.

the load is applied and released in one cycle ($\sigma_{min} = 0$), or it can be similar to a fully reversed load but with nonzero mean stress, as shown in Figure 4.4c. The most general and complicated load is random, as shown in Figure 4.4d. This kind of load appears in many applications; e.g., suspension components of a ground vehicle that is driven over a bumpy road.

It is important to understand the effect of loads on fatigue life calculations. First, load magnitude affects stress magnitudes. Any inaccuracy in stress values amplifies inaccuracy in fatigue life prediction. Second, most of the experimental data obtained from fatigue tests are obtained under uniaxial loads, which lead to uniaxial stress in the structure, such as uniaxial normal stress or shear stress. In practice, structural components are usually under multiaxial loads, yielding a combination of normal and shear stresses. Blindly applying theory developed for uniaxial stress cases to multiaxis stresses can result in erroneous fatigue prediction that contributes to incorrect design decisions.

In Sections 4.3 through 4.5 we assume cyclic loads while discussing basic fatigue and fracture theories. Methods discussed in Section 4.6 assume general random loads.

4.3 The Stress-Life Approach

For crack initiation calculations, there are stress-based and strain-based approaches. Stress-based methods are in general applicable to high-cycle fatigue, where fatigue life is usually between 10^3 and 10^6, which is the endurance limit of the material. The endurance limit is a material property, beyond which the structure is deemed to be of little or no fatigue concern. Theoretically, fatigue life is infinite for such structures. The endurance limit varies for different materials, although 10^6 is the most common case and is often assumed if no better data are at hand. This limit varies according to factors (Juvinall and Marshek 2005) including surface finish, cross-sectional shape of the structural component, type of load, and so forth. This limit has to be corrected from published data, obtained from laboratory tests under a controlled environment, to a lower value by incorporating these factors.

In this section we discuss the S-N diagram for fatigue life prediction of high-cycle fatigue cases. The basic S-N diagram assumes uniaxial stress under fully reversed cyclic load, which is relatively limited in applications. Methods that extend the S-N for applications other than fully reversed cyclic loads are also included in this subsection. Note that stress-life analysis is the simplest and most widely used approach, providing a quick and rough estimate of the fatigue life of structural components.

4.3.1 The S-N Diagram

The S-N diagram is the most basic method for fatigue life prediction of high-cycle fatigue, where fatigue life N is $10^3 < N < 10^6$. It is essentially based on experimental data, mostly under uniaxial, fully reversed cyclic loading. The loading can be bending or torsion. Most of the data and graphs available in textbooks and handbooks are for metals, which are homogeneous and isotropic. Some are available for composite materials; however, in addition to fatigue crack, delamination between material layers is a critical issue to be addressed. We assume homogeneous and isotropic materials in this section.

A typical S-N diagram, shown in Figure 4.5, is on a log-log diagram. The vertical axis is the fatigue strength S_f of the material, and the horizontal axis is the fatigue life N. There are two important fatigue strengths on the vertical axis: S_{10^3} and S_e. S_e is the endurance limit of the material, which is probably the most important material property in high-cycle fatigue calculations because, theoretically (this is backed up by experimental data), if the maximum operating stress of the structural component is less than the endurance limit, the component is said to have infinite fatigue life. That is, fatigue life N is greater than 10^6.

As designers, we prefer a stress level lower than the endurance limit whenever possible. In general, for steels under pure bending, $S_e = 0.5S_{ut}$, where S_{ut} is the ultimate tensile strength. The endurance limit can be different for stress under loads other than pure bending. For steels under axial or torsion stress, the endurance limit is $S_e = 0.45S_{ut}$ and $S_e = 0.29S_{ut}$, respectively. This relation is true for a controlled environment in laboratory tests. It must be corrected by incorporating factors such as size and load, to support the specific operating scenarios for the subject applications. The reader should refer to Juvinall and Marshek (2005) for more details. We assume that the notation S_e represents the corrected endurance limit.

The other property, S_{10^3}, is the fatigue strength corresponding to $N = 10^3$. This is the lower limit of the valid range of high-cycle fatigue. In general, for steels under bending, $S_{10^3} = 0.9S_{ut}$. Under axial and torsion stress, the fatigue strength is $S_{10^3} = 0.75S_{ut}$ and $S_{10^3} = 0.72S_{ut}$, respectively. If the maximum stress of a structural component is between these two strength properties (i.e., $S_{10^3} < \sigma_{max} < S_e$), the fatigue life of the component is finite—$10^3 < N < 10^6$. In this case, the fatigue life corresponding to the maximum stress σ_{max} can be obtained by

$$N = (10^{-\overline{C}}\sigma_{max})^{\frac{1}{b_s}} \tag{4.2a}$$

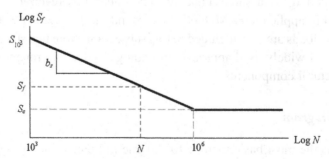

Figure 4.5: Standard S-N diagram.

where

b_s is the slope of the fatigue line (straight line) in the S-N diagram shown in Figure 4.5, called the Basquin exponent.

$\overline{C}$ is the intersect of the fatigue line with the vertical axis.

Both b_s and $\overline{C}$ can be obtained, respectively, by

$$\overline{C} = \log \frac{(S_{10^3})^2}{S_e} \tag{4.2b}$$

and

$$b_s = -\frac{1}{3}\log \frac{S_{10^3}}{S_e} \tag{4.2c}$$

One common mistake made in using these equations, especially Eq. 4.2b, is mixed stress and strength units. If MPa is used for stress and strength, its use must be consistent. If MPa is mixed with Pa (e.g., MPa for S_{10^3} and Pa for S_e), $\overline{C}$ and b_s will be incorrect, which affects the fatigue life calculation in Eq. 4.2a.

Equation 4.2a allows prediction of fatigue life of the structural component when the maximum stress has been calculated. This is often referred to as an analysis problem. On some occasions, we may want to find the allowable stress for a required fatigue life. The allowable stress σ_{all} can be obtained by

$$\sigma_{all} = \frac{S_f}{n}$$

where n is the safety factor and

$$S_f = 10^{\overline{C}} N^{b_s} \tag{4.3b}$$

and N is the desired or required fatigue life. This is a design problem.

The value of b_s, for a particular *S-N* slope, provides a good indication of how accurate the estimate of stress needs to be to give a reliable life prediction. If b_s is 10 then a 7% error in stress causes a 100% error in fatigue life. Even if the stress value is calculated accurately using FEA, if the load magnitude is slightly inaccurate, the stress is affected proportionally (assuming linear elastic), which impacts the life estimate exponentially. Therefore, load must be accurately captured to carry out a reliable fatigue life analysis.

Example 4.1

A steel has a tensile strength of $S_{ut} = 90,000$ psi and a corrected endurance limit of $S_e = 33,000$ psi for a machined surface. Assume that the fatigue strength at $N = 10^3$ is $S_{10^3} = 0.9 S_{ut}$.

(a) Draw an S-N diagram with all needed data marked.
(b) Calculate coefficients $\overline{C}$ and b_s for the equation of the finite life and fatigue strength (i.e., Eq. 4.3b).

(c)   Calculate the fatigue strength for $N = 10,000$.

(d)   If the maximum stress is 34.2 ksi, calculate the corresponding fatigue life N.

Solutions

(a) The S-N diagram can be drawn as shown below, with fatique strength for $N = 1,000$ (i.e., S_{10^3}) and endurance limit S_e calculated below.

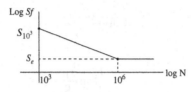

$$S_{10^3} = 0.9\, S_{ut}$$
$$= 0.9(90,000) = 81,000 \text{ psi} = 81 \text{ksi}$$
$$S_e = 33,000 \text{ psi} = 33 \text{ ksi}$$

(b) b_s and $\overline{C}$ can be calculated using Eqs. 4.2b and 4.2c, respectively, as follows

$$b_s = -\frac{1}{3} \log \frac{S_{10^3}}{S_e} = -\frac{1}{3} \log \frac{81}{33} = -0.1300$$

$$\overline{C} = \log \frac{[S_{10^3}]^2}{S_e} = \log \frac{(81)^2}{33} = 2.298 \text{ (in ksi)}$$

(c)   The fatigue strength for $N = 10,000$ can be calculated using Eq. 4.3b as

$$S_f = 10^{\overline{C}}(N)^{b_s} = 10^{2.298}(10,000)^{-0.1300} = 59.98 \text{ ksi}$$

(d)   The fatigue life for a given stress can be obtained using Eq. 4.2a as

$$N = \left(\frac{S_f}{10^{\overline{C}}}\right)^{\frac{1}{b_s}} = \left(\frac{\sigma}{10^{\overline{C}}}\right)^{\frac{1}{b_s}} = \left(\frac{34.2}{10^{2.298}}\right)^{\frac{1}{-0.1300}} = 7.29 \times 10^5 \text{cycles}$$

4.3.2 Nonfully Reversed Cyclic Loads

S-N diagrams assume fully reversed cyclic load or stress. For cases of nonfully reversed stress, where the mean stress σ_m is nonzero, the S-N diagram approach must be modified. There are four commonly employed design criteria to determine fatigue life by incorporating the damage caused by mean stress. They are the Goodman line, Soderberg, Gerber, and yield line criteria, as shown in Figure 4.6. Note that, in the figure, if the stress point (alternating

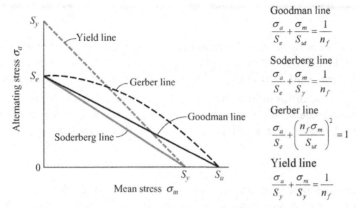

Figure 4.6: Fatigue design criteria for nonzero mean stress. *(Source: Chapter 7; Bernard J. Hamrock, Steven R. Schmid, Bo O. Jacobson, Fundamentals of Machine Elements, 2nd Ed., McGraw-Hill, 2005).*

stress σ_a and mean stress σ_m) that represents the maximum stress of a component is located below the respective failure line, the component is said to be safe and the fatigue life is infinite (more practically, it is greater than 10^6). The safety factor n_f can be calculated, for example, using the Goodman line criterion as

$$\frac{\sigma_a}{S_e} + \frac{\sigma_m}{S_{ut}} = \frac{1}{n_f} \tag{4.4}$$

Comparing these four criteria, it is obvious that the Soderberg line is the most conservative.

On the other hand, when the stress point is above the failure line, the fatigue life of the component is finite and can be calculated by first converting the mean and alternating stresses into equivalent alternating stress σ_A using one of the design criterion, and then calculating the fatigue life using S-N diagram (i.e., Eq. 4.2a) as a fully-reversed stress case. For example, if using the Goodman line criterion, the equivalent alternating stress is

$$\sigma_A = \frac{S_{ut}\sigma_a}{S_{ut} - \sigma_m} \tag{4.5a}$$

For the Soderberg criterion, the equivalent alternating stress is

$$\sigma_A = \frac{S_y\sigma_a}{S_y - \sigma_m} \tag{4.5b}$$

The modified Goodman line that combines the Goodman line with the yield line provides the best match with experimental data and has been the most popular criterion for fatigue design. In Figure 4.6, the Goodman line and yield line intersect at $\sigma_m = [S_{ut}(S_y - S_e)/(S_{ut} - S_e), \sigma_a = S_e(S_{ut} - S_y)/(S_{ut} - S_e)]$. Therefore, when the mean stress σ_m is less than $S_{ut}(S_y - S_e)/(S_{ut} - S_e)$, the Goodman line is employed as the design criterion. If the mean stress is $S_{ut}(S_y - S_e)/(S_{ut} - S_e) < \sigma_m < S_y$, failure is assumed due to yield.

Example 4.2

A straight cantilever rotating beam of solid circular cross section with a 30 mm diameter and 1 m length has an axial load of 50,000 N applied at the end and a stationary transverse (vertical) load of 600 N. The material is AISI 1040 ($S_{ut} = 520$ MPa, $S_y = 320$ MPa). The beam is unnotched.

(a) Find the safety factor for infinite life using the Soderberg line criterion.
(b) What is the equivalent alternating stress σ_A? What is the fatigue life of the beam?
(c) What beam diameter gives a safety factor that exactly equals 1.0?

Solutions

First we calculate endurance limit, alternating stress due to bending and mean stress due to the axial force. Note that for a more conservative design, we use $S_e = 0.45\ S_{ut}$, although both bending and axial loads are present.

$$S_e = 0.45\ S_{ut} = (0.45 \times 520) = 234\ \text{MPa}$$

$$\sigma_a = \frac{Mc}{I} = \frac{(600)(1000)(30/2)}{\frac{\pi}{64}(30)^4} = 226\ \text{MPa}$$

$$\sigma_m = \frac{P}{A} = \frac{50,000}{\frac{\pi}{4}(30)^2} = 70.7\ \text{MPa}$$

(a) Safety factor n_f can be obtained using Eq. 4.4

$$\frac{\sigma_a}{S_e} + \frac{\sigma_m}{S_y} = \frac{1}{n_f}$$

$$\frac{226}{234} + \frac{70.7}{320} = \frac{1}{n_f} \Rightarrow n_f = 0.840$$

(b) Equivalental alternating stress σ_A

$$\sigma_A = \frac{\sigma_a}{1 - \dfrac{\sigma_m}{S_y}} = \frac{226}{1 - \dfrac{70.7}{320}} = 290.1\ \text{MPa}$$

We use S-N approach to calculate fatigue life using the equivalent alternating stress

$$S_{10^3} = 0.75\ S_{ut} = 0.75(520) = 390\ \text{MPa}$$

$$\bar{C} = \log \frac{S_{10^3}^2}{S_e} = \log \frac{390^2}{234} = 2.813$$

$$b_s = -\frac{1}{3}\log \frac{S_{10^3}}{S_e} = -\frac{1}{3}\log \frac{390}{234} = -0.07395$$

$$N = (10^{-\bar{C}}\sigma_A)^{\frac{1}{b_s}}$$

$$= (10^{-2.813} \times 290.1)^{\frac{1}{-0.07395}}$$

$$= 55,090$$

(c)   beam diameter for $n_f = 1$

$$\frac{\sigma_a}{S_e} + \frac{\sigma_m}{S_y} = 1$$

$$\frac{\frac{32M}{\pi d^3}}{S_e} + \frac{\frac{4P}{\pi d^2}}{S_y} = 1$$

$$\frac{\frac{32(600 \times 1000)}{\pi d^3}}{234} + \frac{\frac{4(50,000)}{\pi d^2}}{320} = 1$$

$$\frac{26,120}{d^3} + \frac{198.9}{d^2} = 1$$

$$\Rightarrow d = 32 \, \text{mm}$$

4.3.3 In-Phase Bending and Torsion

S-N diagrams and their associated fatigue strengths assume stresses under uniaxial load, either normal stress under axial or bending load or shear stress under torsion load. Most machine element applications encounter more complicated loading conditions. Two special cases are important. The first situation exists where the applied stresses are synchronous and in phase—that is, at the same frequency (synchronous) with a zero phase angle (in phase). This is called simple multiaxial stress. For example, a cylindrical pressure vessel that is periodically pressurized has a hoop and an axial stress that are both directly related to the pressure variation, so they are subject to their maximum and minimum values at the same time. When the forces and torques are not synchronous or in phase, this situation is called complex multiaxial stress, which is discussed in the next subsection.

For simple multiaxial stress, the principal direction of the stress is unchanged. If the stress is fully reversed, an equivalent von Mises stress is calculated as

$$\sigma_a' = \sqrt{\frac{\left(\sigma_{1_a} - \sigma_{2_a}\right)^2 + \left(\sigma_{2_a} - \sigma_{3_a}\right)^2 + \left(\sigma_{3_a} - \sigma_{1_a}\right)^2}{2}} \tag{4.6}$$

where σ_{1_a}, σ_{2_a}, and σ_{3_a} are the alternating principal stresses. The safety factor n_f can thus be calculated using

$$n_f = \frac{S_e}{\sigma_a'} \tag{4.7}$$

for infinite fatigue life. Or, if $\sigma_a' > S_e$, we insert the equivalent stress σ_a' into Eq. 4.2a to estimate fatigue life for $N < 10^6$.

For nonfully reversed stresses, equivalent von Mises stresses are calculated respectively for the alternating stresses (using Eq. 4.6) and mean stress using

$$\sigma_m' = \sqrt{\frac{\left(\sigma_{1_m} - \sigma_{2_m}\right)^2 + \left(\sigma_{2_m} - \sigma_{3_m}\right)^2 + \left(\sigma_{3_m} - \sigma_{1_m}\right)^2}{2}} \qquad (4.8)$$

where σ_{1_m}, σ_{2_m}, and σ_{3_m} are the mean principal stresses. Then, design criteria such as the modified Goodman line are employed for safety factor calculation using Eq. 4.4. If the stress point falls outside of the safe region, then Eq. 4.5 may be used to calculate equivalent alternating stress and then to calculate fatigue life N using Eq. 4.2a.

For a more specific load, such as a fully reversed normal stress, caused by either axial force or bending moment ($\sigma_m = 0$, $\sigma_a \neq 0$) and steady shear stress due to a constant torque ($\tau_m \neq 0$, $\tau_a = 0$), the safety factor can be obtained by

$$\left(\frac{\sigma_a}{S_e}\right)^2 + \left(\frac{\tau_m}{S_{sy}}\right)^2 = \frac{1}{n_f} \qquad (4.9a)$$

where S_{sy} is the modified shear yield strength and $S_{sy} = S_y/\sqrt{3}$. For a fully reversed bending/axial ($\sigma_m = 0$) with fully reversed torque ($\tau_m = 0$), the safety factor can be obtained by

$$\left(\frac{\sigma_a}{S_e}\right)^2 + \left(\frac{\tau_a}{S_{se}}\right)^2 = \frac{1}{n_f} \qquad (4.9b)$$

where S_{se} is the endurance limit of shear strength and $S_{se} = S_e/\sqrt{3}$. The relation shown in Eqs. 4.9a and 4.9b represents the well-known Gough ellipse, which is backed up by test results (Spotts and Shoup 1998). Equations 4.9a and 4.9b are derived based on the Soderberg fatigue failure criterion and distortion energy theory. They are especially useful for shaft design.

4.3.4 Complex Multiaxial Stress

In many practical applications, loads are simultaneously applied in several different directions, producing stresses with no particular direction bias. The principal direction changes from cycle to cycle. These are complex multiaxial stresses or, simply, multiaxial stresses.

Fatigue occurs on the surface where one of the principal stresses is usually zero. On the surface, the stress state is one of plane stress (i.e., a biaxial stress state) so that there are only three stress components σ_x, σ_y, and τ_{xy}. All other stresses are zero. As a result, multiaxial fatigue problems are usually biaxial in nature.

Complex multiaxial stress for fatigue life calculation is still under investigation by many researchers. Many specific cases of complex multiaxial stress have been analyzed, but no overall design approach applicable to all situations has yet been developed. For the common biaxial stress case of combined bending and torsion, such as occurs in shafts, several approaches have been proposed. One of them, SEQA, based on the ASME Boiler Code, is discussed briefly. SEQA is an equivalent or effective stress that combines the normal and shear stresses and the phase relationship between them into an effective-stress value that can be compared to the ductile materials' fatigue and static strength on a modified-Goodman diagram. It is calculated from

$$SEQA = \frac{\sigma}{\sqrt{2}} \left[1 + \frac{3}{4}Q^2 + \sqrt{1 + \frac{3}{2}Q^2\cos2\phi + \frac{9}{16}Q^4} \right]^{1/2} \tag{4.10}$$

where

> σ is the bending stress amplitude, $Q = 2\tau/\sigma$.
> τ is the torsional stress amplitude.
> ϕ is the phase angle between bending and torsion.

SEQA can be computed for both mean and alternating stresses. This approach is valid for loads that are synchronous with a predictable phase relationship.

In recent years, criteria based on the critical plane approach for multiaxial fatigue evaluation have been gaining popularity for high-cycle fatigue (You and Lee 1996). Here, fatigue evaluation is performed on one plane across a critical location in the component. This plane is called the critical plane, which is usually different for different fatigue models. One of the models proposed by Matake (1977) uses a damage parameter based on the linear combination of shear stress amplitude and maximum normal stress acting on the critical plane. The orientation of this plane is described by the spherical coordinates (ϕ_c, θ_c) of its unit normal vector. The critical plane is defined as the plane on which the shear stress amplitude achieves the maximum value:

$$\begin{cases} (\phi_c, \theta_c) = \max_{(\phi,\theta)}[\tau_a(\phi, \theta)] \\ \tau_a(\phi_c, \theta_c) + k\sigma_{\max}(\phi_c, \theta_c) = \xi \end{cases} \tag{4.11}$$

where the subscript c refers to the critical plane. The material constants k and ξ are given as

$$k = \frac{2S_{se}}{S_e} - 1, \quad \text{and} \quad \xi = S_{se} \tag{4.12}$$

Again, S_e and S_{se} are the endurance limits (also called fatigue limits) in fully reversed bending and torsion, respectively. Socie (1993) proposed a stress-based approach for HCF, in which fatigue life is calculated on the critical plane where the total damage is maximum.

$$\tau_a + k_2\sigma_{max} = \tau'_f(2N)^{b_0} \tag{4.13}$$

The right side of the equation is the elastic part of the strain life. The left side represents the damage parameters defined on the critical plane experiencing the largest range of cyclic shear stress. k_2, τ'_f, and b_0 are the materials parameters.

4.3.5 Cumulative Damage

When a structural component is loaded with multiple stress magnitudes in a set of cyclic loads for their respective number of cycles, the fatigue damage to the component is accumulated from these loading sets. The simplest and most popular approach for calculating cumulative damage is Miner's rule, also referred to as the *Palmgren–Miner linear damage hypothesis*, which states that where there are k different stress magnitudes in a set of cyclic loads, σ_i $(1 \leq i \leq k)$, each contributing $n_i(\sigma_i)$ cycles, then if $N_i(\sigma_i)$ is the number of cycles to failure of the respective stress magnitudes σ_i, failure occurs when

$$\sum_{i=1}^{k} \frac{n_i}{N_i} = C \tag{4.14}$$

where C is experimentally found to be between 0.7 and 2.2. Usually for design purposes, C is assumed to be 1.

Though Miner's rule is a useful approximation in many circumstances, it has major limitations. There is sometimes an effect in the order of the reversals. In some circumstances, cycles of low stress followed by high stress cause more damage than predicted by the rule. This rule does not consider the effect of overload or high stress that may result in a compressive residual stress. High stress followed by low stress may create less damage because of the presence of compressive residual stress.

Although linear damage accumulation has been criticized, it is still the most common approach to fatigue analysis because of its simplicity and effectiveness, and because its accuracy is sufficient in many applications.

Example 4.3

A steel bar is under fully reversed bending stresses of $\sigma_{max} = 300$ MPa and $\sigma_{min} = -300$ MPa applied for 10,000 cycles. The load is changed to ±210 MPa and is applied for 20,000 cycles. Finally, the load is changed to ±350 MPa. How many cycles of operation can be expected at this stress level? Material properties of the part are $S_e = 200$ MPa (modified), $S_y = 490$ MPa, and $S_{ut} = 590$ MPa.

Solutions

We first calculate coefficients for the S-N equation based on the given material properties, then use the Miner's rule to calculate the cumulative fatigue life due to the multiple loads.

$$\overline{C} = \log\frac{S_\ell}{S_e} = \log\frac{531^2}{200} = 3.15$$

$$b_s = -\frac{1}{3}\log\frac{S_\ell}{S_e} = -\frac{1}{3}\log\frac{531}{200} = -0.141$$

$$N_1 = (\sigma_1 10^{-\overline{c}})^{\frac{1}{b_s}} = (300 \times 10^{-3.15})^{\frac{1}{-0.141}} = 59,200$$

$$N_2 = (\sigma_2 10^{-\overline{c}})^{\frac{1}{b_s}} = (210 \times 10^{-3.15})^{\frac{1}{-0.141}} = 743,000$$

$$N_3 = (\sigma_3 10^{-\overline{c}})^{\frac{1}{b_s}} = (350 \times 10^{-3.15})^{\frac{1}{-0.141}} = 19,800$$

Now, use the Miner's rule

$$\frac{n_1}{N_1} + \frac{n_2}{N_2} + \frac{n_3}{N_3} = 1$$

$$n_3 = N_3\left(1 - \frac{n_1}{N_1} - \frac{n_2}{N_2}\right) = 19,800\left(1 - \frac{12,000}{59,200} - \frac{20,000}{743,000}\right) = 15,900$$

4.4 The Strain-Based Approach

Low-cycle fatigue, where fatigue life N is less than 10^3, is usually attributed to stress that is high enough for local plastic deformation to occur. For plastic deformation, the account in terms of stress is less useful and the strain in the material offers a simpler description. Therefore, the strain-based approach is more widely accepted for life estimates in low-cycle fatigue cases.

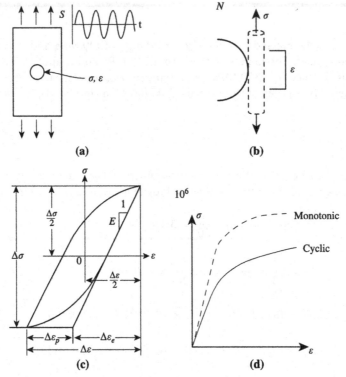

Figure 4.7: Strain-based fatigue life prediction: (a) specimen of center crack under cyclic loading, (b) stress and strain around the crack tip, (c) load path in one cycle, and (d) stress-strain curve.

4.4.1 The Manson–Coffin Equation

Cyclic stress-strain curves are generally obtained from uniaxial stress tests. For a plate with a center hole under normal cyclic loading, as shown in Figure 4.7a, the material near the crack tip experiences large local stress and strain as shown in Figure 4.7b. The typical load path of the materials under a complete load cycle can be seen in Figure 4.7c, where the normal stress is both tensile and compression. In reality cyclic hardening or softening of material alters the shape of the load path. The stress-strain curve shown in Figure 4.7d is usually expressed by a Ramberg–Osgood expression (Ramberg and Osgood 1943):

$$\varepsilon = \frac{\sigma}{E} + \left(\frac{\sigma}{K}\right)^{1/n} \tag{4.15}$$

where E is the elastic modulus, and K and n are fitted to the material test data for a particular material. The first term on the right of Eq. 4.15, σ/E, is equal to the elastic part of the strain, while the second term accounts for the plastic part.

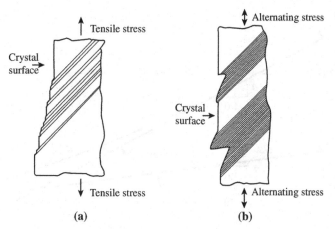

Figure 4.8: Slip band observed in elastic material: (a) monotonic loading, and (b) cyclic loading.

As shown in Figure 4.7d the stress-strain curves for monotonic and cyclic loads are different. This is because cyclic loading causes relatively larger strain given the same stress. When the strain is monotonic, the slip steps that appear on a crystal surface have a relatively simple topology, as shown in Figure 4.8a. On the other hand, under cyclic loading, the slip bands tend to group into packets or striations. The surface topology of these striations is more complex and is indicated schematically in Figure 4.8b. Note that both ridges and crevices tend to be formed. There is good evidence that the crevices are also closely associated with the initiation of cracks.

The first equations proposed for calculating the fatigue life of components correlated the applied stress amplitude with the number of cycles for crack initiation. Basquin's rule (Hertzberg 1983), presented as Eq. 4.16, was discussed in Section 4.3 (in a slightly different form—for example, Eq. 4.3b), where σ_a, σ_f', a, and $2N_f$ are, respectively, the stress amplitude, the fatigue strength coefficient, the fatigue strength exponent, and the number of reversals to initiation.

$$\sigma_a = \frac{\Delta\sigma}{2} = \sigma_f'(2N_f)^a \qquad (4.16)$$

The relation proposed by Basquin offers good agreement for high-cycle fatigue domains, where elastic deformations are present. For low-cycle fatigue domains, Coffin and Manson (published independently by L.F. Coffin in 1954 and S.S. Manson in 1953) suggest an alternative rule, presented in Eq. 4.17, where $\Delta\varepsilon_p$, ε_f', and α are, respectively, the plastic strain range in the notch root, the fatigue ductility coefficient, and the fatigue ductility exponent.

$$\frac{\Delta\varepsilon_p}{2} = \varepsilon_f'(2N_f)^\alpha \qquad (4.17)$$

The previous two expressions are complementary and can be matched, giving a more general expression valid in any fatigue domain, both high- and low-cycle. The result is

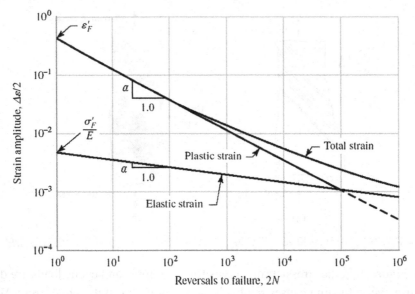

Figure 4.9: Plastic and elastic strain-life curve. *Source: From Shigley 2004, p. 318.*

represented in Eq. 4.18 as well as in Figure 4.9, where $\Delta\varepsilon$ it is the total strain amplitude at the notch root.

$$\frac{\Delta\varepsilon}{2} = \frac{\sigma_f'}{E}(2N_f)^a + \varepsilon_f'(2N_f)^\alpha \qquad (4.18)$$

The first and second terms on the right represent the elastic and plastic portions of the deformation, respectively.

Note that Eq. 4.18 assumes fully reversed stress cases. Morrow (1965) modified Eq. 4.18 to take into account the mean stress effect, resulting in Eq. 4.19.

$$\frac{\Delta\varepsilon}{2} = \frac{(\sigma_f' - \sigma_m)}{E}(2N_f)^a + \varepsilon_f'(2N_f)^\alpha \qquad (4.19)$$

Variable-amplitude loads with blocks of constant amplitude can be analyzed using Miner's linear damage accumulation rule, which is similar to that of HCF, as discussed in Section 4.3.5.

Implementing the strain-based approach for fatigue life prediction requires stress and strain calculations in the plastic range. Using the finite element method for stress and strain calculations requires nonlinear plastic analysis, which can be very expensive. Furthermore, the nominal response of the structure is elastic except around a small region at the crack tip. Therefore, a widely accepted approach is to assume a nominally elastic response and then make use of Neuber's (1961) equation to relate local stress and strain to nominal stress and strain at a stress concentration location near the crack tip. If there is no general yielding over the entire structure—that is, if yielding is confined to the local region of the notch—Neuber's rule takes the form

$$\sigma\varepsilon = \frac{(K_t\sigma_0)^2}{E} = \frac{\sigma_e^2}{E} = \sigma_e\varepsilon_e \tag{4.20}$$

The quantities σ and ε are the local stress and strain (in the plastic range) near the crack tip, respectively; σ_0 is a nominal or average stress, and K_t is an elastic stress concentration factor consistently defined with σ_0. After carrying out a linear elastic analysis using the finite element method, stress concentration factor K_t is incorporated, so $\sigma_e = K_t\sigma_0$ in Eq. 4.20. ε_e is the elastic strain.

While the cyclic stress-strain has the form of Eq. 4.15, combining this with Eq. 4.20 gives

$$\sigma_e^2 = \sigma^2 + E\sigma\left(\frac{\sigma}{K}\right)^{1/n} = \sigma^2 + E\sigma^{1+1/n}\left(\frac{1}{K}\right)^{1/n} \tag{4.21a}$$

or

$$\varepsilon^2 = \left(\frac{\sigma_e}{E}\right)^2 + \varepsilon\left(\frac{\sigma_e^2}{E\varepsilon K}\right)^{1/n} = \left(\frac{\sigma_e}{E}\right)^2 + \varepsilon^{1-1/n}\left(\frac{\sigma_e^2}{EK}\right)^{1/n} \tag{4.21b}$$

Thus, for a stress σ_e obtained using linear elastic FEA, Eq. 4.21a or 4.21b can be solved iteratively for the local stress amplitude σ (assuming fully reversed cyclic load) or local strain ε, respectively, as illustrated in Figure 4.10. The half-strain range $\Delta\varepsilon/2 = \varepsilon$ can be used with the strain-life curve, such as in Eq. 4.18, to obtain the estimated number of cycles N_f that arrives at crack initiation.

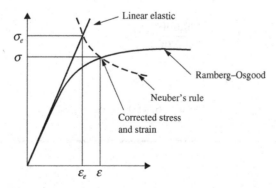

Figure 4.10: Stress correction using Neuber's rule.

Example 4.4

A notched bar made of steel is under a cyclic tensile load $P = 20,000$ lb. The rectangular net cross-section of the bar at the root of the notch is $A = 0.5$ in^2. At the notch the fatigue stress concentration factor is obtained as $K_t = 1.6$. The material properties of the bar that are relevant to our calculation are $K = 155,000$ psi, $n = 0.15$, $\varepsilon_f' = 0.48$, $\sigma_f' = 290,000$ psi, $a = -0.091$, and $\alpha = -0.60$. How many cycles does it take to initiate a fatigue crack at the notch root?

Solution

The nominal stress σ can be calculated as

$$\sigma = \frac{P}{A} = \frac{20,000}{0.5} = 40,000 \text{ psi}$$

Thus, the total stress with stress concentration factor K_t is $\sigma_e = 1.6(40,000) = 64,000$ psi. Using Eq. 4.20,

$$\sigma\varepsilon = \frac{(K_t\sigma_e)^2}{E} = \frac{(1.6 \times 40,000)^2}{30 \times 10^6} = 136.5 \text{ psi}$$

From Eq. 4.21b we have

$$\varepsilon^2 = \left(\frac{\sigma_e}{E}\right)^2 + \varepsilon^{1-1/n}\left(\frac{\sigma_e^2}{EK}\right)^{1/n} = \left(\frac{64,000}{30 \times 10^6}\right)^2 + \varepsilon^{1-1/0.15}\left(\frac{64,000}{30 \times 10^6 \times 155,000}\right)^{1/0.15}$$

$$= 4.551 \times 10^{-6} + \varepsilon^{-0.667}(9.176 \times 10^{-8})$$

which can be solved iteratively, or with tools such as MATLAB®, as $\varepsilon = 0.002996$ in./in. Then, from Eq. 4.18,

$$0.002996 = \frac{290,000}{30,000,000}(2N_f)^{0.091} + 0.48(2N_f)^{0.60}$$

which can be solved iteratively as

$$N_f = 6,150 \text{ cycles}$$

4.4.2 Multiaxial Analysis

The strain-based approach discussed so far assumes stress and strain under uniaxial loads. As discussed in Section 4.3.3, most machine element applications encounter more complicated loading conditions. Simple multiaxial cases assume that stress or strain is synchronous and in phase. Otherwise, we have complex multiaxial stress and strain.

Similar to the stress-based approach, for simple multiaxial cases the principal direction of the strain is unchanged. An equivalent stress-strain approach is suitable for such cases, such as von Mises equivalent strain or the ASME Boiler Code. The von Mises equivalent strain is further illustrated in this subsection. If the strain is fully reversed, based on the definition of an equivalent stress parameter proposed by the von Mises yield criterion, the equivalent "uniaxial" strain amplitude parameter $\Delta \varepsilon_{\text{eff}}/2$ can be defined as (Chu et al. 1993)

$$
\frac{\Delta \varepsilon_{\text{eff}}}{2} = \frac{1}{\sqrt{2(1 + \nu_{\text{eff}})}} \left[\left(\frac{\Delta \varepsilon_1}{2} - \frac{\Delta \varepsilon_2}{2} \right)^2 + \left(\frac{\Delta \varepsilon_2}{2} - \frac{\Delta \varepsilon_3}{2} \right)^2 + \left(\frac{\Delta \varepsilon_1}{2} - \frac{\Delta \varepsilon_3}{2} \right)^2 \right.
$$
$$
\left. + 6 \left(\frac{\Delta \varepsilon_{12}}{2} + \frac{\Delta \varepsilon_{23}}{2} + \frac{\Delta \varepsilon_{13}}{2} \right)^2 \right]^{\frac{1}{2}}
\tag{4.22}
$$

where

$$
\nu_{\text{eff}} = 0.5 - (0.5 - \nu) \frac{E_{\text{eff}}}{E}
\tag{4.23}
$$

where E_{eff} is the effective secant modulus, which is the effective modulus at 2% strain.

As with stress-based approaches, complex multiaxial strain for fatigue life calculation is still under investigation by many researchers. Criteria based on the critical plane approach has gained popularity. Examples include the Fatemi–Socie model (Fatemi and Socie 1988) and the Smith, Watson, and Topper model (SWT) (Smith et al. 1970).

The Fatemi–Socie model employs the following equation:

$$
\gamma_a \left(1 + k_1 \frac{\sigma_n}{S_y} \right) = \frac{\tau_f'}{G} (2N_f)^a + \gamma_f' (2N_f)^\alpha
\tag{4.24}
$$

where

γ_a is the shear strain amplitude on the critical plane during a cycle.
σ_n is the maximum normal stress on the critical plane during a cycle.
S_y is the material yield strength.
G is the shear modulus.
k_1, γ_f', and τ_f' are materials parameters for fatigue.

The right side of Eq. 4.24 is the description of the strain-life Manson–Coffin curve in torsion. The term on the left side represents the damage parameter on the plane experiencing the largest range of shear strain (critical plane). The fatigue is calculated on the critical plane where the total damage is maximum.

Smith, Watson, and Topper's (SWT) model assumes that the amount of fatigue damage in a cycle is determined by $\sigma_{max}\varepsilon_a$, where σ_{max} is the maximum tensile stress and ε_a is the strain amplitude on the maximum normal strain plane (critical plane). The fatigue life can be calculated using

$$\varepsilon_a \sigma_{max} = \frac{\sigma'_f}{E}(2N_f)^a + \sigma'_f \varepsilon'_f (2N_f)^{a+\alpha} \tag{4.25}$$

4.5 Fracture Mechanics

The mere presence of a crack does not mean that a structure is unsafe. In fact, the damage-tolerant design and analysis approach takes into account the presence of flaws and predicts a component's useful remaining service life (residual life). It is common practice to subject critical structural components to periodic inspections to identify the presence of cracks, and then to monitor crack growth at certain intervals. If the geometric shape of the structure, flaw shape and size, material, and loading, are known, in many cases it is possible to predict the period of subcritical crack growth using crack propagation analysis techniques. Note that a crack undergoing subcritical crack growth propagates under a remotely applied stress (away from the crack tip) well under material yield or ultimate strength.

Crack nucleation can be predicted using the methods discussed in Sections 4.2 and 4.3. The crack grows to a certain size under repetitive or cyclic loads until the stress intensity factor reaches the fracture toughness of the material, at which point the crack grows in an unpredictable manner that leads to component and system failure.

In this section we focus on stage 2 fatigue, crack propagation. The question that is often asked by designers is how the crack grows in a stable fashion, including rate and direction. This question can often be answered by fracture mechanics. We start with a discussion of an infinite plate with a center crack under uniaxial normal load, where both the energy method and the stress intensity factor approaches are touched on. Then we introduce linear elastic fracture mechanics (LEFM), which is the most popular and widely applied method, and follow with a mixed-mode fracture where both normal and shear loads are incorporated. We also briefly discuss the newly developed extended finite element method (XFEM) that successfully alleviates the troublesome mesh generation problem by using regular FEA for crack propagation simulation. In this section, we assume cyclic load. For crack propagation under random loads, peak-valley editing together with rain-flow counting, as to be discussed in Section 4.6, can be employed to convert the random load into a pseudocyclic load. The theory and the computation method introduced are still applicable.

4.5.1 Basic Approaches

There are two basic approaches to fracture analysis: energy criterion and stress intensity. These are equivalent under certain circumstances.

The energy approach states that fracture occurs when the energy available for crack growth is sufficient to overcome the resistance of the material, which is referred to as fracture toughness—an important material property. Griffith (1921) was the first to propose the energy criterion for fracture, but Irwin (1957) is primarily responsible for developing the approach involving energy release rate, G, which is defined as the rate of change in potential energy with crack area for a linear elastic material. When $G = G_c$, the critical energy release rate is reached, which is a measure of fracture toughness.

For a through crack of length $2a$ in an infinite plate subject to a remote tensile stress, as shown in Figure 4.11a, the energy release rate is given by

$$G = \frac{\pi\sigma^2 a}{E} \tag{4.26}$$

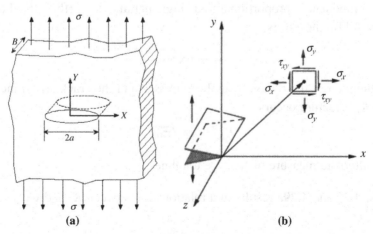

Figure 4.11: Infinite plate with center crack of size $2a$: (a) under tensile load, and (b) stress element near the crack tip.

where

    E is Young's modulus.

    σ is the remotely applied stress.

    a is the half-crack length.

When G reaches a critical value G_c, due to an increase in stress σ or in crack size a, or both, fracture failure occurs, where

$$G_c = \frac{\pi \sigma_f^2 a_c}{E} \qquad (4.27)$$

The energy release rate G is the driving force for fracture. Note that the remotely applied stress σ_f is usually smaller than the material yield strength (ductile material) or ultimate tensile strength (brittle material). Therefore, whenever cracks appear in the structure, the structure's strength in withstanding operating load deteriorates.

The stress intensity approach employs stress intensity factor (SIF) K_I as the driving force for fracture. For a stress element near the crack tip, the stress components σ_x, σ_y, and τ_{xy}, as shown in Figure 4.11, can be calculated using the equations

$$\begin{cases} \sigma_x = \dfrac{K_I}{\sqrt{2\pi r}} \cos \dfrac{\theta}{2} \left(1 - \sin \dfrac{\theta}{2} \sin \dfrac{3\theta}{2} \right) \\[3mm] \sigma_y = \dfrac{K_I}{\sqrt{2\pi r}} \cos \dfrac{\theta}{2} \left(1 + \sin \dfrac{\theta}{2} \sin \dfrac{3\theta}{2} \right) \\[3mm] \tau_{xy} = \dfrac{K_I}{\sqrt{2\pi r}} \cos \dfrac{\theta}{2} \sin \dfrac{\theta}{2} \cos \dfrac{3\theta}{2} \end{cases} \qquad (4.28)$$

Note that each component is proportional to a single parameter K_I (the SIF). For the plate shown in Figure 4.11a, the SIF is

$$K_I = \sigma \sqrt{\pi a} \qquad (4.29a)$$

When the combination of the remotely applied stress σ and the crack size a increases to a critical value K_{Ic}, fracture occurs:

$$K_{Ic} = \sigma_f \sqrt{\pi a_c} \qquad (4.29b)$$

Thus, K_{Ic} is an alternate measure of fracture toughness.

Comparing Eqs. 4.26 and 4.29a results in a relationship between K_I and G:

$$G = \frac{K_I^2}{E} \qquad (4.30)$$

Thus the energy and stress intensity approaches to fracture mechanics are equivalent for linear elastic materials. Note that Eq. 4.30 holds for plane stress problems. For plane strain problems, the denominator on the right becomes $E/(1-v)$, where v is Poisson's ratio.

4.5.2 Linear Elastic Fracture Mechanics

Linear elastic fracture mechanics is a branch of fracture mechanics that deals with problems in which the size of the plastic zone around the crack tip is very small in comparison to the domain size. LEFM holds well for the brittle mode of fracture, which governs the fracture until the SIFs are greater than the material fracture toughness. Fracture in metals after this point is usually accompanied by significant plastic yielding, and since LEFM is unable to analyze crack growth in such cases, it predicts a conservative estimate of life. However, for a majority of common structural applications, fracture toughness is still considered the failure criterion. Because of this fact, and because of its simplicity, LEFM is widely used.

A major achievement in the theoretical foundation of LEFM was the introduction of SIF K as a parameter for the intensity of stresses close to the crack tip. The SIF shown in Eq. 4.28 is only applicable to the center crack of an infinite thin plate under tensile load, as shown in Figure 4.11a. We must note that the expression for K_I in Eq. 4.28 is different for geometries other than the center-cracked infinite plate. Consequently, it is necessary to introduce a dimensionless geometric factor, Y, to characterize the component geometry, initial crack geometry (e.g., edge crack or center crack), and loads (tensile or shear). We thus have

$$K_I = Y\sigma\sqrt{\pi a} \tag{4.31}$$

where Y is a function of crack length a and width w of the sheet, given by

$$Y\left(\frac{a}{W}\right) = \sqrt{\sec\left(\frac{\pi a}{W}\right)} \tag{4.32}$$

for a plate of finite width W containing a through-thickness crack of length $2a$ at the center, or

$$Y\left(\frac{a}{W}\right) = 1.12 - \frac{0.41}{\sqrt{\pi}}\frac{a}{W} + \frac{18.7}{\sqrt{\pi}}\left(\frac{a}{W}\right)^2 - \dots \tag{4.33}$$

for a plate of finite width W containing a through-thickness edge crack of length a.

The geometric factor Y can be obtained analytically using the stress function (as discussed in Chapter 2) for components of simple geometry such as rectangular plates of center or edge cracks, or circular cross-section of a pipe with edge cracks, of which, mostly for planar problems. Note that the famous Westergaard stress function (Anderson 1994) of complex variables yields Eq. 4.28.

A more general approach for calculating SIF is the *J*-integral, which is basically a path-independent contour integral around the crack tip. Its use as a fracture parameter was introduced by Rice (1968). The *J*-integral is given by

$$J = \int_{\Gamma} \left[W dy - T_i \frac{\partial z_i}{\partial x} ds \right] = \int_{\Gamma} \left[W \, \delta_{1j} - \sigma_{ij} \frac{\partial z_i}{\partial x} \right] n_j d\Gamma \qquad (4.34)$$

where

σ_{ij} is the Cauchy stress tensor.

z_i is the *i*th component of the displacement.

$\boldsymbol{n}$ is the outward normal vector to an arbitrary contour around the crack tip (as shown in Figure 4.12).

W is the strain energy, defined as

$$W = \int_{0}^{\varepsilon_{ij}} \sigma_{ij} d\varepsilon_{ij} \qquad (4.35)$$

For linear elastic materials, $W = \frac{1}{2}\sigma_{ij}\varepsilon_{ij}$.

Physically, the *J*-integral may be interpreted as the energy flowing through the contour Γ per unit crack advance. Under elastic conditions, it is equivalent to Griffith's energy release rate

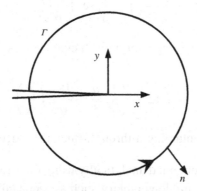

Figure 4.12: Path-independent closed contour around the crack tip.

(Anderson, 1994), and its relation to the stress intensity factors for a planar structure in uniaxial cases is given by

$$G = J = \frac{K_I^2}{E^*} \tag{4.36}$$

where $E^* = E/(1 - v^2)$ for plane strain and $E^* = E$ for plane stress. Note that for numerical implementation using the finite element method, the J-integral is usually converted to a domain form by applying the divergence theorem for more accurate results.

4.5.3 Mixed Mode

Up to this point, we have assumed that the structure is under uniaxial loads. In reality, loads are applied to a structural component in different directions. For a structure under general loads, three independent movements of the upper and lower crack surfaces with respect to each other (corresponding to three independent cases of loading) define the three crack opening modes, as shown in Figure 4.13. SIFs for the three modes are denoted by K_I, K_{II}, and K_{III}. In a general case, more than one loading mode may be present, and the crack-tip fields can be produced by an appropriate linear combination of those corresponding to these three modes. Such problems are referred to as mixed-mode.

For a planar structural component, mixed-mode problems consist of a crack-opening mode (1) and a shearing mode (2) due to in-plane normal N and shear loads S, respectively, as shown in Figure 4.14a. According to LEFM theory, the displacements and stresses near the crack tip are given respectively by (Anderson 1994)

$$\begin{cases} u_r = \frac{1}{\mu} \sqrt{\frac{r}{2\mu}} \left[K_I \left(\kappa - \cos\frac{\theta}{2} + \sin\theta\sin\frac{\theta}{2} \right) + K_{II} \left(\kappa + \sin\frac{\theta}{2} + \sin\theta\cos\frac{\theta}{2} \right) \right] \\ u_\theta = \frac{1}{\mu} \sqrt{\frac{r}{2\mu}} \left[K_I \left(\kappa + \sin\frac{\theta}{2} + \sin\theta\cos\frac{\theta}{2} \right) + K_{II} \left(\kappa - \cos\frac{\theta}{2} - \sin\theta\sin\frac{\theta}{2} \right) \right] \end{cases} \tag{4.37}$$

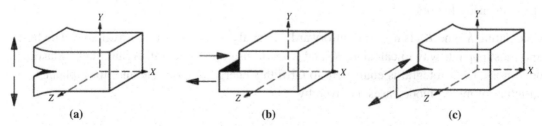

Figure 4.13: Modes of crack-tip opening: (a) mode 1: crack opening, (b) mode 2: crack shearing, and (c) mode 3: crack tearing.

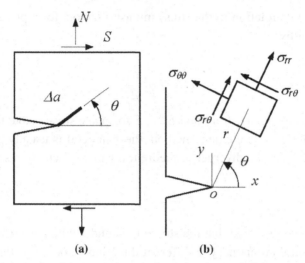

Figure 4.14: Mixed mode: (a) normal and shear loads, and (b) stress element near the crack tip.

and

$$
\begin{cases}
\sigma_{rr} = \dfrac{1}{\sqrt{2\pi r}} \left[\dfrac{K_I}{4}\left(5\cos\dfrac{\theta}{2} - \cos\dfrac{3\theta}{2} \right) + \dfrac{K_{II}}{4}\left(-5\sin\dfrac{\theta}{2} + 3\sin\dfrac{3\theta}{2} \right) \right] \\[3mm]
\sigma_{\theta\theta} = \dfrac{1}{\sqrt{2\pi r}} \left[\dfrac{K_I}{4}\left(3\cos\dfrac{\theta}{2} + \cos\dfrac{3\theta}{2} \right) + \dfrac{K_{II}}{4}\left(-3\sin\dfrac{\theta}{2} - 3\sin\dfrac{3\theta}{2} \right) \right] \\[3mm]
\sigma_{r\theta} = \dfrac{1}{\sqrt{2\pi r}} \left[\dfrac{K_I}{4}\left(\sin\dfrac{\theta}{2} + \sin\dfrac{3\theta}{2} \right) + \dfrac{K_{II}}{4}\left(\cos\dfrac{3\theta}{2} + 3\cos\dfrac{3\theta}{2} \right) \right]
\end{cases}
\tag{4.38}
$$

where K_I and K_{II} are the SIFs of mode 1 and mode 2 crack, respectively. In these expressions, (r, θ) define the location of a point in a local crack tip coordinate system (shown in Figure 4.14b), μ is the shear modulus, ν is Poisson's ratio, and κ is the Kolosov coefficient, whose value is $\kappa = (3 - \nu)/(1 + \nu)$ for plane stress and $\kappa = 3 - 4\nu$ for plane strain. These equations show that the near-tip stresses and displacements are completely determined by the stress intensity factors.

Calculating K_I and K_{II} is not straightforward. No analytical expression is available. The best and most popular way to calculate K_I and K_{II} is to use the J-integral. Again, under elastic conditions, the J-integral is equivalent to Griffith's energy release rate and, for a planar structure of mixed-mode cases, is given by

$$
G = J = \frac{1}{E^*} (K_I^2 + K_{II}^2)
\tag{4.39}
$$

As shown in Eq. 4.39, the evaluation of the J-integral does not give separate values of mode 1 and mode 2 stress intensity factors. Yau et al. (1980) proposed an interaction integral technique in which two kinematically admissible states of a body are superimposed to extract the mixed-mode SIFs.

Consider two independent equilibrium states of a cracked body. State 1 is defined as the actual state for the given boundary conditions, while State 2 is an auxiliary state. The J-integral for the two superposed states is

$$J^{(1+2)} = J^{(1)} + J^{(2)} + M^{(1+2)} \tag{4.40}$$

where $J^{(1)}$ and $J^{(2)}$ are the J-integrals for actual state and auxiliary state, respectively, and $M^{(1+2)}$ is the interaction integral for the two equilibrium states:

$$M^{(1+2)} = \int_{\Gamma} \left[W^{(1+2)} \delta_{1j} - \sigma_{ij}^{(1)} \frac{\partial z_i^{(2)}}{\partial x} - \sigma_{ij}^{(2)} \frac{\partial z_i^{(1)}}{\partial x} \right] n_j d\Gamma \tag{4.41}$$

where $W^{(1+2)}$ is the interaction strain energy density:

$$W^{(1+2)} = \frac{1}{2} \left(\sigma_{ij}^{(1)} \varepsilon_{ij}^{(2)} + \sigma_{ij}^{(2)} \varepsilon_{ij}^{(1)} \right) = \sigma_{ij}^{(1)} \varepsilon_{ij}^{(2)} = \sigma_{ij}^{(2)} \varepsilon_{ij}^{(1)} \tag{4.42}$$

On the other hand, Eq. 4.39 for the superposed state can be written as

$$
\begin{aligned}
J^{(1+2)} &= \frac{1}{E^*} \left[\left(K_I^{(1+2)} \right)^2 + \left(K_{II}^{(1+2)} \right)^2 \right] = \frac{1}{E^*} \left[\left(K_I^{(1)} + K_I^{(2)} \right)^2 + \left(K_{II}^{(1)} + K_{II}^{(2)} \right)^2 \right] \\
&= J^{(1)} + J^{(2)} + \frac{2}{E^*} \left(K_I^{(1)} K_I^{(2)} + K_{II}^{(1)} K_{II}^{(2)} \right)
\end{aligned}
\tag{4.43}
$$

Thus,

$$M^{(1+2)} = \frac{2}{E^*} \left(K_I^{(1)} K_I^{(2)} + K_{II}^{(1)} K_{II}^{(2)} \right) \tag{4.44}$$

To obtain mode 1 stress intensity factor, the auxiliary state is chosen to be the pure mode 1 asymptotic condition with $K_I^{(2a)} = 1$ and $K_{II}^{(2a)} = 0$. Substituting this in Eq. 4.44 gives

$$K_I^{(1)} = \frac{E^*}{2} M^{(1+2a)} = \frac{E^*}{2} \int_{\Gamma} \left[W^{(1+2a)} \delta_{1j} - \sigma_{ij}^{(1)} \frac{\partial z_i^{(2a)}}{\partial x} - \sigma_{ij}^{(2a)} \frac{\partial z_i^{(1)}}{\partial x} \right] n_j d\Gamma \tag{4.45}$$

where $z_i^{(2a)}$ and $\sigma_{ij}^{(2a)}$ can be obtained by plugging $K_I^{(2a)} = 1$ and $K_{II}^{(2a)} = 0$ into Eqs. 4.37 and 4.38, respectively. Then $\sigma_{ij}^{(2a)}$ and $\varepsilon_{ij}^{(2a)}$ (obtained from $z_i^{(2a)}$) are inserted into Eq. 4.45 for $M^{(1+2a)}$.

Similarly, mode 2 stress intensity factor can be obtained by choosing the auxiliary state to be the pure mode 2 asymptotic condition with $K_I^{(2b)} = 0$ and $K_{II}^{(2b)} = 1$. Substituting this in Eq. 4.44 gives

$$K_{II}^{(1)} = \frac{E^*}{2}M^{(1+2b)} = \int_\Gamma \left[W^{(1+2b)}\delta_{1j} - \sigma_{ij}^{(1)}\frac{\partial z_i^{(2b)}}{\partial x} - \sigma_{ij}^{(2b)}\frac{\partial z_i^{(1)}}{\partial x} \right]n_j d\Gamma \qquad (4.46)$$

Note that the asymptotic fields used for the auxiliary state are valid for LEFM only, and hence the interaction integral method presented here is also applicable only to LEFM.

4.5.4 Quasistatic Crack Growth

Laboratory experiments show that the rate of crack growth is a function of the stress intensity factor. In Regime B of the crack shown in Figure 4.15 (i.e., the stable crack propagation stage, also called the Paris regime), the crack growth rate can be calculated using the Paris power law (Paris et al. 1961) or Forman's equation (Forman et al. 1967).

The Paris power law relates the crack propagation rate under fatigue loading to stress intensity factors. For a given fatigue loading, assume that $\Delta K = K_{max} - K_{min}$ is the SIF range, where K_{max} and K_{min} correspond to the SIFs of maximum and minimum stresses in a load cycle, respectively. Suppose that the crack grows by an amount Δa in ΔN cycles. Crack growth rate is related to ΔK as

$$\frac{\Delta a}{\Delta N} \approx \frac{da}{dN} = C(\Delta K)^m \qquad (4.47a)$$

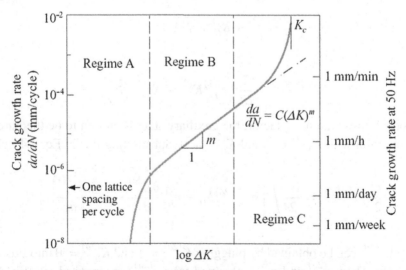

Figure 4.15: Rate of crack growth. *(Source: Chapter 7; Bernard J. Hamrock, Steven R. Schmid, Bo O. Jacobson, Fundamentals of Machine Elements, 2nd Ed., McGraw-Hill, 2005)*

where C and m are material constants. For mode 1 loading, $\Delta K = K_{I\max} - K_{I\min}$. Forman's equation refines the Paris law by incorporating stress ratio R and fracture toughness K_c into the growth rate calculation:

$$\frac{\Delta a}{\Delta N} \approx \frac{da}{dN} = \frac{C(\Delta K)^p}{K_c(1-R) - \Delta K} \tag{4.47b}$$

where C, m, and p are material constants for crack propagation.

For general mixed-mode loading, ΔK is replaced by an equivalent SIF range, ΔK_{eq}, given by

$$\Delta K_{eq} = \sqrt{\Delta K_I^2 + \Delta K_{II}^2} \tag{4.48}$$

Usually, the crack growth per cycle Δa is very small—almost on the order of 10^{-8} in. Hence, in numerical implementation, instead of computing crack growth in every load cycle, usually the value of Δa is predetermined. As a rule of thumb, $\Delta a = a_i/10$, where a_i is the initial crack length. Once Δa is fixed, the number of corresponding cycles is computed using the following equation if the Paris law is employed:

$$\Delta N = \frac{\Delta a}{C(\Delta K_{eq})^m} \tag{4.49}$$

Along with crack propagation rate, crack growth angle θ_c is a necessary parameter in modeling crack propagation. Among the many criteria, the maximum hoop stress criterion is often employed, which states that the crack propagates in a direction where the hoop stress is maximum. Based on the maximum hoop stress criterion, the expression for θ_c is given as

$$\theta_c = 2\tan^{-1}\left(\frac{1}{4}(K_I/K_{II}) \pm \sqrt{(K_I/K_{II})^2 + 8}\right) \tag{4.50}$$

It can be seen from Eqs. 4.49 and 4.50 that ΔN and θ_c are functions of K_I and K_{II}.

4.5.5 The Extended Finite Element Method

Crack propagation calculations are usually performed using finite element method. Since the crack tip is singular where the stress field is infinitely high, as shown in Eq. 4.28 or 4.38 according to LEFM theory, a very much refined mesh must be created around it to capture the surrounding high stress, as shown in Figure 4.16a. In addition, when the crack propagates, the structure has to be remeshed to conform to the newly created boundary edges (or faces for 3D) at the crack tip, as illustrated in Figure 4.16b. Very often a large finite element model is required for crack propagation calculation. However, it is more troublesome for mesh, generation around the crack tip for structures with complex geometry, especially for 3D structural components.

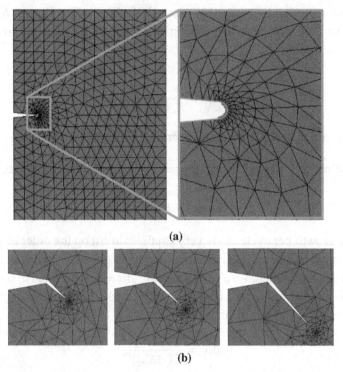

(a)

(b)

Figure 4.16: Finite element mesh for crack growth calculation: (a) very much refined mesh at the crack tip, and (b) mesh refinement following the evolving crack tip.

Instead of the regular finite element method, the newly developed extended finite element method alleviates problems of mesh refinement around the crack tip and remesh as crack grows. XFEM was developed in late 1990s by Belytschko and Black (1999) to deal with the shortcomings of the finite element method in solving discontinuous problems. One of its key advantages is that in such problems the finite element mesh does not need to be updated to track the crack path or the interface movement.

XFEM is a computational technique in which special enrichment functions are used to incorporate the discontinuity caused by the crack surfaces and crack-tip fields into a regular finite element approximation. The XFEM displacement approximation for a vector-valued function $u(x)$: $\mathbf{R}^2 \rightarrow \mathbf{R}^2$ is typically given as

$$u^h(\boldsymbol{x}, t) = \sum_{i \in I} u_i(t) N_i(\boldsymbol{x}) + \sum_{j \in J} b_j(t) N_i(\boldsymbol{x}) H(\psi(\boldsymbol{x}, t)) + \sum_{k \in K} N_k(\mathbf{x}) \left(\sum_{\ell=1}^{4} a_k^\ell(t) B_\ell(r, \theta) \right)$$

(4.51)

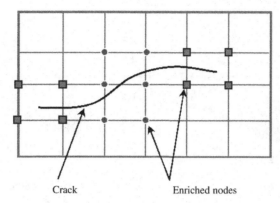

Crack   Enriched nodes

Figure 4.17: Nodal enrichment in XFEM.

where

$N_i(x)$ is the shape function associated with the node i.

t is a monotonically increasing time parameter that represents the load cycles.

J is the set of all nodes whose support is bisected by the crack (shown by filled circles in Figure 4.17).

The set K contains all nodes of the elements containing the crack tip (shown by filled squares in Figure 4.17).

The first term in Eq. 4.51 is the regular finite element shape function; the second term represents the Heaviside step function $H(\psi(x,t))$ employed to model discontinuity due to crack; and the last term incorporates the near-tip asymptotic displacement fields using branch functions, B_ℓ (shown in Figure 4.18), which are defined by

$$B_\ell(r, \theta) = \left\{ \sqrt{r} \sin \frac{\theta}{2}, \sqrt{r} \cos \frac{\theta}{2}, \sqrt{r} \sin \frac{\theta}{2} \sin \theta, \sqrt{r} \cos \frac{\theta}{2} \sin \theta \right\} \qquad (4.52)$$

where (r,θ) are defined in a polar coordinate system at the crack tip and $\theta = 0$ is tangent to the crack. Because of the branch functions, a relatively coarse mesh can be used near the crack-tip region.

The level set method (LSM) is a numerical technique used to track the motion of interfaces. In LSM the crack is modeled using signed-distance functions ϕ and ψ (shown in Figure 4.19), which are stored at nodes. Thus, it is always possible to know which elements are cut by the crack and which elements contain the crack tip. LSM couples naturally with XFEM and facilitates selection of nodes for enrichment. These enrichment functions appear in the form of extra degrees of freedom in the finite element stiffness matrix. At the end of a crack growth cycle, the signed-distance functions are updated to account for changes in crack geometry,

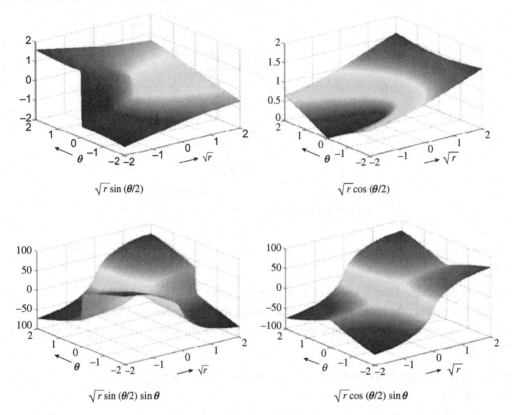

$\sqrt{r}\sin(\theta/2)$ $\quad$ $\sqrt{r}\cos(\theta/2)$

$\sqrt{r}\sin(\theta/2)\sin\theta$ $\quad$ $\sqrt{r}\cos(\theta/2)\sin\theta$

Figure 4.18: Branch functions.

therefore, no remeshing is required. Thus XFEM and LSM provide an elegant scheme for crack growth simulation.

The superiority of XFEM and LSM in solving problems with discontinuities and singularities has been demonstrated and widely recognized. Figure 4.20 demonstrates an application of the method to crack propagation in an engine connecting rod (Edke and

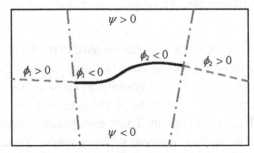

Figure 4.19: Crack representation using level sets.

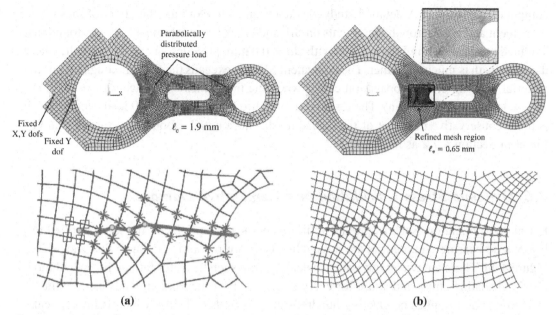

Figure 4.20: Two-dimensional engine connecting rod for simulating crack propagation using XFEM and LSM: (a) $\Delta a = 1.5$ mm and 8554 dof; (b) $\Delta a = 0.8$ mm and 25,186 dof.

Chang 2011), where the finite element mesh does not have to be updated when the crack tip advances.

In Figure 4.20a, an initial crack of $a = 7$ mm (an arbitrary choice) is introduced at the left side of the semicircular arc of the center slot, as shown. A crack propagation analysis is conducted with a crack growth increment $\Delta a = 1.5$ mm. Figure 4.20a shows the crack propagation path until failure—that is, until the equivalent stress intensity factor K_{eq} exceeds the fracture toughness of the material.

For mixed-mode cases, generally the crack propagation path is curvilinear. In this case, however, the path appears to be zigzag because of the alternating positive and negative signs of the crack propagation angle θ_c. A large crack growth increment and large element size are the main reasons behind these oscillations. The equivalent stress intensity factor K_{eq} exceeds the fracture toughness of the material after four crack propagation cycles, when the final crack size reaches $a_f = 13$ mm ($a + 4\Delta a = 7 + 4 \times 1.5$ mm). When a large value of Δa (such as 1.5 mm) is selected, the analysis may not accurately predict crack path and residual life, as indicated in this coarse mesh case. Note that the estimated residual service life N is about 194,000 load cycles.

To improve analysis accuracy and reduce the zigzag crack path, a smaller Δa must be employed. A very small value of Δa requires very fine mesh, thereby greatly increasing the computational burden. A detailed study was undertaken to examine the effect of mesh refinement and crack growth increments on SIF and crack propagation path. It was found that a refined mesh, shown in Figure 4.20b with $\Delta a = 0.8$ mm, provides an excellent result, where the crack path is much smoother. The equivalent SIF K_{eq} exceeds the fracture toughness of the material after six crack propagation cycles when the final crack size reaches $a_f = 11.8$ mm ($a + 4\Delta a = 7 + 6 \times 0.8$ mm). The residual service life N is about 129,000 load cycles, which is more conservative than that of the coarse mesh. More details about this example can be found in Section 4.7.2 as a case study.

4.6 Dynamic Stress Calculation and Cumulative Damage

For many applications such as ground vehicles and heavy equipment, the loads applied to heavy load bearing components are not cyclic. They are random, similar to that shown in Figure 4.4d. There are two major issues in dealing with such loads for fatigue life calculation. First, calculating stress and strain for every single point of the random loads using finite element method implies conducting hundreds or thousands of FEAs, which is impractical. Second, the principal directions of the stresses at critical locations of the component vary in time.

The first issue is well addressed using quasistatic FEA for dynamic stress calculations. Various fatigue calculation methods have been proposed to address the second issue (see Sections 4.3 and 4.4). The stress history is then employed to predict the fatigue life of the component using either a strain- or stress-based crack initiation life prediction method and linear elastic fracture mechanics for crack propagation life. The overall process is shown in Figure 4.21.

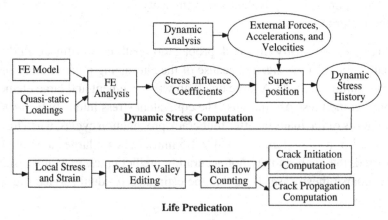

Figure 4.21: Computation process for fatigue life.

4.6.1 Dynamic Stress Calculations

Dynamic stress can be obtained either from experiment (mounting sensors or transducers on a physical component) or from simulation. Using simulation, a representative load history, including inertia forces and external forces (such as joint reaction forces and torques), for accurate dynamic stress computation must first be generated. Multibody dynamic analysis methods (discussed in Chapter 3), which have typically been used for dynamic motion analysis, can be used for dynamic load analysis of mechanical systems. All bodies of the dynamic model are usually assumed to be rigid. For the suspension components of a vehicle, the rigid-body assumption usually yields reasonably accurate analysis results to support structural design for durability.

Quasistatic FEA are then performed to obtain stress influence coefficients for the structural components. The stress influence coefficients are superposed with the dynamic analysis results, including external forces, accelerations, and angular velocities to compute dynamic stress history. Sanders and Tesar (1978) showed that the quasistatic deformation evaluation is a valid form of approximation for most industrial mechanisms that are stiff and operate substantially below their natural frequencies. Note they assumed that deformation caused by applied external and inertia forces is small compared with the geometry of the structural component. They further assumed that the material from which the component is fabricated behaves in a linear elastic fashion.

The finite element model of the component of interest corresponds to a body in the multibody dynamic model. Since dynamic stress histories contain very large amounts of data, it is generally necessary to reduce or condense the amount of data by, for example, peak-valley editing before computing crack initiation and propagation life. These values are then used to perform a cycle-counting procedure to transform variable-amplitude stress or strain histories into a number of constant-amplitude stress or strain histories. These histories are used to compute the component's crack initiation life as well as the crack propagation life.

For a component subject to external forces (including joint reaction forces and torques) and inertia forces obtained from multibody dynamic analysis, the quasistatic equation in a matrix form of the finite element method can be written as

$$Kz = F_e(t) - F_i(t) \tag{4.53}$$

where

K is the stiffness matrix.

z is a vector of nodal displacements.

$F_e(t)$ and $F_i(t)$ are vectors of external and inertia force histories, respectively, obtained from dynamic analysis.

Since the loading condition can vary with time in a dynamic system, dynamic stress can be calculated as follows:

$$\sigma(t) = DBK^{-1}[F_e(t) - F_i(t)] \tag{4.54}$$

where

D is the elasticity matrix.

B is the strain-displacement matrix.

The quasistatic method separates the external forces and inertia forces acting on the component into two parts: time dependent (external and inertia force histories) and time independent (quasistatic loading), and treats the quasistatic loading as static forces. The stress influence coefficients are obtained by performing FEA for each quasistatic loading separately. The dynamic stresses can be calculated using the superposition principle; that is, external and inertial force histories are multiplied by the corresponding stress influence coefficients.

A set of unit loads is used to calculate the stress influence coefficients corresponding to the joint reaction forces and torques. The unit loads are applied at a given point x in all degrees of freedom where joint reaction forces and torques act. For example, if a set of joint reaction forces and torques acts on the kth finite element node, the corresponding quasistatic loads q^k are three unit forces and three unit torques in the body reference frame of the body x_1-x_2-x_3 applied to the kth node as six loading cases. Therefore, the stress influence coefficients σ^k_{SIC} can be obtained using FEA:

$$\sigma^k_{SIC} = DBK^{-1}q^k \tag{4.55}$$

The inertia body force applied to a point x in the component due to accelerations, angular velocities, and angular accelerations, as shown in Figure 4.22, can be expressed as

$$f_i(x) = f_i^a(x) + f_i^r(x) + f_i^t(x) = -\rho(x)a_i - \rho(x)a_i^r + \rho(x)a_i^t \tag{4.56}$$

where

$\rho(x)$ is mass density.

$f_i(x)$ is the x_i-component of the inertia body force per unit mass.

$f_i^a(x), f_i^r(x),$ and $f_i^t(x)$ are inertia body forces per unit mass in the translational, radial, and tangential directions, respectively.

a_i is the instantaneous translational acceleration and is independent of the location of point x.

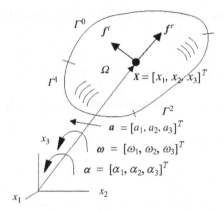

Figure 4.22: Inertia forces applied to a component.

a_i^r is the centripetal acceleration toward the instantaneous axis of the rotation and is perpendicular to it.

a_i^t is the tangential acceleration.

The radial and tangential accelerations $a_i^r(x)$ and $a_i^t(x)$ at point x can be written as

$$a_i^r(x) = \omega_{ij}\omega_{jk}x_k \tag{4.57}$$

and

$$a_i^t(x) = \alpha_{ij}x_j \tag{4.58}$$

where x_k is the kth coordinate of point x, ω_{ij} is the instantaneous angular velocity, and α_{ij} is the instantaneous angular acceleration. Hence, the inertia body force at point x is

$$f_i(x) = \rho(x)(-a_i - \alpha_{ij}x_j + \omega_{ij}\omega_{jk}x_k) \tag{4.59}$$

It can be seen from Eq. 4.59 that the inertia force, $f_i(x)$, is linearly dependent on components of the acceleration a and the angular acceleration α. However, the inertia force is not linearly dependent on components of the angular velocities ω. Instead, it depends linearly on the combinations of components of the angular velocities ω, including six terms $\omega_1\omega_2$, $\omega_2\omega_3$, $\omega_2\omega_3$, $\omega_2^2 + \omega_3^2$, $\omega_1^2 + \omega_3^2$, and $\omega_1^2 + \omega_2^2$. Therefore, to compute the quasistatic loading of inertia forces, 12 loading cases are assumed.

Note that the stress influence coefficients of the first six quasistatic loads can be obtained by applying unit accelerations to perform FEA directly, using commercial FEA codes such as ANSYS®. However, equivalent nodal forces corresponding to the last six quasistatic loads involve angular velocities, which can be applied to the finite element model as external nodal forces. The stress influence coefficients σ_{SIC}^{ine} due to inertia forces can be obtained using FEA:

$$\sigma_{SIC}^{ine} = DBK^{-1}q_i^{ine}, \quad i = 1, ..., 12 \tag{4.60}$$

The dynamic stress is calculated using the superposition principle as

$$\sigma(t) = \sigma^{ine}(t) + \sigma^{ext}(t) \tag{4.61}$$

where

$$
\begin{aligned}
\sigma^{ine}(t) = & \sum_{i=1}^{3} \sigma_{SICi}^{ine} a_i(t) + \sum_{i=1}^{3} \sigma_{SIC(i+3)}^{ine} \alpha_i(t) \\
& + \sigma_{SIC7}^{ine}\omega_1(t)\omega_2(t) + \sigma_{SIC8}^{ine}\omega_2(t)\omega_3(t) + \sigma_{SIC9}^{ine}\omega_3(t)\omega_1(t) \\
& + \sigma_{SIC10}^{ine}\left(\omega_2^2(t) + \omega_3^2(t)\right) + \sigma_{SIC11}^{ine}\left(\omega_1^2(t) + \omega_3^2(t)\right) + \sigma_{SIC12}^{ine}\left(\omega_1^2(t) + \omega_2^2(t)\right)
\end{aligned}
\tag{4.62}
$$

in which σ_{SIC}^{ine} is obtained from Eq. 4.60, and

$$\sigma^{ext}(t) = \sum_{k=1}^{n} \sigma_{SIC}^{k}F_i^{k}(t) \tag{4.63}$$

where σ_{SIC}^{k} can be obtained from Eq. 4.55, and n is the number of nodes at which external forces $F^k(t)$ are applied.

4.6.2 Peak-Valley Editing

As discussed earlier, variable-amplitude fatigue life prediction usually requires either strain or stress histories as input. Often these strain or stress histories are in the form of time histories obtained from multibody dynamic analysis (or from experiment by mounting sensors or transducers on a physical component for data collection). These time histories contain stress or strain values collected or computed at preset time intervals. Depending on the time step used and the length of the analysis, these types of stress-time or strain-time histories can often contain very large amounts of data. For the purpose of fatigue life computation, it is generally necessary to reduce or condense the amount of data in them. The cycle-counting process, such as rain-flow counting, requires that a variable-amplitude time history be put into reduced form (peak-valley edited) before life calculations can be performed. Also, fewer data are easier to manipulate.

A common method for reducing the number of points in a variable-amplitude time history is to perform peak-valley editing on it (Downing and Socie 1982). Peak-valley editing has two main purposes. The first is to produce a history that contains only sequential peak and valley reversal points. Peaks and valleys are points in the time history where a change in loading direction occurs. For example, data between points A and B in Figure 4.23a do not contain any stress reversals; hence, they are all removed. The second purpose is to remove any ranges with magnitudes less than a prescribed minimum allowable value. For example, if the minimum range of stress variation is 20 ksi, all data points between points B and C are removed, as shown in Figure 4.23b.

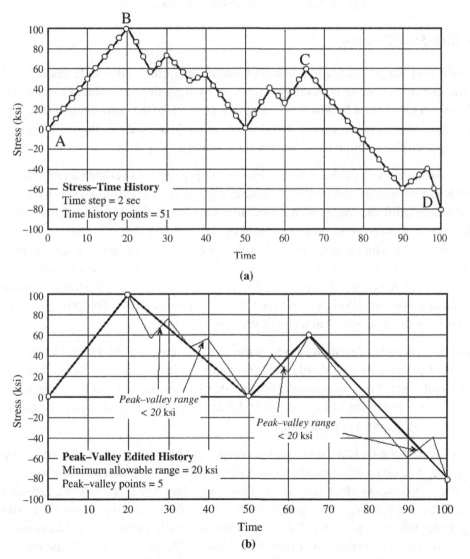

Figure 4.23: Peak-valley editing: (a) original stress-time history, and (b) edited stress-time history.

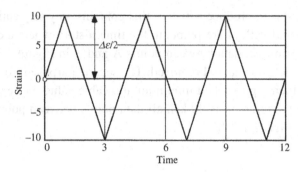

Figure 4.24: Constant-amplitude strain history.

4.6.3 Rain-Flow Counting

Most traditional fatigue life prediction models require a constant-amplitude loading as input. Constant-amplitude loading is usually represented by repeating sinusoidal or triangular waveforms. In the case of fatigue crack initiation prediction, the conventional strain-life relationship is used to relate a constant-amplitude strain loading $\Delta\varepsilon/2$, shown in Figure 4.24, to the fatigue crack initiation life $2N_f$, using, for example, Eq. 4.18.

However, components of mechanical systems usually experience a variable-amplitude loading that yields a stress or strain history with variable amplitude, such as the one shown in Figure 4.23. To predict fatigue life of a component subject to a variable-amplitude loading history, the variable-amplitude stress or strain history must be converted into several constant-amplitude cycles. The procedure for this conversion is referred to as cycle counting procedure. A number of such procedures have been proposed over the years, but all of them must use some rules to decide when or how a cycle is defined from a variable-amplitude history. A well-accepted procedure, called rain-flow counting (Downing and Socie 1982), is discussed next. This procedure attempts to define cycles that correspond to a closed stress-strain hysteresis loop.

The rain-flow counting algorithm is used in the analysis of fatigue data to reduce a spectrum of varying stress into a set of simple stress reversals. Its importance is that it allows the application of Miner's rule to assess the fatigue life of a structure subject to complex loading. The algorithm was developed by Endo and Matsuiski in 1968; they describe the process in terms of rain falling off a pagoda roof.

Figure 4.25a shows a typical stress history, composed of repetitive blocks. A stress block (stress points A to G) is identified and rotated 90 deg. clockwise, as shown at the top of Figure 4.25b. The rotated stress block (or loading history) (points A—G) resembles a Japanese pagoda. The corresponding stress and strain history is plotted directly below the loading history. In the lower stress-strain plot, three cycles are easily identified: one large overall cycle (A—D—G), one intermediate cycle in the center of the plot (C—B—C), and one smaller cycle (E—F—E). Each cycle has its own strain range and mean stress.

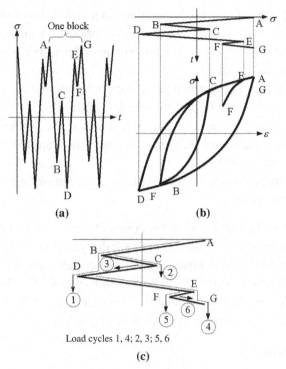

Figure 4.25: Rain-flow cycle counting: (a) stress history, (b) hysteresis loops, and (c) rain-flow counting.

From a deformation viewpoint, the process proceeds as follows. Start at A, the maximum strain, and load the material to B in compression. Then reload to point C and compress to D. When the material reaches the strain at point B during the loading from C to D, the material remembers its prior deformation and deforms along a path from A to D as if event C–B never happened. This is better illustrated in the next part of the loading. Load from D to E and unload to F. Now load from F to G. When the material reaches the strain at point E during the loading from F to G, the material remembers its prior deformation and deforms along a path from D to G as if the event E–F never happened. As a result, rain-flow counting identifies three cycles: A–D–G, B–C–B, and E–F–E.

Several rules are imposed on raindrops falling down these sloping roofs so that the rain flow may be used to define cycles and half-cycles of fluctuating stress in the spectrum. Rain flow is initiated by placing raindrops successively at the inside of each peak (maximum) and valley (minimum). The rules are as follows:

1. The rain is allowed to fall on the roof and drip down to the next slope. For example, the rain flow begins at point A, drips to point B, then to point D. After point D, there are no more roofs to drip to, so stress points A–D are counted as a half-cycle, cycle 1.

2. Step 1 is true except that, if the rain drop initiates at a valley (for example, point B shown in Figure 4.25c), it must be terminated when it comes opposite a valley more negative (in this case point D) than the valley from which it initiated. Therefore, stress points B—C are counted as a half-cycle, cycle 2.
3. Similarly, if the rain flow initiates at a peak, it must be terminated when it comes opposite a peak more positive than the peak from which it initiated.
4. The rain flow must stop if it meets the rain from a roof above. For example, the rain begins at point C and ends beneath cycle 1, right under point B. Therefore, stress points C—B are counted as a half-cycle, cycle 3.
5. Half-cycles of identical magnitude (but opposite sense) must be paired up to count the number of complete cycles.

Following the rules, from Figure 4.25c, stress points D—E—G are counted as a half-cycle (rule 1), cycle 4. Stress points E—F are counted as half-cycle 5 (rule 1), and points F—E are counted as half-cycle 6, as shown in Figure 4.25c. Half-cycles 1 and 4 form a complete cycle (rule 5); hence, the hysteresis loop A—D—G shown in Figure 4.25b. Similarly, half-cycles 2 and 3 form the second cycle, the smaller loop B—C in Figure 4.25b. Finally, half-cycles 5 and 6 form the third cycle, the smallest loop F—G in Figure 4.25b.

4.6.4 Blocks to Failure

Obtaining the fatigue crack initiation life $2N_f$ for each rain-flow cycle does not directly yield the required "blocks to failure" result for the fatigue crack initiation life prediction involved with the variable stress or strain amplitudes. These individual lives must be combined through the use of an appropriate damage summation routine to obtain the required "blocks to failure" result.

The most popular and widely used damage summation is the Palmgren—Miner linear damage rule, or Miner's rule, discussed in Section 4.2.5. Miner's rule simply states that the failure occurs when the summation of the individual damage values caused by each rain-flow cycle reaches a value of 1; that is,

$$\sum_{i=1}^{k} D_i \geq 1 \tag{4.64}$$

where

    k is the total number of cycles defined from the variable-amplitude history.
    D_i is the damage for the ith defined cycle.

Miner's rule defines the damage per individual rain-flow cycle as

$$D_i = \frac{2}{(2N_f)_i} = \frac{1}{(N_f)_i} \tag{4.65}$$

where

$(2N_f)_i$ is the fatigue crack initiation life (reversals to failure) for the ith defined cycle obtained by solving the nonlinear strain-life relation, Eq. 4.18 or 4.19.

$(N_f)_i$ represents the corresponding fatigue life expressed in cycles to failure.

From the definition of damage per cycle given in Eq. 4.65, it can be seen that a given rain-flow cycle consumes $1/(N_f)_i$ of the total fatigue, where $(N_f)_i$ is the fatigue life in cycles to failure for the given ith cycle.

Defining the total accumulated damage for one loading block as D_{block}, Miner's rule predicts failure when

$$D_{block} \geq 1 \qquad (4.66)$$

Using Eqs. 4.65 and 4.66, Miner's rule can be expressed in a form that directly yields the variable-amplitude fatigue life in "blocks to failure" as

$$B_f = \frac{1}{D_{block}} = \frac{1}{\sum\limits_{i=1}^{k} D_i} = \frac{1}{\sum\limits_{i=1}^{k} \frac{2}{(2N_f)_i}} \qquad (4.67)$$

where B_f represents the blocks to failure due to the given variable-amplitude loading.

This can be thought of as assessing what proportion of life is consumed by stress reversal at each magnitude and then forming a linear combination of their cumulative effects.

4.7 Fatigue and Fracture Simulation Software

Three types of fatigue software are included in this section: the general-purpose codes for crack initiation, non-FEA crack propagation, and FEA crack propagation.

4.7.1 General-Purpose Codes for Crack Initiation

Several general-purpose codes have been developed for predicting crack initiation life, supporting both high- and low-cycle fatigue. Some support basic stress-life and strain-life estimates of uniaxial stresses. These include the fatigue computation of Pro/MECHANICA Structure and SolidWorks Simulation. Some codes are implemented as postprocessors or modules of commercial FEA software, including ANSYS, MSC Fatigue®, and ABAQUS®.

nCode DesignLife™, an ANSYS module, provides advanced fatigue analysis in the ANSYS Workbench 11 SP1 environment. Results and materials data from Workbench simulations can be directly accessed by DesignLife, which was developed by HBM-nCode, Inc. Both stress-life (S-N) and strain-life (ε-N) fatigue life estimations are implemented in this program. It also implements both Goodman and Gerber mean stress correction. In strain-life calculations, DesignLife uses Neuber (1961) and Hoffmann and Seeger (1985) notch corrections and mean stress correction by Morrow (1965) or Smith-Watson-Topper (1970). When temperature difference is important in a model, the software provides difference curves for different temperatures and interpolates between them. For multiaxial problems, the Dang Van (Dang Van and Papadopoulos 1987, Dang Van et al. 1989) method is used. In addition, the software has spot and seam weld tools and allows temperature variations to be included in simulations.

MSC.Fatigue is an advanced fatigue life estimation software package for use with finite element analysis results. It provides state-of-the-art fatigue design tools that can be used to optimize product life. MSC.Fatigue was developed by nCode International Ltd. in conjunction with MSC Software Corporation. It is integrated with MSC/Patran® for its excellent meshing and visualization capabilities, as well as with MSC Adams® for dynamic simulation and dynamic stress calculations for fatigue life estimates.

Safe Technology Ltd. has developed a set of durability calculation programs collected under the name fe-safe™. In normal stress-life fatigue calculation, fe-safe implements Goodman (Dowling 2007), Gerber (Dowling 2007), Buch (1997), or user-defined mean stress corrections. Morrow (1961), SWT (1970), or user-defined mean stress correction is available for estimating life with the strain-life method. For multiaxial analysis the software gives the option of Dang Van (Dang Van and Papadopoulous 1987, Dang Van et al. 1989) for stress-based analysis and Brown-Miller (Brown and Miller 1973) for combining normal and shear strain in strain-based analysis. All of these modules are collected in a standalone product called safe4fatigue™, a suite that can be used for fatigue calculation. fe-safe has been integrated into ABAQUS as a postprocessor for fatigue life estimates.

FEMFAT was created by the Engineering Center Steyr in Austria. The software is a fatigue postprocessor utilizing finite element method to predict fatigue life in components. Finite element modeling is carried out in a separate program, and FEMFAT is compatible with many of the most commonly used CAE programs.

In addition to commercial-based fatigue programs, there are a number of online free fatigue postprocessors. One of these was developed by Darrell F. Socie, a professor of mechanical engineering at the University of Illinois, and is called Fatigue Calculator. The software is opened in a web browser with no need for installation.

4.7.2 Non-FEA-Based Crack Propagation

To date, there are two mainstream approaches available for the calculation of fracture parameters—the closed-form solution and the finite element method (including the boundary element method). In the closed-form approach a known solution provides fracture parameters as a function of geometry, external loads, and crack size. The known solutions may be derived from analytical expressions or more usually, interpolated from results obtained using FEM. These predictions generally assume that a crack grows under single prescribed mode, mostly mode 1 (crack opening) conditions in a single plane (normal to the principal stress direction for some limiting load conditions). As a result of these simplifications, a constant principal stress orientation must be assumed. However, if it can be used, this approach provides a potentially fast method to determine crack propagation. A number of software packages use this method—for example, AFGROW and NASGRO®—because it is efficient, but it is limited since it does not support structures of complex geometry and general loading conditions.

AFGROW is a fatigue analysis software program dedicated to crack propagation calculation. Its development began in the early 1980s under the name ASDGRO. Since then it has been in constant development and has been rewritten multiple times. It was first used in crack growth analysis for the Sikorsky H-53 helicopter. In aerodynamic situations, the stress field is often arbitrary so the software has been modified to simulate stress intensity solutions for cracks under these conditions. After a while, the Navy took interest in the project and, when it started funding development, ASDGRO became AFGROW, which is currently being further developed and is being used by the U.S. Air Force.

For crack growth, AFGROW can calculate a single crack, multiple cracks, and single or multiple asymmetric cracks. Also, it takes residual stresses into account, and contains multiple crack growth laws for the user to choose from. In addition to pure crack growth AFGROW allows inclusion of temperature, retardation, and repair (if the crack has a bonded repair patch) effects on the crack. To support these calculations, a library with material data is included that can be modified, and in which new materials can be added or old ones edited. The methods for calculating crack growth are the Forman equation (Forman et al. 1967), the Harter T-method (Harter 1994), the NASGRO equation (Forman and Mettu 1992), the Walker equation (Walker 1970), and tabular lookup, which follows the Harter T-method and the Walker equation, interpolating between curves for different stress ratios.

NASGRO is a crack propagation program with roots in the 1980s. It was created by NASA for fracture control analysis of space hardware. It was further developed by NASA in collaboration with SwRI (Southwest Research Institute) for aging aircraft, and the software was expanded for use in the broader aerospace/aircraft industry. NASGRO is entirely a crack growth program. The equation is a further development of the Paris law to give better crack growth prediction. Because of this equation, many new constants, depending on the material, are required and it is therefore implemented in a large database of material data. Addition of new user-specified material data is also possible. When loading the component, the software uses spectrums of load data called blocks. Several blocks can be added together to create the total load. NASGRO was developed around its own crack growth law, which is a more complex derivation of the Paris law or, more specifically, an improvement of it, by Forman and Mettu (1992).

4.7.3 FEA-Based Crack Propagation

The finite element method (FEM) solves partial differential equations (PDEs) for stress and strain by discretizing the domain of structural components. The primary advantage of FEM is its flexibility in terms of types of analysis (such as nonlinear), geometry complexity, and general loading conditions. Many commercial FEM packages are available for fracture mechanics applications. To support crack propagation, add-in software modules such as ZENCRACK provide commercial FEM packages with a 3D crack propagation capability, including ABAQUS MSC.Marc®, and ANSYS. ZENCRACK uses the *J*-integral for calculating fracture parameters. A similar approach is employed in Franc2D and Franc3D, which use displacements to calculate stress intensity factors and crack growth. Also, MSC.Fatigue incorporates nCode®, developed, by nCode International Ltd. (Fjeldstad and Wormsen 2006) for fracture analysis. As for boundary element methods, BEASY™ is the best known commercial code with crack propagation capabilities. It provides a linear solution with *J*-integral capabilities for 2D applications. In 3D the linear solution uses displacements to generate stress intensity factors.

All of these tools require regenerating the finite element mesh as the crack advances, in addition to creating very refined mesh around the crack tip to capture high stress in the area. This requirement leads to extremely large finite element models and requires extensive computation time. Essentially, a successful crack growth simulation is at the mercy of the automatic mesh generator offered by these tools.

The newly developed XFEM with LSM alleviates the mesh problem. The first version of XFEM was recently implemented in ABAQUS 6.9.

4.8 Case Studies and Tutorial Example

Two case studies, a tracked vehicle roadarm for crack initiation and 2D engine connecting rod for crack propagation using XFEM, are included in this section. In addition, one tutorial example is presented.

4.8.1 Case Study: Tracked Vehicle Roadarm

A roadarm of a tracked vehicle shown in Figure 4.26 (Chang et al. 1997) is employed as a case study to demonstrate the crack initiation life prediction discussed in this chapter. The multibody dynamic model of the tracked vehicle and its simulation environment are

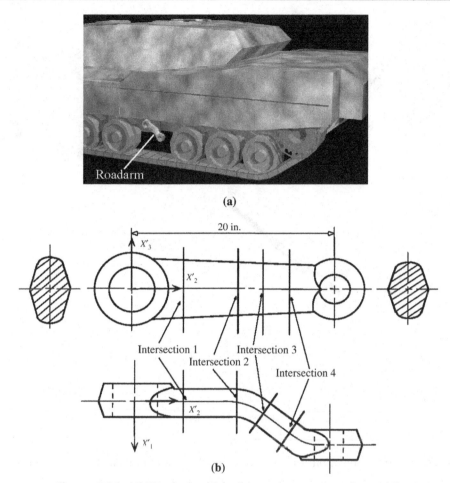

Figure 4.26: (a) Tracked vehicle (b) roadarm geometric model.

described first. Then a structural finite element model of the roadarm is presented with contours of crack initiation life and von Mises stress at the peak load of the simulation period.

A 17-body dynamics model, shown in Figure 4.27a, was generated to carry out vehicle dynamic analysis for the tracked vehicle on Aberdeen Proving Ground 4 (APG4), as shown in Figure 4.27b, at a constant speed of 20 miles per hour forward (positive X_2-direction in Figure 4.27a). The road was fairly uneven, featuring a few bumps and washboard-like terrain, as shown in Figure 4.27c.

Dynamic analysis and design systems (DADS) was used to generate the dynamics model and to perform dynamic analysis. A 20-sec dynamic simulation was performed at a maximum integration time step of 0.05 sec. An output interval of 0.05 sec was predefined for this analysis, and a total of 400 sets of results were generated. The joint reaction forces applied at

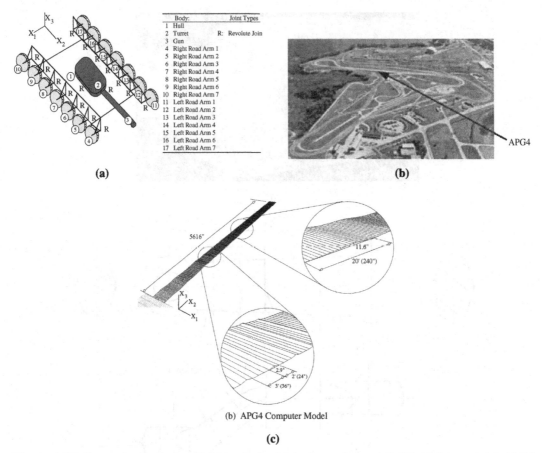

(a)

(b)

(b) APG4 Computer Model

(c)

Figure 4.27: Dynamic simulation: (a) Tracked vehicle dynamic model, (b) APG4, and (c) APG4 road profile.

the wheel end of the roadarm, accelerations, angular velocities, and angular accelerations of the roadarm were obtained from the analysis. A time history of joint reaction forces at the wheel end is shown in Figure 4.28.

Four beam elements, STIF4, and 310 ANSYS 20-node isoparametric finite elements, STIF95, were used for the roadarm FEM model. A number of rigid beams were created to connect nodes at the inner surface of the two holes to end nodes of beam elements to simulate the roadwheel shaft and torsion bar, respectively. Displacement constraints were defined at the end nodes of the beam elements that simulated the torsion bar, and joint reaction forces and torque were applied at the end node of the other beam element that simulated the shaft of the roadwheel, as shown in Figure 4.29. The roadarm was made of S4340 steel, with material properties of Young's modulus $E = 3.0 \times 10^7$ psi and Poisson's ratio $\nu = 0.3$. Note that the FEM's coordinate systems were identical to the body reference frame of the roadarm in the

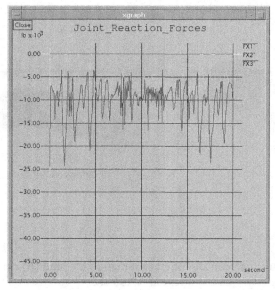

Figure 4.28: Joint reaction forces applied to the roadarm.

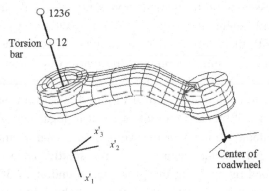

Figure 4.29: Roadarm finite element model.

dynamic model. Therefore, the loading history generated from the dynamic analysis could be used without transformation.

Finite element analysis was first performed to obtain stress influence coefficients for the roadarm using ANSYS. Eighteen quasistatic loads were applied, as discussed in Section 4.6. Among the loads, the first 6 that corresponded to external joint forces were three unit forces and three unit torques applied at the center of the roadwheel, in the x_1' -, x_2' -, and x_3' -directions, and the remaining 12 that corresponded to inertia forces were unit accelerations, unit angular accelerations, and unit combinations of angular velocities, as

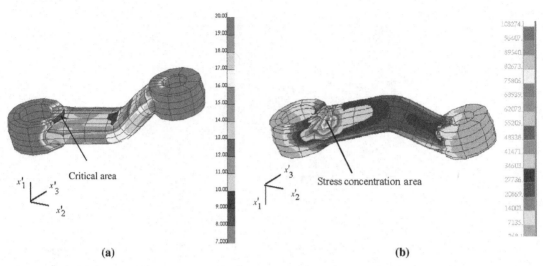

Figure 4.30: Analysis results: (a) contour of crack initiation life, and (b) contour of static von Mises stress at 17.35 sec.

also discussed in Section 4.6. The stress influence coefficients obtained from analyses were 6 component stresses at finite element nodes in the x'_1, x'_2, and x'_3 coordinates. Dynamic stresses at finite element nodes were then calculated by superposing stress influence coefficients with their corresponding external forces and accelerations and velocities in the time domain. To compute the multiaxial crack initiation life of the roadarm, the equivalent von Mises stress approach discussed in Section 4.3.3 was employed. The fatigue life contour is given in Figure 4.30a. Note that the spectrum in Figure 4.30a is the number of blocks to initiate a crack in logarithm. A static stress contour shown in Figure 4.30b demonstrates that the worst-case scenario employed for durability design using stresses as performance measures is problematic. The stress contour shown in Figure 4.30b was obtained by applying the peak load found at 17.35 sec of the 20-sec simulation, including 6 joint reaction forces at the roadwheel end, and accelerations, angular accelerations, and angular velocities of the roadarm. Note that, from Figures 4.30a and 4.30b, the stress concentration area identified as the worst case did not conform to the critical areas where the crack was first initiated. Design based on maximum stress might not address the more critical issue of fatigue.

4.8.2 Case Study: Engine Connecting Rod

Consider an engine connecting rod (Section 4.5.5). The thickness was assumed to be 1 mm. Material properties for the rod were as follows: Young's modulus $E = 210$ GPa, yield strength $S_y = 210$ MPa, Poisson's ratio $\nu = 0.3$. The Paris constants were $C = 5.6 \times 10^{-12}$ mm/cycle and $m = 3.5$, and the fracture toughness was $K_c = 100$ MPa$\sqrt{\text{mm}}$. The load acting on the connecting rod in terms of the rotation angle θ is given by Eq. 4.68 (Hwang et al. 1997).

$$T_F = \begin{cases} 43.794\,\theta^2 + 30.19 \text{ at left inner circle } \left(\dfrac{-40}{180}\pi \leq l\right) \\ \\ 9.54\,\theta^2 - 42.97 \text{ at right inner circle } \left(\dfrac{140}{180}\pi \leq \theta\right) \end{cases} \tag{4.68}$$

Figure 4.31 shows the FEM model of the rod with 6,058 dof, which is considered adequate for stress analysis in general; in particular, the mesh was refined at the high-stress areas. The average element length ℓ_e in the crack region for this mesh was about 1.9 mm. This, of course, varied from element to element and is only given here to provide a rough idea of element size. The maximum principal stress distribution in the connecting rod is shown in Figure 4.31b. Although the maximum stress appears to be at the fixed node on the left side, it is merely an artificial stress concentration due to displacement constraints. The real maximum stress of $\sigma_{\max} = 124$ MPa occurs on the left semicircular edge of the slot. An initial crack of $a = 7$ mm

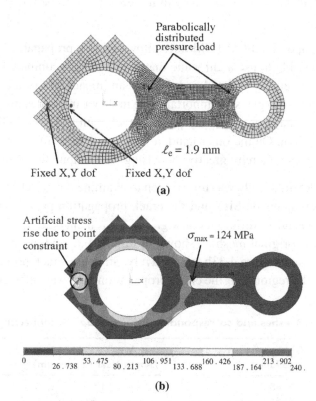

Figure 4.31: Connecting rod: (a) finite element model, and (b) contour of maximum principal stress.

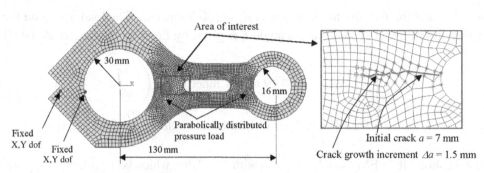

Figure 4.32: Finite element model of the connecting rod.

was introduced in this location and crack propagation analysis was conducted with $\Delta a = 1.5$ mm. Figure 4.32 shows the crack propagation path until failure—that is, until K_{eq} exceeded the fracture toughness of the material (Edke and Chang 2010).

For mixed-mode cases, generally the crack propagation path is curvilinear. In this case, however, the path appeared to be zigzag due to alternating positive and negative signs of θ_c. As discussed in Section 4.5.4, the crack growth increment had to be reduced to minimize these oscillations.

From an analysis standpoint, XFEM-LSM does allow crack to propagate within an element. However it is not advisable to use a Δa value that is quite small compared to element size. Thus, Δa and mesh size are interrelated and do have an impact on the accuracy of analysis results. This relationship also plays an important role in downstream design studies. If a large value Δa is selected, it may not accurately predict crack path and service life, leading to erroneous optimum design. On the other hand, a very small value of Δa requires correspondingly fine mesh, thereby greatly increasing the computational burden.

For these reasons, a detailed study was undertaken to examine the effect of mesh refinement and crack growth increments on SIFs and the crack propagation path. Three different meshes (see Table 4.1) and 12 different crack growth increment sizes ($\Delta a = 0.1 - 1.2$ mm) were used. mesh 1, the original mesh, is shown in Figure 4.32 and mesh 2 and mesh 3 are shown in Figure 4.33a and Figure 4.33b, respectively. Since the crack propagation path was roughly known, only the region near the crack propagation path was selected for refinement.

Table 4.1: Meshes and corresponding values of Δa for connecting rod.

Variable	Degrees of Freedom	Average Element Length Near Crack (mm)	Range of Δa Values (mm)
Mesh 1	6058	1.9	0.9–1.2
Mesh 2	8554	0.65	0.3–1.2
Mesh 3	25,186	0.22	0.1–1.2

Because of the relation between mesh size and Δa size, not all values of Δa could be used for all meshes. In these cases, crack propagation analysis was conducted. Figures 4.34a, b, and c show the crack propagation path corresponding to different Δa values for the three meshes. The scale on both axes has been modified to clearly illustrate the effect. It can be observed that for mesh 1 decreasing Δa value did not have any noticeable impact on the oscillations. However, oscillations were significantly reduced from mesh 1 to mesh 2, especially for smaller values of Δa. For mesh 3, there were virtually no oscillations, and this was true for all values of Δa. Thus, the oscillations seemed to be more sensitive to mesh than Δa size. Since Δa sizes of 0.9 to 1.2 mm were common across all meshes, these are used to illustrate the effect of mesh refinement in Figures 4.35.

Mesh 1 exhibited significant oscillation and hence was not suitable for accurate prediction of crack path. Mesh 2 did exhibit some oscillations, especially in the first few crack growth cycles, but the crack path predicted was very close to the oscillation-free crack path predicted by mesh 3, the finest mesh. It can thus be concluded that the additional computational burden

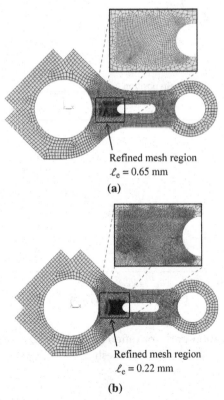

Refined mesh region
$\ell_e = 0.65$ mm

(a)

Refined mesh region
$\ell_e = 0.22$ mm

(b)

Figure 4.33: Finite element model of the connecting rod, (a) mesh 2: intermediate mesh, and (b) mesh 3: fine mesh.

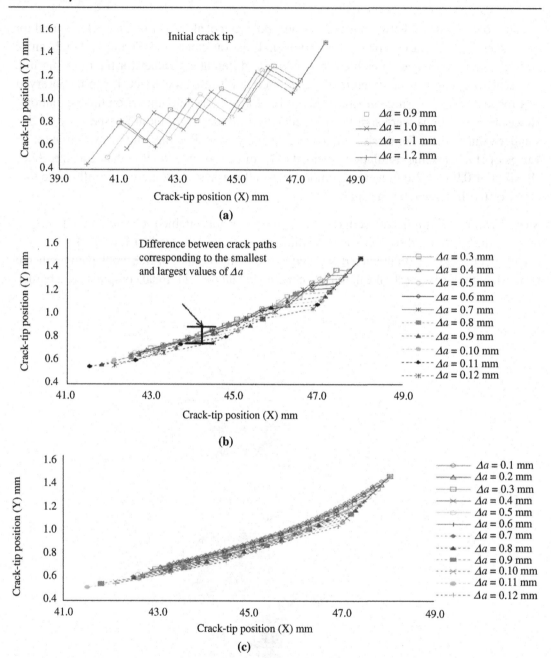

Figure 4.34: Crack propagation path for different values of Δa: (a) mesh 1, (b) mesh 2, and (c) mesh 3.

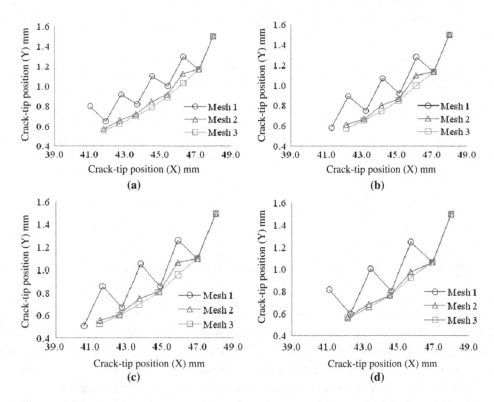

Figure 4.35: Cross-mesh comparison of crack propagation path: (a) $\Delta a = 0.9$ mm, (b) $\Delta a = 1.0$ mm, (c) $\Delta a = 1.1$ mm, and (d) $\Delta a = 1.2$ mm.

associated with mesh 3 is not justified considering this minor difference in crack paths and that mesh 2 can be used for further studies.

Since no analytical solution was available for this problem, the best available case— $\Delta a = 0.1$ mm for mesh 3—was taken as the basis for comparison. The crack trajectory was then approximated as a second-order curve and second-order polynomials were fitted through the crack-tip coordinates for different values of Δa for mesh 2 and mesh 3. While all Δa combinations for mesh 2 and mesh 3 were very close to each other, the curve for mesh 2 and the $\Delta a = 0.8$ mm case matched most closely with the best case. This can be observed from Figure 4.36, which also shows coefficients for the curve-fit polynomials and their respective R^2 values. It was also observed that the SIF values for these two cases were within 1% and that the service life values were within 9% of each other. Hence, the combination of mesh 2 and $\Delta a = 0.8$ mm is adequate for further studies.

Fatigue lives of these three meshes with different Δa were also compared. As shown in Table 4.2, residual life estimated using different meshes and crack size increment Δa varied

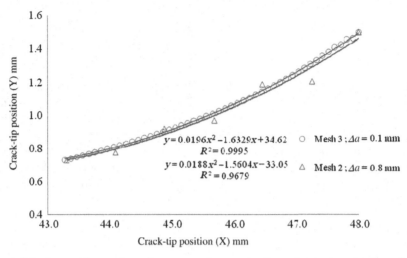

Figure 4.36: Comparison of crack propagation paths for reference and selected cases.

significantly. With the same mesh, a smaller Δa led to a less residual life, which tended to lead to more conservative designs. For a given Δa value, refined mesh again provided less residual life. The residual life of mesh 1 with $\Delta a = 1.2$ mm was almost twice that of mesh 3 with $\Delta a = 0.1$ mm, which was considered significant. In general, a safety factor of less than 2 is employed for mechanical components, such as those in ground vehicle suspensions. It is important to note that the large variation in residual life found in this example may not be adequately addressed in design using a safety factor approach.

For the selected mesh (mesh 2) and Δa size (0.8 mm), the crack propagation path and stress distribution at failure are shown in Figures 4.37a and 4.37b, respectively. The crack growth path was quite smooth, as seen in Figure 4.37. The peculiar plastic zone shape for the plane

Table 4.2: Residual life results for different meshes and corresponding values of Δa.

Variable	Degrees of Freedom	Average Element Length Near crack (mm)	Δa Value (mm)	Residual Life (cycles)
Mesh 1	6058	1.9	0.9	192,425
			1.2	213,170
Mesh 2	8554	0.65	0.3	124,137
			0.7	141,916
			0.8	143,640
			1.2	154,408
Mesh 3	25,186	0.22	0.1	116,650
			0.4	127,079
			0.8	139,643
			1.2	152,702

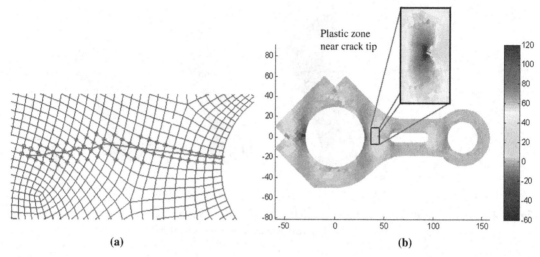

Figure 4.37: Analysis results of mesh 2 with $\Delta a = 0.8$ mm, (a) crack propagation path, and (b) Y-stress distributions (MPa).

strain condition could be observed near the crack-tip region, as shown in Figure 4.37b. The plastic zone size was quite small compared to the overall size of the connecting rod and hence the LEFM assumption was valid for this example.

4.8.3 Tutorial Example: Crankshaft

A crankshaft, which is a load-bearing component in a slider-crank mechanism and is shown in Figure 4.38a, is employed as a tutorial example for HCF calculation using SolidWorks Simulation. The major geometric dimensions of the crankshaft are shown in Figure 4.38b. Note that two small fillets (radius 0.05 in.) were added to the intersections between the two cylinders at the end and the crank body. A bearing load of 250 lb was added to the outer cylindrical surface of the shaft (the cylinder on the top end), and the outer cylindrical surface

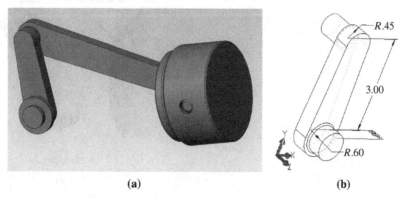

Figure 4.38: (a) Slider-crank mechanism, and (b) crankshaft.

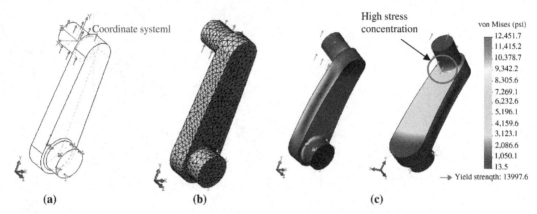

Figure 4.39: Crankshaft finite element model: (a) load and boundary condition, (b) finite element mesh, and (c) von Mises stress fringe plot.

of the lower cylinder was fixed, as shown in Figure 4.39a. The crankshaft was meshed with 11,867 tetrahedral finite elements (Figure 4.39b). A static analysis was carried out, and the von Mises stress plot, shown in Figure 4.39c, had a stress concentration with a stress level of 12,450 psi. Note that this stress was lower than the material yield strength of AL2014, $S_y = 13,998$ psi, provided in the SolidWorks material library.

An HCF analysis was carried out, assuming that the 250 lb bearing load was fully reversed. The equivalent von Moses stress method was chosen for fatigue calculation. For AL2014, the S-N diagram provided by SolidWorks Simulation is shown in Figure 4.40a. The crack initiation fatigue life fringe plot is shown in Figure 4.40b, in which the lowest life is located at the area where the maximum von Mises stress occurs, as often expected. The lowest life is

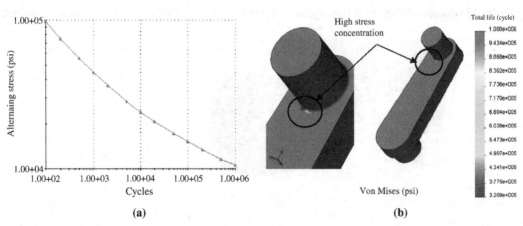

Figure 4.40: Fatigue life of the crankshaft: (a) S-N diagram, and (b) fatigue life fringe plot.

about 320,000 cycles. This result can be verified by drawing a horizontal line for the maximum stress 12,450 psi on the S-N diagram, intersecting the stress line with the S-N curve, and projecting the corresponding fatigue life on the horizontal axis.

4.9 Summary

In this chapter, we discussed methods for fatigue analysis, both crack initiation and crack propagation. For crack initiation, there are stress-based and strain-based approaches for high- and low-cycle fatigue analyses, respectively. The stress-based approach is simple and easy to calculate, but it offers only a very rough estimate of fatigue life since it uses elastic stress as the input while physical fatigue initiation is driven by local plastic strains. Therefore, the stress-based approach is applicable to components with minimum and limited local plasticity areas. The strain-based approach considers local plastic strain and is more suitable for fatigue life calculations, especially when local strains are significant in structural components. We have to examine carefully the stress and strain of the structural components and choose an adequate method that provides reliable results for design decision making.

We also discussed fracture mechanics for crack propagation and facture analysis. This discussion included the powerful *J*-integral for stress intensity factor calculation as well as mixed modes for 2D applications. In addition, we briefly covered a recently developed method that supports crack propagation analysis, XFEM. This method alleviates the need for remeshing when crack propagates, which is a huge advantage in tackling problems with complex geometry. The XFEM capability in ABAQUS offer superior capabilities in for crack propagation computation.

We briefly reviewed software tools that offer fatigue calculation capabilities, and provided case studies that give a general idea of the kind of applications that are possible in simulating fatigue life computation in general. We hope this chapter provided enough information to increase the reader's familiarity with fatigue analysis to address structural durability in engineering design. With more practice will come more confidence and competence in the use of software tools for carrying out crack initiation and crack propagation computations.

Questions and Exercises

4.1. In Section 4.1 we discussed several famous incidents that involved fatigue failures. Please find and review three more incidents that involved failure caused by structural fatigue. Describe the incidents, identify the nature of the failures, and offer opinions in terms of preventing similar incidents.

4.2. A 20-mm-diameter shaft transmits a variable torque of ± 400 N-m. The frequency of the torque variation is $0.1 \sec^{-1}$. The shaft is made of high-carbon steel (AISI 1080, $S_{ut} = 615$ MPa). Find the fatigue life of the shaft (in hours).

4.3. For the shaft in Exercise 4.2, we assume that the loading is one of completely reversed torsion. During a typical 30 sec of operation under overload conditions, the nominal stress was calculated to be as shown in the figure below. Estimate the life of the shaft when it is operating continually under these conditions.

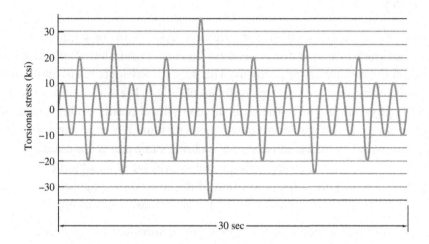

4.4. A flood-protection dam gate is supposed to operate only once per week for 120 years, but after 40 years of use it needs to be operated twice per day (each time the high tide comes in). Determine how much lower the bending stress must be from then on to still give a total life of 120 years. The material being fatigued is medium-carbon steel (AISI 1040, $S_{ut} = 520$ MPa).

4.5. A plate with two fillets is loaded with a cyclic tensile load $P = 1000$ N, as shown in the following figure. The material properties of the bar that are relevant to the calculation have been found to be $K = 155,000$ psi, $n = 0.15$, $\varepsilon_f' = 0.48$, $\sigma_f' = 290,000$ psi, $a = -0.091$, and $\alpha = -0.60$. How many cycles would it take to initiate a fatigue crack at the fillet root?

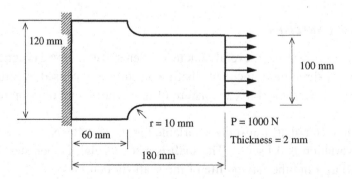

References

Anderson, T.L., 1994. Fracture Mechanics: Fundamentals and Applications, second ed. CRC Press.

Bannantine, J.A., Comer, J.J., Handrock, J.L., 1990. Fundamentals of Metal Fatigue Analysis. Prentice Hall.

Bannerman, D.B., Young, R.T., 1946. Some improvements resulting from studies of welded ship failures. Welding Journal 25 (3), 223–236.

Belytschko, T., Black, T., 1999. Elastic crack growth in finite elements with minimal remeshing. International Journal for Numerical Methods in Engineering 45 (5), 601–620.

Brown, M.W., Miller, K.J., 1973. A theory for fatigue failure under multiaxial stress-strain conditions. Proceedings, Institution of Mechanical Engineering 187 (1), 745–755.

Buch, A., May-June 1997. Prediction of the comparative fatigue performance for realistic loading distributions. Progress in Aerospace Sciences 33 (5–6), 391–430.

Chang, K.H., Yu, X., Choi, K.K., 1997. Shape Design Sensitivity Analysis and Optimization For Structural Durability. International Journal of Numerical Methods in Engineering 40, 1719–1743.

Chu, C.C., Conle, F.A., Bonnen, J.J.F., 1993. Multiaxial stress-strain modeling and fatigue life prediction of sae axle shafts, Advances in Multiaxial Fatigue. ASTM STP 1191. In: McDowell, E. (Ed.), ASTM, Philadelphia, PA, 1993, pp. 37–54.

Coffin, L.F., 1954. A study of the effects of cyclic thermal stresses on a ductile metal. Transaction of ASME. 76, 931–950.

Collins, J.A., 1993. Failure of Materials in Mechanical Design: Analysis, Prediction, Prevention, second ed. John Wiley.

Dang Van, K., Griveau, B., Message, O., 1989. On a new multiaxial fatigue limit criterion: theory and application, biaxial and multiaxial Fatigue. Mechanical Engineering Publications, London. EGF 3479–3496.

Dang Van, K., Papadopoulos, Y.V., 1987. Multiaxial fatigue failure criterion: a new approach, Proceedings of the Third International Conference on Fatigue and Fatigue Thresholds, Fatigue 87. University of Virginia, Charlottesville.

Dowling, N.E., 2007. Mechanical Behavior of Materials, Engineering Methods for Deformation, Fracture and Fatigue, third ed. Pearson Education.

Downing, S.D., Socie, D.F., 1982. Simple rainflow counting algorithm. International Journal of Fatigue 4 (1), 31–40.

Edke, M., Chang, K.H., 2010. Shape Sensitivity Analysis For 2-D Mixed Mode Fractures Using Extended FEM (XFEM) And Level Set Method (LSM), Mechanics Based Design of Structures and Machines. submitted, accepted for publications in June 2009 38 (03), 328–347. 2010.

Edke, M., Chang, K.H., 2011. Shape Optimization For 2-D Mixed Mode Fractures Using Extended FEM (XFEM) And Level Set Method (LSM). Structural and Multidisciplinary Optimization 44 (2), 165–181. http://dx.doi.org/10.1007/s00158-010-0616-5.

Endo, T., Matsuiski, M., March 1968. Fatigue of metals subjected to varying stress. Paper presented at the Kyushu District Meeting of the Japan Society of Mechanical Engineers, No. 68-2. Fukuoka Japan, pp. 37–40, in Japanese.

Fatemi, A., Socie, D., 1988. A critical plane approach to multiaxial fatigue damage including out-of-phase loading. Fatigue and Fracture of Engineering Materials and Structures 11 (3), 149–165.

Fjeldstad, A., Wormsen, G., 2006. Härkegård Simulation of fatigue crack growth in components with random defects.

Forman, R.G., Hearney, V.E., Engle, R.M., 1967. Numerical analysis of crack propagation in cyclic-loaded structures. Journal of Basic Engineering Trans of ASME 89, 459–464.

Forman, R.G., Mettu, S.R., 1992. Behavior of surface and corner cracks subjected to tensile and bending loads in TI-6AL-4 V alloy, Fracture Mechanics 22nd Symposium. In: Ernst, H.A., Saxena, A., McDowell, D.L. (Eds.), American Society for Testing, Materials.

Griffith, A.A., 1921. The phenomena of rupture and flow in solids. Philosophical Transactions of the Royal Society of London A 221, 163–198.

Halfpenny A. A practical discussion on fatigue. HBM nCode, www.ncode.com.

Harter, J.A., 1994. MODGRO Version 1.2, Technical Memorandum AFWALTM-88-157-FIBE. Wright-Patterson AFB, AFWAL Flight Dynamics Laboratory.

Hertzberg, R.W. 1983. Deformation and fracture mechanics of engineering materials. Wiley

Hoffmann, M., Seeger, T., 1985. A Generalized Method for Estimating Multiaxial Elastic-Plastic Notch Stresses and Strains, Part 1: Theory, Journal of Engineering Materials and Technology. October 1985 Vol 107 (4), 250−254.

Hwang, H-Y., Choi, K.K., Chang, K.H., 1997. Shape Design Sensitivity Analysis and Optimization Using p-Version Finite Element Analysis. Mechanics of Structures and Machines vol. 25. (No. 1), 103−137.

Irwin, G., 1957. Analysis of stresses and strains near the end of a crack traversing a plate. Journal of Applied Mechanics 24, 361−364.

Juvinall, R.C., Marshek, K.M., 2005. Fundamentals of Machine Component Design, fourth ed. Wiley.

Manson, S.S. 1953. Behaviour of Materials Under Conditions of Thermal Stress, NACA TN-2933.

Matake, T., 1977. An explanation on fatigue limit under combined stress. Bulletin JSME 20, 257−263.

Morrow, J., 1965. Cyclic plastic strain energy and fatigue of metals. Internal Friction Damping and Cyclic Plasticity, ASTMSTP378, July, 45−84.

National Bureau of Standards, 1983. The Economic Effects of Fracture in the United States. Department of Commerce., U.S.

National Transportation Safety Board, 1989. Aircraft Accident Report: Aloha Airlines, Flight 243, Boeing 737−200. N73711 near Maui, Hawaii. 28 April 1988.

National Transportation Safety Board, 1990. United Airlines Flight 232 McDonnell Douglas DC-10-10 Sioux Gateway Airport. Iowa, Sioux City.

Neuber, H., 1961. Theory of stress concentration for shear-strained prismatical bodies with arbitrary nonlinear stress-strain laws. Journal of Applied Mechanics E28, 544.

Paris, P.C., Gomez, M.P., Anderson, W.E., 1961. A rational analytic theory of fatigue. The Trend in Engineering 13, 9−14.

Ramberg, W., Osgood, W.R., 1943. Description of stress-strain curves by three parameters. Technical Note No. 902, National Advisory Committee for Aeronautics.

Rice, J.R., 1968. A path independent integral and the approximate analysis of strain concentration by notches and cracks. Journal of Applied Mechanics 35, 379−386.

Sanders, J.R., Tesar, D., 1978. The analytical and experimental evaluation of vibration oscillations in realistically proportioned mechanisms. ASME Paper No. 78-DE-1.

Shigley, J.E., Mischke, C.R., Budynas, R.G., 2004. Mechanical Engineering Design, seventh ed. McGraw-Hill.

Smith, J., Watson, P., Topper, T., 1970. A stress strain function for fatigue of metals. Journal of Materials 4 (5), 293−298.

Socie, D., 1993. Critical plane approaches for multiaxial fatigue damage assessment. In: McDowell, D.L., Ellis, R. (Eds.), Advances in Multiaxial Fatigue. ASTM International, pp. 7−36.

Spotts, M.F., Shoup, T.E., 1998. Design of Machine Elements, seventh ed. Prentice Hall.

Walker, K., 1970. The effect of stress ratio during crack propagation and fatigue for 2024-T3 and 7075-T6 aluminum. Paper ASTM STP 462. Proceedings of the American Society for Testing and Materials.

Yau, J.F., Wang, S.S., Corten, H.T., 1980. A mixed-mode crack analysis of isotropic solids using conservation laws of elasticity. Journal of Applied Mechanics 47, 335−341.

You, B.R., Lee, S.B., 1996. A critical review on multiaxial fatigue assessments of metals. International Journal of Fatigue 18 (4), 235−244.

Sources

Abaqus: *www.3ds.com*
AFGROW: *www.afgrow.net*
ANSYS: *www.ansys.com*
BEASY: *www.beasy.com*

FEMFAT: *www.femfat.com*
fe-safe: *www.safetechnology.com*
Franc2D, Franc3D: *www.cfg.cornell.edu/index.htm*
MSC Fatigue, MSC.Marc: *www.mscsoftware.com*
NASGRO: *www.swri.org*
nCode DesignLife: *www.ncodeinc.com*
Pro/MECHANICA Structure: *www.ptc.com*
safe4fatigue: *www.safetechnology.com*
SolidWorks Simulation: *www.solidworks.com*
ZENCRACK: *www.zentech.co.uk*

Reliability Analysis

Chapter Outline

We live in a world of uncertainties. When we roll a die, the outcome is uncertain. We know only that there is a 50% probability we will get an even (or odd) number. When we walk to work or school, we are often uncertain if it is going to rain later in the day. Meteorologists can tell us only that there is a 30% chance of rain, and they will never be wrong giving us such a probabilistic prediction.

In engineering design, there are uncertainties. Take a simple cantilever beam as an example. Uncertainties or variabilities exist in loading, material properties, geometric size, and material strength. However, when we calculate the maximum bending stress, we usually assume that all of the numbers we insert into the bending stress equation are deterministic; that is, there is no uncertainty in these quantities. When we are given a safety factor to determine if the design is safe, we compare the bending stress of the cantilever beam with its material strength to see if the ratio of strength to stress is greater than the given safety factor. We are usually satisfied with the design if the strength-to-stress ratio is greater than the given safety factor. However, is the design verified by the safety factor approach truly safe? Is the design verified truly reliable? Does a design with a safety factor greater than 1 never fail? Not likely? What if we were to increase the safety factor to 2 or greater? How safe is the design in reality? How reliable can a design be? How do we tell?

These questions can be answered only through a probabilistic approach. A deterministic approach using a safety factor or a worst-case scenario is not sufficient to address the safety or reliability of a product design. A safe and reliable product can be ensured only by considering probabilities or statistics in the design process.

Mechanical engineers must understand the importance of the probabilistic aspect of product design, and must be able to apply adequate reliability analysis methods to engineering problems. This chapter is devoted to the subject of reliability analysis. We also touch on design from a probabilistic perspective, although a more in-depth discussion of reliability-based design is provided in *Design Theory and Methods using CAD/CAE*, a book in The Computer Aided Engineering Design Series.

Although reliability analysis is the focus of this chapter, the topic is no doubt a substantial one—usually an entire book is necessary for a comprehensive treatment. Because providing all of the details of reliability analysis in just one chapter is not feasible, this chapter is organized to focus on introducing the most popular and powerful of its methods. Therefore, the first-order reliability analysis method (FORM), the second-order reliability method

(SORM), Monte Carlo simulation, importance sampling, and the response surface method are introduced with sufficient detail to help the reader understand them and apply them to engineering applications. See Choi et al. (2006) for comprehensive discussion of the subject.

One important assumption made in this chapter is that readers are familiar with the basics of engineering statistics, such as random variables, the probability density function, the cumulative distribution function, and so forth. These are briefly reviewed in Section 5.3. Readers are encouraged to review this section before attempting to grasp concepts and mathematic formulations for the reliability analysis methods introduced in the following sections.

Another point worth mentioning is that we assume that all uncertainties are irreducible. In general, uncertainties may be classified as reducible or irreducible. Reducible uncertainties are normally caused by lack of data, modeling simplifications, human errors, and the like, and can usually be handled by, among other things, collecting more data, improving analysis models with a better understanding of the problem, and implementing stricter quality control. Irreducible uncertainties, on the other hand, are caused by phenomena related to the stochastic nature of the physical problem and cannot be reduced by more knowledge or data. Addressing reducible uncertainties for reliability analysis is beyond the scope of this chapter. We discuss only irreducible uncertainties.

In addition to reliability analysis methods, this chapter discusses general-purpose software tools for carrying them out. A practical example, a high-mobility multipurpose wheeled vehicle (HMMWV) roadarm, is included as a case study to illustrate and demonstrate the reliability analysis methods discussed. Almost all examples included in this chapter are structural problems to maintain focus. However, theory and methods introduced are applicable to other types of engineering problems.

Overall the objectives of this chapter are as follows:

- To provide basic probabilistic theory and reliability analysis methods using simple examples.
- To explain the importance of the probabilistic nature of engineering design.
- To enable the reader to apply the analysis methods to basic engineering problems.
- To provide a basic knowledge of reliability analysis software.

5.1 Introduction

In engineering design, the traditional deterministic approach has been successfully applied to systematically reduce cost and improve product performance. However, the existing uncertainties or variabilities in physical quantities, such as loads, materials, manufacturing tolerance, and so forth, are unavoidable in engineering design. These uncertainties must be considered in the product design process.

The traditional way of dealing with uncertainties is to use conservative values of the uncertain quantities as well as safety factors in the framework of deterministic design. Conservative values lead to a product that is often overdesigned and therefore heavy and inefficient. A safety factor approach offers a safety measure that is only relative. Although the use of a safety factor is satisfactory with most design applications, determining an adequate safety factor for a given design problem is uncertain. A larger safety factor usually makes the product "safer"; however, if unnecessarily large it too often yields an overdesigned product, which is not desirable. What level of safety factor is considered to be just right? How safe is the design obtained using a safety factor? This safety factor approach does not provide design engineers with a full picture of the reliability level of the product. The question we should ask is not if the design will fail but instead what the probability is that it will fail. This question can only be answered from a probabilistic perspective. Reliability analysis is key in addressing the safety and reliability in product design.

Reliability analysis deals primarily with the effects of random variability on the performance of an engineering component or system during the design phase. To carry out a reliability analysis, a failure mode must first be identified. A failure mode describes how the product or component fails; it indicates that response exceeds the component's or system's design limit. For example, when the maximum stress of a cantilever beam exceeds its material strength, the beam fails. In this case, the maximum stress performance of the beam is defined as the failure mode.

After a failure mode is defined, variabilities of the physical parameters or manufacturing process that affect stress and strength must be identified. For example, the length and cross-section dimensions of the beam can vary depending on the tolerance requirements in its manufacture. Material properties, such as elasticity modules or yield strength, are uncertain and depend on the manufacturing process of the raw material. Certainly, the load applied to the beam can vary according to how the beam is loaded and how the load is controlled. Reliability analysis takes these uncertainties into consideration and estimates a failure probability that predicts the percentage of incidents in which the maximum stress of the beam exceeds its strength. There is no need to determine a safety factor. The result offered by reliability analysis is far more precise and effective than that provided by a safety factor.

A number of methods have been developed to support reliability analysis in product design. In particular, for probabilistic structural design, Monte Carlo simulation, the first-order reliability method, the second-order reliability method, importance sampling, and the response surface method, among others, have been applied to solve practical applications on a broad basis. Our emphasis in this chapter is on structural problems. To stay focused we use mostly structural examples to illustrate concept, theory, and methods, which, however, are applicable to other engineering problems that involve different failure modes and associated performance measures.

In this chapter, we start by introducing the basic concept of probability of failure using simple examples. Here we compare deterministic and probabilistic approaches to demonstrate that reliability is indispensable in product design. A brief review of essential topics in engineering statistics relevant to our discussion is also provided in Section 5.3. Readers are encouraged to review this section before moving to later sections, which are more involved with statistics theory and require a good prior understanding. The key part of this chapter is Section 5.4, in which reliability analysis methods are introduced. We also briefly discuss system reliability and present failure probability prediction for a series system using FORM. We discuss reliability analysis software in Section 5.6. Example problems modeled and solved using some of the methods discussed are presented. The chapter wraps up with a case study.

5.2 Probability of Failure—Basic Concepts

We start our discussion with the basic concepts of failure probability. The goal is to introduce key ideas in reliability analysis using a simple cantilever beam, avoiding sophisticated math and theory at the beginning. We first recall the safety factor approach that we are familiar with in deterministic design, we will point out a few of its shortcomings. We then see how a probabilistic approach can be applied to the same beam example, in which the adequacy of the approach is demonstrated. We also briefly touch on probabilistic design by adjusting the beam dimension in an attempt to reduce its failure probability. We compare the results with that of the worst-case approach commonly employed in deterministic design.

5.2.1 Deterministic Design versus Probabilistic Prediction

In engineering analysis and design using a deterministic approach, we assume the physical parameters to be deterministic. When we say the length of a cantilever beam is 10 in., we assume that every single beam manufactured is exactly 10 in. long. In fact, not a single beam is exactly 10 in. long; in fact, 10 in. is a nominal value we use. In statistics, this value often refers to an average or a mean of the length parameter, which can be obtained by measuring the length of a bulk of cantilever beams that were manufactured following the same design specifications. Similarly, material yield strength, which can be found in a strength of materials textbook, is a mean value. In deterministic design, we are essentially employing mean values of the physical parameters for our calculations.

For a steel cantilever beam of solid circular cross-section, shown in Figure 5.1, the length and diameter are given as 10 in. and 1 in., respectively. The load applied at the tip is 393 lb. With these values, the maximum bending stress can be calculated as

$$s = \frac{32M}{\pi d^3} = \frac{32(p\ell)}{\pi d^3} = \frac{32(393)(10)}{\pi(1)^3} = 40{,}000 \text{ psi} = 40 \text{ ksi} \tag{5.1}$$

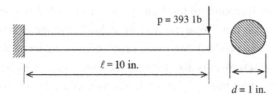

FIGURE 5.1: Cantilever beam of solid circular cross-section.

If the yield strength of the steel is $S_y = 50$ ksi, the safety factor can be calculated as $n = S_y/s = 50/40 = 1.25$. This is simple and almost effortless. But the question is whether we are satisfied with the design. Usually we are since the safety factor is greater than 1. As a result, we are under the impression that the beam will not fail; that is, the bending stress will not be greater than the yield strength.

Is this true? Are we sure? Is the design really safe? How safe? Is safety factor $n = 1.25$ good enough? Will we see any failure; in other words, is there any possibility that the bending stress will be greater than the yield strength if hundreds or thousands of the cantilever beams are manufactured and tested? If the beam fails, how often does failure occur? If we increase the safety factor from 1.25 to 1.5 by, for example, using a stronger material with a higher yield strength, how much improvement in terms of reducing the possibility of failure can we expect?

These are valid questions and the very ones we should ask as design engineers. We have to answer them by calculating probability of failure using reliability analysis. Mathematically, the probability of failure of this stress failure mode for the cantilever beam can be defined as

$$P_f = P(s > S_y) = P(S_y - s \leq 0) \tag{5.2}$$

where $P(\cdot)$ is the probability of the failure mode $\cdot$, and the failure mode in this case is defined as $S_y - s \leq 0$. Equation 5.2 is referred to as a simple strength-load (or strength-stress) approach for reliability analysis, and the equation $S_y - s = 0$ is referred to as the limit state function.

Next, we assume both stress s and strength S_y to be random variables of normal distribution; that is, both have a distribution like the bell-shape curve shown in Figure 5.2. The stochastic characteristics of the stress and strength are usually obtained from experiments. We assume that the mean value and standard deviation for the yield strength are $\mu_{S_y} = 50$ ksi and $\sigma_{S_y} = 6$ ksi, respectively. Notice that the bending stress is affected by load and geometric dimensions. Variabilities in these parameters affect its stochastic characteristics. For the time being, we simply assume that the mean value and standard deviation of the bending stress are $\mu_s = 40$ ksi and $\sigma_s = 8$ ksi, respectively.

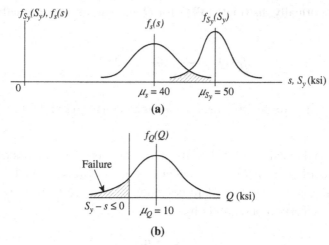

FIGURE 5.2: Probability distributions of (a) bending stress s and yield strength S_y and (b) failure mode $Q = S_y - s$.

The distribution functions $f_{S_y}(S_y)$ and $f_s(s)$ shown in Figure 5.2a are the respective probability density functions (PDFs) of the random variables S_y and s. A probability density function describes the relative likelihood for this random variable to take on a given value. The area underneath each curve is 1, which represents the 100% probability of the entire sample space. The overlapped area underneath the two intersecting curves represents beam failure $S_y - s \leq 0$; that is, bending stress s is greater than yield strength S_y, meaning that the overlapped area represents the probability of failure. The question is how we calculate the probability of failure.

We first rewrite the failure mode as

$$Q = S_y - s \tag{5.3}$$

where Q is a random variable with normal distribution since both S_y and s are assumed normally distributed. The probability density function $f_Q(Q)$ of the random variable Q is shown in Figure 5.2b, in which the area underneath the curve to the left of the origin represents the beam failure. Therefore, the probability of failure P_f can be calculated as

$$P_f = \int_{-\infty}^{0} f_Q(Q)dQ \tag{5.4}$$

where $f_Q(Q)$ is the normally distributed PDF for Q and can be written mathematically as

$$f_Q(Q) = \frac{1}{\sigma_Q\sqrt{2\pi}} e^{-\frac{1}{2}\left(\frac{Q-\mu_Q}{\sigma_Q}\right)^2} \tag{5.5}$$

where μ_Q and σ_Q are the respective mean value and standard deviation of the random variable Q.

Instead of using Eq. 5.4 to calculate P_f directly, it is easier and more common to transform the PDF of a normal distribution $f_Q(Q)$ to a standard (or normalized) normal distribution $\phi(z)$, which has mean value $\mu_z = 0$ and standard deviation $\sigma_z = 1$, as depicted in Figure 5.3. The transformation can be carried out simply by

$$z = \frac{Q - \mu_Q}{\sigma_Q} \tag{5.6}$$

and the standard normal distribution function is

$$\phi(z) = \frac{1}{\sqrt{2\pi}} e^{-\frac{1}{2}z^2} \tag{5.7}$$

Thus, failure probability can be calculated as

$$P_f = \int_{-\infty}^{0} f_Q(Q)dQ = \int_{-\infty}^{z} \frac{1}{\sqrt{2\pi}} e^{-\frac{1}{2}t^2} dt = \Phi(z) \tag{5.8}$$

where $\Phi(z)$ is the cumulative distribution function (CDF) of $\phi(z)$. Note that $\Phi(z)$ for a given z can be found in engineering statistics textbook. Likewise, the function value for $\Phi(z)$ can be found using any engineering software with statistics capabilities, such as MATLAB. The MATLAB function for the value of the normal cumulative distribution function is `normcdf(z)`. For example, `normcdf(0)` $= 0.5$.

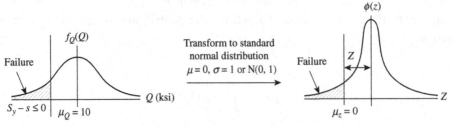

FIGURE 5.3: Transformation of a random variable of normal distribution into one of standard normal distribution.

Now let us go back to the cantilever beam. The mean value and standard deviation of the random variable Q can be calculated, respectively, as

$$\mu_Q = \mu_{S_y} - \mu_s = 50 - 40 = 10 \text{ ksi} \tag{5.9a}$$

$$\sigma_Q = \sqrt{\sigma_{S_y}{}^2 + \sigma_s{}^2} = \sqrt{6^2 + 8^2} = 10 \text{ ksi} \tag{5.9b}$$

Thus,

$$z = \frac{Q - \mu_Q}{\sigma_Q} = \frac{0 - 10}{10} = -1$$

and the probability of failure can be calculated using Eq. 5.8 as

$$P_f = \Phi(-1) = 0.159 = 15.9\%$$

The result shows that there is a 15.9% probability that the beam will fail, in which its maximum bending stress exceeds its yield strength. Note that Eqs. 5.9a and 5.9b are simply the results of subtracting two random variables. For a combination other than subtraction, refer to Table 5.1. Derivations of these equations can be found in a statistics textbook.

It is apparent that the probability of failure calculated by reliability analysis can more precisely describe the reliability of the cantilever beam. The result shows that a safety factor $n = 1.25$ does not guarantee that the beam will never fail. A safety factor can only go so far in providing a relative measure in terms of the reliability of a design.

Now we know that the beam will fail with a quite high probability. The question is how we reduce the failure probability, for example, using a stronger material. If we use a stronger steel with yield strength $S_y = 60$ ksi, the corresponding safety factor can be calculated as

Table 5.1: Mean value and standard deviation of combined independent random variables x and y.

Functions	μ_Q	σ_Q
$Q = c$	c	0
$Q = cx$	$c\mu_x$	$c\sigma_x$
$Q = x + c$	$\mu_x + c$	σ_x
$Q = x \pm y$	$\mu_x + \mu_y$	$\sqrt{\sigma_x^2 + \sigma_y^2}$
$Q = xy$	$\mu_x\mu_y$	$\sqrt{\mu_x^2\sigma_y^2 + \mu_y^2\sigma_x^2}$
$Q = x/y$	μ_x/μ_y	$\dfrac{\sqrt{\mu_x^2\sigma_y^2 + \mu_y^2\sigma_x^2}}{\mu_y^2}$
$Q = \dfrac{1}{x}$	$\dfrac{1}{\mu_x}$	$\dfrac{\sigma_x}{\mu_x^2}$

$n = 60/40 = 1.5$, increased from 1.25 if we keep the same beam dimensions. In this case, the failure probability can be calculated as follows. First the mean value of the random variable Q increases to 20 ksi as in the following equation.

$$\mu_Q = \mu_{S_y} - \mu_s = 60 - 40 = 20 \text{ ksi}$$

Note that σ_Q stays the same since we are not changing any of the standard deviation of the stress and strength. Hence,

$$z = \frac{-20}{10} = -2$$

and

$$P_f = \Phi(-2) = 0.0228 = 2.3\%$$

That is, only 2.3% of the beams will fail, which is a significant improvement over the previous 15.9% failure rate. Is using stronger steel the only way to reduce failure probability? What if we tighten the tolerance in the manufacturing process? If we do so, let us assume that the standard deviation of the bending stress is reduced from 8 to 4 ksi (recall that the bending stress is a function of beam dimensions). We assume the same steel of yield strength $S_y = 40$ ksi as before, so $\mu_Q = 10$ ksi and

$$\sigma_Q = \sqrt{\sigma_{S_y}{}^2 + \sigma_s{}^2} = \sqrt{6^2 + 4^2} = 7.21 \text{ ksi} \tag{5.10}$$

Therefore, $z = \dfrac{-10}{7.21} = -1.39$, and the probability of failure is

$$P_f = \Phi(-1.39) = 0.0808 = 8.08\%$$

which is lower than the previous 15.9%.

Failure probability is reduced when manufacturing quality is improved by tightening tolerance. Note that this improvement has nothing to do with the safety factor (which is still $n = 1.25$) and has nothing to do with stronger material (yield strength is the same, $S_y = 40$ ksi).

There are several possible scenarios that can help us reduce the standard deviation of the bending stress from 8 to 4 ksi. To close out this discussion, we mention one possibility that involves manufacturing tolerance. We assume that beam length is a random variable of normal distribution with mean value and standard deviation $\mu_\ell = 10$ in. and $\sigma_\ell = 2$ in. respectively. We also assume that the load and diameter of the cross-section of the beam are deterministic. The bending stress is thus

$$s = \frac{32(p\ell)}{\pi d^3} = \frac{32(393)\ell}{\pi(1)^3} = 4000\ell \text{ psi} = 4\ell \text{ ksi}$$

If $\mu_\ell = 10$ in. and $\sigma_\ell = 2$ in. the mean value and standard deviation of the bending stress are $\mu_s = 4\mu_\ell = 40$ in. and $\sigma_s = 4\sigma_\ell = 8$ in. respectively. If we tighten the manufacturing tolerance that reduces the standard deviation of the beam length from 2 to 1 in. the standard deviation of the bending stress becomes $\sigma_s = 4\sigma_\ell = 4$ in. which gives $\sigma_Q = 7.21$ ksi as shown in Eq. 5.10.

What if two or more parameters in the bending stress equation are random? Their uncertainties affect the stochastic characteristics of the bending stress. How do we calculate failure probability of the beam when we have more random variables? This issue is formally addressed in Section 5.4, where we introduce reliability analysis methods.

5.2.2 Probabilistic Design

In this subsection we touch a bit on the subject of probabilistic design, in which effects of random variability on the performance of an engineering system or component are considered in product design. Instead of offering a comprehensive discussion of this subject, we simply bring you some ideas about engineering design considering uncertainties. A more in-depth discussion, commonly referred to as reliability-based design, can be found in *Design Theory and Methods using CAD/CAE*, a book in The Computer Aided Engineering Design Series.

Previously, we demonstrated that reducing the standard deviation of the beam length by tightening the manufacturing tolerance reduces failure probability. From the perspective of design, we often vary geometric dimensions to achieve better product performance. As mentioned before, usually the mean values of the respective dimensions are employed for deterministic design. In this subsection we use the same cantilever beam to illustrate the concept of probabilistic design.

We assume that the same random variable of yield strength: normal distribution with mean value and standard deviation $\mu_{S_y} = 50$ ksi and $\sigma_{S_y} = 6$ ksi, respectively. As before, we assume that the load and diameter of the beam cross-section are deterministic. Thus, the only random variable that affects the bending stress is the beam length, which has normal distribution with mean value and standard deviation μ_ℓ and σ_ℓ, respectively. We further assume that $\sigma_\ell = 0.1\mu_\ell$; as a result, we are left with one design variable, μ_ℓ.

Following the previous discussion we have

$$\mu_Q = \mu_{S_y} - \mu_s = 50 - 4\mu_\ell \tag{5.11a}$$

$$\sigma_Q = \sqrt{\sigma_{S_y}^2 + \sigma_s^2} = \sqrt{6^2 + (4\sigma_\ell)^2} = \sqrt{6^2 + (0.4\mu_\ell)^2} \tag{5.11b}$$

and

$$z = \frac{-(50 - 4\mu_\ell)}{\sqrt{6^2 + (0.4\mu_\ell)^2}} \qquad (5.11c)$$

If the design requirement is to have a failure probability no greater than $P_f = 0.001$, the corresponding z value is

$$z = \Phi^{-1}(P_f) = \Phi^{-1}(0.001) = -3.09 \qquad (5.12)$$

which again can be obtained from any statistics textbook or reference manual. If you use MATLAB, the function to call is norminv(p), where the input parameter p is the failure probability P_f. For example, norminv(0.001) $= -3.09$.

If we equate Eqs. 5.11c and 5.12, the mean value can be calculated as $\mu_\ell = 7.34$ in. If the failure probability is reduced 10 times to $P_f = 0.0001$, $z = \Phi^{-1}(0.0001) = -3.72$. Then the mean value becomes $\mu_\ell = 6.43$ in. This result shows that a reduction of 0.91 in. (from 7.34 to 6.34) in beam length reduces the failure probability 10 times, from $P_f = 0.001$ to $P_f = 0.0001$. We know that the safety factor and the failure probability are roughly related. As seen before, increasing the safety factor reduces failure probability. This relation can be more clearly illustrated for the cantilever beam as follows.

As mentioned earlier, mean values are employed for deterministic design. Therefore, the safety factor of the cantilever beam can be also written as

$$n = \frac{S_y}{s} = \frac{\mu_{Sy}}{\mu_s} = \frac{50}{4\mu_\ell} \qquad (5.13)$$

Thus,

$$\mu_\ell = \frac{50}{4n} = \frac{12.5}{n} \qquad (5.14a)$$

and

$$z = \frac{-50(1 - 1/n)}{\sqrt{6^2 + (5/n)^2}} \qquad (5.14b)$$

$$P_f = \Phi(z) = \Phi\left(\frac{-50(1 - 1/n)}{\sqrt{6^2 + (5/n)^2}}\right) \qquad (5.14c)$$

The mean value and failure probability can be graphed for a range of safety factors—for example, between 1 and 2, as shown in Figure 5.4, which clearly shows that increasing the safety factor reduces failure probability drastically. However, no matter how large the safety factor is, failure probability will never become zero. For example, when the safety factor increases to $n = 2$, the failure probability is 0.0059%, which is very small but not zero. It is important to keep in mind that reliability analysis offers much more precise information in failure probability estimates. The question is not if a safety factor $n = 2$ is sufficient; instead, the question is if a 0.005% failure rate is acceptable or if the rate is so small that the design can be relaxed a bit. It is evident that the probabilistic approach supports more informative design decision making than the deterministic approach in engineering design.

One last topic to discuss in this section is the absolute-worst-case method that we commonly employ for design following deterministic approach. We use the same cantilever beam to illustrate the approach and point out a few important points and shortcomings of the method.

With the absolute-worst-case method we assume that that the tolerances of yield strength and diameter of the beam cross-section are, respectively,

$$S_y = \mu_{S_y} \pm 3\sigma_{S_y} = 50 \pm 3(6) \text{ ksi}$$

and

$$d = \mu_d \pm 0.25\sigma_d = 1 \pm 0.25(0.1) \text{ in.}$$

in which a 3σ and a 0.25σ tolerance is assumed respectively for these two variables.

The absolute-worst-case design calls for a worst-case scenario, which in this case implies determining the largest possible bending stress and smallest possible material strength, and using both to carry out design of the beam. In this example, we assume that the length of the beam is to be determined using the worst-case scenario. As before, the standard deviation of

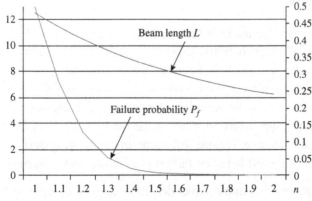

FIGURE 5.4: Influence of the safety factor on beam length and failure probability.

the beam length is assumed to be 0.1 of its mean value, $\sigma_\ell = 0.1\mu_\ell$. The worst-case scenario for the beam can be identified as

$$S_y = \mu_{S_y} - 3\sigma_{S_y} = 50 - 3(6) = 32 \text{ ksi}$$

and

$$d = \mu_d + 0.25\sigma_d = 1 + 0.25(0.1) = 1.025 \text{ in.}$$

The beam length can be determined by equating the largest stress with the smaller strength:

$$s = \frac{32(p\ell)}{\pi d^3} = S_y$$

Therefore, the beam length can be calculated as

$$\ell = \frac{\pi d^3 S_y}{32p} = \frac{\pi 1.025^3(32{,}000)}{32(393)} = 8.609 \text{ in.}$$

From Eqs. 5.11c and 5.12, we have for length $\ell = 8.609$ in.

$$z = \frac{-(50 - 4\mu l)}{\sqrt{6^2 + (0.4\mu l)^2}} = \frac{-(50 - 4(8.609))}{\sqrt{6^2 + (0.4(8.609))^2}} = -2.25$$

and the failure probability $P_f = \Phi(-2.25) = 0.0122 = 1.22\%$. That is, 1.22% of the beam will fail even if we impose a worst-case design scenario. After all, the absolute-worst-case design is not so absolute. On the other hand, if a failure rate greater than 1.22% can be tolerated, for example, if the required minimum failure rate is 5%, the worst-case design is in fact yielding an overdesigned cantilever beam.

5.2.3 Short Summary

Several important points were made in the cantilever beam example. First, increasing the safety factor reduces failure probability. However, employing the safety factor approach for design does not guarantee zero failure. Second, failure probability is also reduced by tightening manufacturing tolerance, which narrows the variation of geometric dimensions. Third, in addition to narrowing the variation of random variables, desired failure probability can be achieved by adjusting the mean value of the random variable. Finally, the absolute-worst-case design is not absolute. Failure will occur. In engineering design, the main objective is often carrying out a trade-off between failure probability and a reasonable product cost.

It is critical to bring probability into product design to adequately address the uncertainties in physical parameters that affect product performance. Design engineers must understand the probabilistic nature of physical problems and be able to take uncertainties into consideration in the various design phases.

In the following sections, we introduce the subject of reliability analysis systematically and with more rigor. A short review of the fundamentals of engineering statistics are provided next for readers to sharpen their statistics skills before turning to the more sophisticated reliability analysis methods in Section 5.4.

5.3 Basics of Statistics and Probabilistic Theory

Before discussing reliability analysis methods, we briefly review a few basic subjects in statistics and probabilistic theory. This is not a thorough review; instead, only the basics and those topics relevant to the reliability analysis methods in Section 5.4 are included.

5.3.1 Events and Basic Probability Rules

In the following paragraphs, basic terms are defined and basic probability rules are described.

A sample space Ω is a set of all possible outcomes of an experiment. An event E is a subset of the sample space of an experiment. Thus, for example, the sample space of rolling a die is $\Omega = \{1, 2, 3, 4, 5, 6\}$. An event of an even number can be defined as $E_1 = \{2, 4, 6\}$. The probability of event E_1 is obviously $P(E_1) = \frac{1}{2}$. Also, as mentioned in the previous section, a failure mode of a structure (or a failure event in statistics) can be modeled as $E_2 = \{S_y \le s\}$, where s and S_y are the maximum stress and yield strength, respectively. The probability of failure is the probability of event E_2; i.e., $P_f = P(E_2) = P(S_y \le s) = P(S_y - s \le 0)$.

If a system is modeled by a number of failure events, failure of the system can be calculated by a union or an intersection of the individual failure events. For a series system, any individual failure event causes system failure. Therefore, the failure of the system is a union of all individual events:

$$E = E_1 \cup E_2 \cup \ldots \cup E_m = \cup_{i=1}^{m} E_i \tag{5.15}$$

where m is the number of individual events. A parallel system does not fail unless all individual events fail simultaneously. Therefore, the failure of the system is an intersection of all individual events:

$$E = E_1 \cap E_2 \cap \ldots \cap E_m = \cap_{i=1}^{m} E_i \tag{5.16}$$

Disjoint, or mutually exclusive, events are defined as

$$E_1 \cap E_2 = \emptyset \tag{5.17}$$

where ø is an empty (or impossible) event. For rolling a die, the two events $E_3 = \{2\}$ and $E_4 = \{6\}$ are disjoint since it is impossible to have both 2 and 6 at the same time when a die is rolled.

There are three fundamental axioms:

> *Axiom 1*: For any event E, $0 \le P(E) \le 1$
> *Axiom 2*: For a sample space Ω, $P(\Omega) = 1$
> *Axiom 3*: For mutually exclusive events, $E_1, E_2, \ldots, E_m$, $P(\cup_{i=1}^{m} E_i) = \sum_{i=1}^{m} P(E_i)$

Note that mutually exclusive events are not independent. *Mutually exclusive* and *independent* are two unrelated terms in statistics.

To determine whether events are independent, we have to first understand conditional probability. Conditional probability of an event E_2 given another event E_1 is defined by

$$P(E_2|E_1) = \frac{P(E_2 \cap E_1)}{P(E_1)} \tag{5.18a}$$

Thus,

$$P(E_2 \cap E_1) = P(E_2|E_1)P(E_1) \tag{5.18b}$$

For example, what is the probability that the total of two dice is greater than 8, given that the first die is a 6?

Let us define event $E_1 = \{6\}$ and for the time being, define event E_2, which is the total of two dice greater than 8 regardless of the first die. Then $E_2 = \{(3,6), (4,5), (4,6), (5,4), (5,5), (5,6), (6,3), (6,4), (6,5), (6,6)\}$. There are overall 36 ($6 \times 6 = 36$) possible outcomes when we roll two dice, so the probability of event E_2 is

$$P(E_2) = \frac{10}{36} = \frac{5}{18}$$

Also, $E_1 \cap E_2 = \{(6,3), (6,4), (6,5), (6,6)\}$, so $P(E_1 \cap E_2) = \frac{4}{36} = \frac{1}{9}$. We know that the answer to this question; i.e., the probability that the total of two dice is greater than 8, given that the first die is a 6, is 2/3 because only when the second die is 3, 4, 5, or 6 (4 out of a possible 6) is the total of two dice greater than 8. Therefore, $P(E_2|E_1) = \frac{4}{6} = \frac{2}{3}$. Let us see if Eq. 5.18a gives us the right answer:

$$P(E_2|E_1) = \frac{P(E_2 \cap E_1)}{P(E_1)} = \frac{1/9}{1/6} = \frac{2}{3}$$

It does.

If event E_2 is statically independent of E_1—that is, $P(E_2) = \dfrac{5}{18}$ regardless if the first die is 6—then

$$P(E_2|E_1) = P(E_2) \tag{5.19}$$

Therefore, if E_1 and E_2 are statically independent, from Eq. 5.18b we have

$$P(E_2 \cap E_1) = P(E_2)P(E_1) \tag{5.20}$$

Again, independent events are not disjoint and vice versa.

The total probability theorem states that if $E_1, E_2,..., E_m$ are mutually exclusive, then, for an event A, we have

$$
\begin{aligned}
P(A) &= P(A|E_1)P(E_1) + P(A|E_2)P(E_2) + ... + P(A|E_m)P(E_m) \\
&= P(A \cap E_1) + P(A \cap E_2) + ... + P(A \cap E_m) \\
&= \sum_{i=1}^{m} P(A|E_i)P(E_i)
\end{aligned}
\tag{5.21}
$$

Since

$$P(A \cap E_i) = P(A|E_i)P(E_i)$$

we have

$$P(E_i|A) = \frac{P(A|E_i)P(E_i)}{P(A)} = \frac{P(A|E_i)P(E_i)}{\sum_{l=1}^{m} P(A|E_i)P(E_i)} \tag{5.22}$$

Equation 5.22 is an extended form of the famous and powerful Bayes' theorem (or law or rule), which is known as the law of total probability.

Bayes' theorem provides the actual probability of an event given the measured test probabilities. It has been applied to many problems in engineering and other disciplines for solving problems encountered in, for example, correction for measurement errors and relating actual probability to measured test probability. The following example illustrates the application of the theorem to a simple engineering problem.

EXAMPLE 5.1

Consider steel beams that are tested before use. Let A denote the event that the beams are supplied by vendor A; let E_1 denote the event that beams pass the test; and let E_2 denote the event that beams fail the test. E_1 and E_2 are mutually exclusive. Let the pass rate be 95%—that is, $P(E_1) = 0.95$, and then $P(E_2) = 0.05$. For those beams that pass the test, 90% are supplied by vendor A (i.e., $P(A|E_1) = 0.9$), and for those that fail, 20% are from vendor A (i.e., $P(A|E_2) = 0.2$). What is the probability of beams supplied by vendor A passing the test? In other words, what is $P(E_1|A)$? Also, what percentage of the beams are supplied by vendor A? In other words, what is $P(A)$?

Solution

The answers to these questions can be found using Bayes' theorem as shown in Eq. 5.20. Before we apply the equation to a solution, we analyze the problem so we have a better understanding of the approach to be taken. The problem can be analyzed using either the rectangle or the tree diagram shown in the figure below.

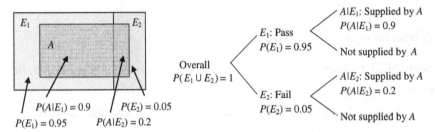

Problem analysis diagrams: (a) rectangle and (b) tree.

From the figure, we see that the percentage of beams that pass the test and are supplied by vendor A is $P(E_1 \cap A) = P(A|E_1)P(E_1) = (0.9)(0.95) = 0.855$. The number that fail the test and are supplied by vendor A is $P(E_2 \cap A) = P(A|E_2)P(E_2) = (0.2)(0.05) = 0.01$. Hence, the percentage of beams supplied by vendor A is $P(E_1 \cap A) + P(E_2 \cap A) = 0.855 + 0.01 = 0.865 = P(A)$. The probability of the beams supplied by vendor A passing the test is $0.855/(0.855 + 0.01) = 0.988$. The above analysis can be summarized in one equation as follows:

$$P(E_1|A) = \frac{P(E_1 \cap A)}{P(E_1 \cap A) + P(E_2 \cap A)} = \frac{P(A|E_1)P(E_1)}{P(A|E_1)P(E_1) + P(A|E_2)P(E_2)}$$

$$= \frac{0.9 \times 0.95}{0.9 \times 0.95 + 0.2 \times 0.05} = 0.988$$

which is exactly the application of Bayes' theorem shown in Eq. 5.22.

5.3.2 Random Variables and Distribution Functions

In this subsection, we discuss random variables and probabilistic distributions that are essential to reliability analysis.

Random Variables

A random variable is one whose value is determined by the outcome of a random experiment. In this chapter, X denotes a random variable and x denotes a value of the random variable in an experiment, which represents an event that is a subset of the sample space. In general, a random variable takes on various values x within the range $-\infty < x < \infty$.

Random variables are of two types: discrete and continuous. A discrete random variable is one whose set of assumed values is countable (i.e., arises from counting). A continuous random variable is one whose set of assumed values is uncountable (i.e., arises from measurement). Most random variables encountered in engineering design are continuous—for example, material strength is obtained from measurement instead of counting. Therefore, the discussion in this chapter focuses on continuous random variables.

Distribution Functions

The function $f_X(x)$ is called a probability density function (PDF) for the continuous random variable where the total area under the PDF curve bounded by the x-axis, as shown in Figure 5.5a, is equal to 1:

$$\int_{-\infty}^{\infty} f_X(x)dx = 1 \qquad (5.23)$$

The cumulative distribution function of a continuous random variable X, shown in Figure 5.5b, is defined as

$$F_X(x) = \int_{-\infty}^{x} f_X(s)ds \qquad (5.24)$$

If $F_X(x)$ is continuous, the probability of X having a value between a and b can be calculated as

$$F_X(b) - F_X(a) = \int_{-\infty}^{b} f_X(x)dx - \int_{-\infty}^{a} f_X(x)dx = \int_{a}^{b} f_X(x)dx \qquad (5.25)$$

Equation 5.25 is illustrated in Figure 5.6, in which $F_X(b) - F_X(a)$ is the area under the curve bounded by the x-axis between a and b. If the random variable X is continuous and if the first derivative of the CDF exists, then the PDF $f_X(x)$ is given by the first derivative of the CDF, $F_X(x)$:

$$f_X(x) = \frac{dF_X(x)}{dx} \qquad (5.26)$$

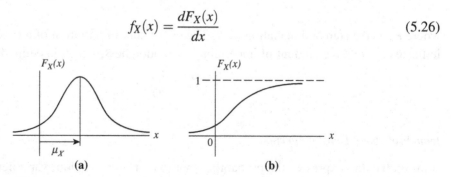

FIGURE 5.5: Distribution functions of a continuous random variable X: (a) PDF and (b) CDF.

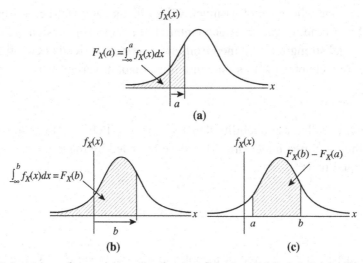

FIGURE 5.6: Value of cumulative distribution function $F_X(x)$: (a) $F_X(a)$, (b) $F_X(b)$, and (c) $F_X(b) - F_X(a)$.

Mean Value and Standard Deviation

The mean value (or expected value or average) of a random variable X with PDF $f_X(x)$ is defined as

$$\mu_X = E(X) = \int_{-\infty}^{\infty} x f_X(x) dx \tag{5.27}$$

The variance σ_X^2 of X, which is a measure of how far a set of numbers is spread out, is defined as

$$\sigma_X^2 = E[(X - \mu_X)^2] = \int_{-\infty}^{\infty} (x - \mu_X)^2 f_X(x) dx \tag{5.28}$$

where σ_X is the standard deviation of X. The coefficient of variation of a random variable X indicates the relative amount of uncertainty or randomness, which is defined as

$$V_X = \frac{\sigma_X}{\mu_X} \tag{5.29}$$

Joint Probability Density Function

Joint probability expresses the probability that two or more random variables will exist simultaneously. In general, if there are n random variables, the outcome is an n-dimensional vector of them. For example, the probability of a two-dimensional case, in which the vector of random variables is $X = [X, Y]^T$, can be calculated as

$$P(a < X < b, c < Y < d) = \int_c^d \int_a^b f_{XY}(x, y)dxdy \qquad (5.30)$$

where $f_{XY}(x, y)$ is the joint probability density function of the random variables X and Y.

If two random variables X and Y are correlated, X can be affected by the value taken by Y. One of the best ways to visualize the possible relationship is to plot the (X,Y) pair that is produced by several trials of the experiment. An example of correlated samples is shown in Figure 5.7, in which the correlation coefficients ρ (to be discussed next) are 0.5 and -0.5, respectively.

For correlated random variables X and Y, the covariance defined in Eq. 5.31a can be used as a measure to describe a linear association between the two random variables:

$$\begin{aligned} C_{XY} = \text{Cov}(X, Y) &= E[(X - \mu_X)(Y - \mu_Y)] \\ &= \int_{-\infty}^{\infty} \int_{-\infty}^{\infty} (x - \mu_X)(y - \mu_Y)f_{XY}(x, y)dxdy \end{aligned} \qquad (5.31a)$$

Cov(X,Y) can also be written as

$$\begin{aligned} \text{Cov}(X, Y) &= E[(X - \mu_X)(Y - \mu_Y)] \\ &= E[XY] - \mu_X\mu_Y \end{aligned} \qquad (5.31b)$$

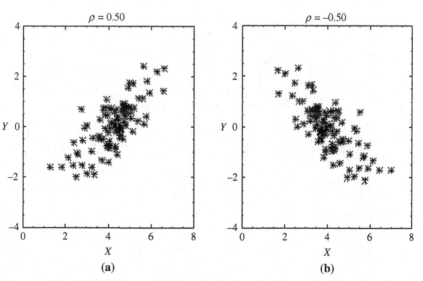

FIGURE 5.7: Correlated random variables X and Y: (a) X and Y correlated with a positive correlation coefficient, $\rho = 0.5$; (b) X and Y correlated with a negative correlation coefficient, $\rho = -0.5$.

The correlation coefficient ρ mentioned previously is a nondimensional measure of the correlation, which is defined as

$$\rho = \frac{C_{XY}}{\sigma_X \sigma_Y} = \frac{\text{Cov}(X, Y)}{\sigma_X \sigma_Y}, -1 \leq \rho \leq 1 \tag{5.32}$$

where σ_X and σ_Y are the standard deviations of the random variables X and Y, respectively.

If X and Y are statistically independent (that is, there is no linear relationship between them), the variables are uncorrelated and the correlation coefficient $\rho = 0$. $\rho > 0$ indicates a positive relationship between X and Y; that is, Y increases as X increases. For example, $\rho = 0.5$, shown in Figure 5.7a, reveals such a correlation. $\rho < 0$ indicates a negative relationship between X and Y; that is, Y decreases as X increases, as depicted in Figure 5.7b. $\rho = 1$ indicates a perfect positive linear relationship between X and Y; Y linearly increases as X increases. $\rho = -1$ gives a perfect negative linear relationship between X and Y (also called anticorrelation); Y linearly decreases as X increases.

Similarly, for a vector of random variables $X = [X_1, X_2, ..., X_n]^T$ with joint PDF $f_X(x)$, the elements in the vectors of expected values and the covariance matrix are, respectively, $\mu_i = E[X_i]$, $i = 1, n$; and $C_{ij} = \text{Cov}(X_i, X_j)$, $i, j = 1, n$, which is defined as

$$C_{ij} = \text{Cov}(X_i, X_j) = E[(X_i - \mu_i)(X_j - \mu_j)]$$
$$= \int\limits_{-\infty}^{\infty} \int\limits_{-\infty}^{\infty} (x_i - \mu_i)(x_j - \mu_j) f_{X_i X_j}(x_i, x_j) dx_i dx_j \tag{5.33a}$$

Equation 5.33a can also be written in a matrix form as

$$C = \begin{bmatrix} E\left[(X_1 - \mu_1)^2\right] & \cdots & \text{Symmetric} \\ E[(X_2 - \mu_2)(X_1 - \mu_1)] & E\left[(X_2 - \mu_2)^2\right] & \cdots \\ & \cdots & \\ E[(X_n - \mu_n)(X_1 - \mu_1)] & \cdots & E\left[(X_n - \mu_n)^2\right] \end{bmatrix}$$

$$= \begin{bmatrix} C_{11} & \cdots & & \text{Symmetric} \\ C_{21} & C_{22} & \cdots & \\ & & \cdots & \\ C_{n1} & & \cdots & C_{nn} \end{bmatrix}_{n \times n} \tag{5.33b}$$

where

$C_{ii} = \sigma_i^2$. σ_i is the standard deviation of the random variable X_i.

The correlation coefficient between X_i and X_j is

$$\rho_{ij} = \frac{C_{ij}}{\sigma_i \sigma_j}; \quad i,j = 1, n; \quad -1 \le \rho_{ij} \le 1 \tag{5.34}$$

The following example illustrates how random variables can be correlated and how the correlation using correlation coefficient ρ_{ij} can be characterized.

EXAMPLE 5.2

Consider a cantilever beam of solid circular cross-section, as shown below. The diameter of the beam is $d = 0.467$ in. The beam is loaded with two random forces F_1 and F_2 at $L_1 = 6$ in. and $L_2 = 10$ in. respectively. The mean values and standard deviations of these two random forces are, respectively, $\mu_1 = 20$ lb, $\sigma_1 = 4$ lb; and $\mu_2 = 10$ lb, $\sigma_2 = 2$ lb. The shear force V and maximum bending stress s at the root of the beam are, respectively,

$$V = F_1 + F_2, \text{ and } s = \frac{32M}{\pi d^3} = \frac{32(F_1 L_1 + F_2 L_2)}{\pi d^3} = 100(F_1 L_1 + F_2 L_2)$$

Calculate the mean value and standard deviation of the maximum bending stress s. If the forces F_1 and F_2 are independent, what is the correlation coefficient between V and s?

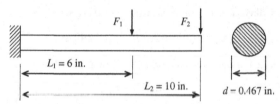

Cantilever beam cross-section.

Solution

The mean value and standard deviation of the shear force are, respectively,

$$\mu_V = \mu_1 + \mu_2 = 20 + 10 = 30 \text{ lb}$$

and

$$
\begin{aligned}
\sigma_V^2 &= E\left[(V - \mu_V)^2\right] \\
&= E\left[((F_1 + F_2) - (\mu_1 + \mu_2))^2\right] \\
&= E\left[((F_1 - \mu_1) + (F_2 - \mu_2))^2\right] \\
&= E\left[(F_1 - \mu_1)^2 + 2(F_1 - \mu_1)(F_2 - \mu_2) + (F_2 - \mu_2)^2\right] \\
&= E\left[(F_1 - \mu_1)^2\right] + E\left[(F_2 - \mu_2)^2\right] \\
&= \sigma_1^2 + \sigma_2^2 \\
&= 4^2 + 2^2 \\
&= 20
\end{aligned}
$$

Therefore, the standard deviation of the shear force is

$$\sigma_V = \sqrt{20} = 4.472 \text{ lb}$$

The mean value and standard deviation of the maximum bending stress are, respectively,

$$\mu_s = 100(\mu_1 L_1 + \mu_2 L_2) = 100(20 \times 6 + 10 \times 10) = 22,000 \text{ psi}$$

and

$$\sigma_s^2 = E\left[(s - \mu_s)^2\right]$$

$$= E\left[(100(F_1 L_1 + F_2 L_2) - 100(\mu_1 L_1 + \mu_2 L_2))^2\right]$$

$$= E\left[(100(L_1(F_1 - \mu_1) + L_2(F_2 - \mu_2)))^2\right]$$

$$= E\left[\left(10,000(L_1^2(F_1 - \mu_1)^2 + 2L_1 L_2(F_1 - \mu_1)(F_2 - \mu_2) + L_2^2(F_2 - \mu_2)^2\right)\right]$$

$$= 10,000\left\{L_1^2 E\left[(F_1 - \mu_1)^2\right] + 2L_1 L_2 E\left[(F_1 - \mu_1)(F_2 - \mu_2)\right] + L_2^2 E\left[(F_2 - \mu_2)^2\right]\right\}$$

$$= 10,000(L_1^2 \sigma_1^2 + 2L_1 L_2 \text{Cov}(F_1, F_2) + L_2^2 \sigma_2^2)$$

$$= 10,000(6^2 \times 4^2 + 0 + 10^2 \times 2^2) = 9,760,000$$

Therefore, the standard deviation of the maximum bending stress is

$$\sigma_s = \sqrt{9,760,000} = 3,124 \text{ psi}$$

The correlation coefficient between V and s can be found as follows:

$$C_{Vs} = \text{Cov}(V, s) = E[(V - \mu_V)(s - \mu_s)] = E[Vs] - \mu_V \mu_s$$

$$= E[(F_1 + F_2)100(F_1 L_1 + F_2 L_2)] - (\mu_1 + \mu_2)100(\mu_1 L_1 + \mu_2 L_2)$$

$$= 100\left\{L_1 E\left[F_1^2\right] + L_2 E\left[F_2^2\right] + (L_1 + L_2)E\left[F_1 + F_2\right] - L_1 \mu_1^2 - L_2 \mu_2^2 - (L_1 + L_2)\mu_1 \mu_2\right\}$$

$$= 100\left\{L_1 E\left[F_1^2\right] + L_2 E\left[F_2^2\right] - L_1 \mu_1^2 - L_2 \mu_2^2\right\}$$

$$= 100\left\{L_1\left(E\left[F_1^2\right] - \mu_1^2\right) + L_2\left(E\left[F_2^2\right] - \mu_2^2\right)\right\}$$

$$= 100\left\{L_1 \sigma_1^2 + L_2 \sigma_2^2\right\}$$

$$= 100\left\{6 \times 4^2 + 10 \times 2^2\right\}$$

$$= 13,600$$

$$\rho_{Vs} = \frac{C_{Vs}}{\sigma_V \sigma_s} = \frac{13,600}{4,472 \times 3,124} = 0.973$$

Hence, the covariance matrix $\mathbf{C}$ is

$$\mathbf{C} = \begin{bmatrix} 1 & 0.973 \\ 0.973 & 1 \end{bmatrix}$$

5.3.3 Probabilistic Distributions

In this subsection, we discuss a few common distribution function types, specifically, probability density functions and cumulative distribution functions. We include normal, lognormal, and Weibull distributions (Weibull is one of the extreme value distributions). These are commonly found in engineering applications and are used in the examples in Section 5.4, where we discuss reliability analysis methods.

Normal Distribution

A normal distribution function, shown in Figure 5.8, reveals a bell-shape curve. The normal distribution of a random variable X with expected value μ_X and standard deviation σ_X is denoted $N(\mu_X, \sigma_X)$ or $X \sim N(\mu_X, \sigma_X)$, and its PDF is defined as

$$f_X(x) = \frac{1}{\sqrt{2\pi}\sigma_X}e^{-\frac{1}{2}\left(\frac{x-\mu_X}{\sigma_X}\right)^2}, -\infty < x < \infty \tag{5.35a}$$

The CDF of X is

$$F_X(x) = \Phi\left(\frac{x-\mu_X}{\sigma_X}\right) = \int_{-\infty}^{x} \frac{1}{\sqrt{2\pi}\sigma_X}e^{\frac{1}{2}\left(\frac{s-\mu_X}{\sigma_X}\right)^2} ds \tag{5.35b}$$

where $\Phi(\bullet)$ is the standard (or normalized) normal distribution function with mean value 0 and standard deviation 1, represented as $N(0,1)$.

Therefore, a random variable X of normal distribution, denoted $X \sim N(\mu_X, \sigma_X)$, can be normalized by a simple transformation $z = \dfrac{x-\mu_X}{\sigma_X}$, in which Z is a random variable of

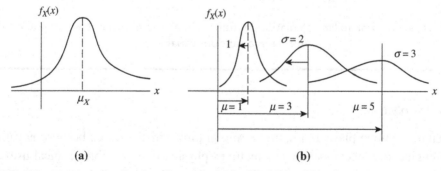

FIGURE 5.8: Normal distribution function $f_X(x)$: (a) $f_X(x)$ with mean value μ_X and standard deviation σ_X; (b) $f_X(x)$ with several mean values and standard deviations.

standard normal distribution (i.e., $Z \sim N(0,1)$). The cumulative distribution function of the random variable Z of standard normal distribution is thus given by

$$F_X(x) = \int_{-\infty}^{x} \frac{1}{\sqrt{2\pi}\sigma_X} e^{-\frac{1}{2}\left(\frac{s-\mu_X}{\sigma_X}\right)^2} ds = \int_{-\infty}^{z} \frac{1}{\sqrt{2\pi}} e^{-\frac{t^2}{2}} dt = \Phi(z) \qquad (5.36)$$

EXAMPLE 5.3

Continue with Example 5.2. If the two random forces F_1 and F_2 are normally distributed, what is the probability that the bending stress s exceeds 24,000 psi?

Solution

From Example 5.2, we have the mean value and standard deviation of the bending stress: $\mu_s = 22,000$ psi and $\sigma_s = 3,124$ psi. First, we calculate z by

$$z = \frac{s - \mu_s}{\sigma_s} = \frac{24,000 - 22,000}{3,124} = 0.640$$

The probability can then be obtained as (see below diagram)

$$P(s > 24,000) = P(z > 0.640) = 1 - \Phi(0.640) = 1 - 0.739 = 0.261 = 26.1\%$$

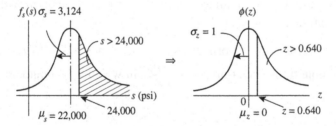

Illustration of the probability calculation by transforming random variable s to standard and normalized random variable z

Lognormal Distribution

Lognormal distribution plays an important role in probabilistic design because negative values of engineering phenomena are sometimes physically impossible. Typical uses of lognormal distribution are found in descriptions of fatigue failure, failure rates, and other phenomena involving a large range of data.

The lognormal distribution of a random variable X with expected value μ_X and standard deviation σ_X is denoted $LN(\mu_X, \sigma_X)$ and is defined as

$$f_X(x) = \frac{1}{\sqrt{2\pi}\sigma_Y} e^{-\frac{1}{2}\left(\frac{\ln(x)-\mu_Y}{\sigma_Y}\right)^2}, 0 < x < \infty \tag{5.37a}$$

in which $f_X(x)$ is the PDF of the random variable X, and

$$\sigma_Y = \sqrt{\ln\left(\left(\frac{\sigma_X}{\mu_X}\right)^2 + 1\right)}$$

and

$$\tag{5.37b}$$

$$\mu_Y = \ln\left(\mu_X\right) - \frac{1}{2}\sigma_Y^2$$

are the standard deviation and expected value for the normal distribution variable $y = \ln(x)$. A few lognormal distribution functions are shown in Figure 5.9. The cumulative distribution function of a lognormal distribution is given as

$$F_X(x) = \int\limits_{-\infty}^{\ln(x)} \frac{1}{\sqrt{2\pi}\sigma_Y} e^{-\frac{1}{2}\left(\frac{s-\mu_Y}{\sigma_Y}\right)^2} ds \tag{5.37c}$$

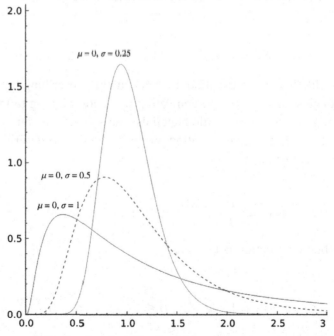

FIGURE 5.9: Lognormal distribution function f_X with several mean values and standard deviations.

Extreme Value Distributions

Extreme value distributions are used to represent the maximum or minimum of a number of samples of various distributions. There are three types, described in the following paragraphs.

Type 1, also called the Gumbel distribution, is a distribution of the maximum or minimum of a number of samples of normally distributed data. A Gumbel distribution function is defined as

$$f_X(x) = a e^{-e^{-a(x-b)}} e^{-a(x-b)}, \quad -\infty < x < \infty, \, a > 0 \tag{5.38a}$$

where a and b are scale and location parameters, respectively. The cumulative distribution function of a Gumbel distribution is given as

$$F_X(x) \, e^{-e^{-a(x-b)}}, \quad -\infty < x < \infty, \, a > 0 \tag{5.38b}$$

Type 2, also called the Frechet distribution, is defined as

$$f_X(x) = \frac{\kappa}{v} \left(\frac{v}{x}\right)^{\kappa+1} e^{-\left(\frac{v}{x}\right)^{\kappa}}, \quad 0 < x < \infty, \, \kappa \geq 2 \tag{5.39a}$$

The cumulative distribution function of a Frechet distribution is given as

$$F_X(x) = e^{-\left(\frac{v}{x}\right)^{\kappa}}, \quad 0 < x < \infty, \, \kappa \geq 2 \tag{5.39b}$$

Type 3, also called the Weibull distribution, is well suited to describing the weakest-link phenomenon, a situation where there are competing flaws contributing to failure. It is often used to describe fatigue, fracture of brittle materials, and strength in composites. The distribution of wind speeds at a given location on earth can also be described with this distribution, which is defined as

$$f_X(x) = \frac{a x^{a-1}}{b^a} e^{-\left(\frac{x}{b}\right)^{a}}, \quad 0 \leq x, \, a > 0, \, b > 0 \tag{5.40a}$$

The CDF of a Weibull distribution is given as

$$F_X(x) = 1 - e^{-\left(\frac{x}{b}\right)^{a}}, \quad 0 \leq x, \, a > 0, \, b > 0 \tag{5.40b}$$

A few Weibull distribution functions are shown in Figure 5.10.

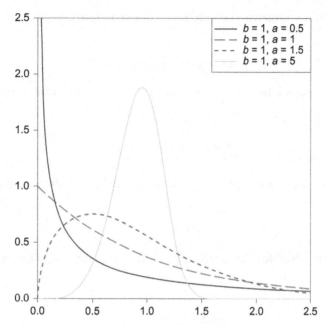

FIGURE 5.10: Weibull distribution function f_X with several mean values and standard deviations.

5.4 Reliability Analysis Methods

We have reached the key topic of this chapter, reliability analysis methods. Our focus is on those that are the most commonly employed: Monte Carlo simulation, the first-order reliability method (FORM), the second order reliability method (SORM), importance sampling, and the response surface method. In addition, we discuss the transformation of random variables into standard normal distribution that is necessary to employ some of these methods. We assume a single failure mode in a product component in this discussion. Multiple failure modes or system failure are discussed in Section 5.5.

5.4.1 The Limit State Function

The first step in reliability analysis is to identify the failure mode in which performance exceeds design limit—that is, how the product or component fails. The mode must be written in a mathematical form that involves random variables. Such a mathematical representation is called the *limit state function*.

A vector of random variables $X = [X_1, X_2, \ldots, X_n]^T$ with joint probability density function $f_X(x)$ can be used to model the uncertainties of a physical problem. For a structural problem, X can be used to model uncertainties in loads, yield strength, geometric dimensions, material properties, and so forth. Realization of $X = [X_1, X_2, \ldots, X_n]^T$ is denoted $x = [x_1, x_2, \ldots, x_n]^T$, which is a point in the n-dimensional space. A limit state function of these variables can be written as

$$g(x) = g(x_1, x_2, \ldots, x_n) = 0 \tag{5.41}$$

which divides the x-space into two regions, safe region R_s and failure region R_f:

$$g(x) = \begin{cases} > 0, x \in R_s \\ \leq 0, x \in R_f \end{cases} \tag{5.42}$$

If in the limit state function x is replaced by random variables X, the so-called safety margin M is defined as

$$M = g(X) \tag{5.43}$$

where M is a random variable. The probability of failure P_f of a structure with this failure mode is thus

$$P_f = P(M \leq 0) = P(g(X) \leq 0) = \int_{g(x) \leq 0} f_X(x) dx = \int_{R_f} f_X(x) dx \tag{5.44}$$

and the reliability of the structure can be simply obtained as $P_R = 1 - P_f$.

The probability integration in Eq. 5.44 is visualized with a two-dimensional case in Figure 5.11a, which shows the joint PDF $f_X(x)$ and its contour projected on the x_1-x_2-plane. All points on the projected contours have the same values of $f_X(x)$ or the same probability density. The limit state function $g(x) = 0$ is also shown. The failure probability P_f is the volume underneath the surface of the joint PDF $f_X(x)$ in the failure region $g(x) \leq 0$. To show the integration more clearly, the contours of the joint PDF $f_X(x)$ and the limit state function $g(x) = 0$ are plotted on the x_1-x_2-plane, as shown in Figure 5.11b.

The direct evaluation of the probability integration of Eq. 5.44 is very difficult if not impossible. First, the integration is multidimensional since often multiple random variables are involved in engineering problems. Second, the joint PDF $f_X(x)$ is in general a nonlinear function. Third, the limit state function $g(x)$ is often nonlinear without an analytical form, in which case a numerical method, such as finite element analysis (FEA), is employed for a solution. For this reason, methods other than direct integration have been developed, which are discussed next.

5.4.2 Monte Carlo Simulation

A powerful statistical analysis tool that has been widely used in both engineering and nonengineering applications—*Monte Carlo simulation*, or, simply, *simulation*—is the simplest and most reliable analysis method among many others. After a set of random

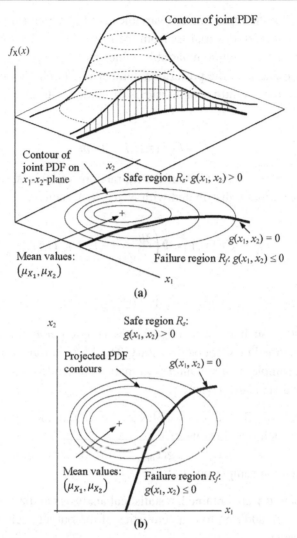

FIGURE 5.11: Probability integration using a two-dimensional example: (a) isometric view and (b) projected view on the x_1-x_2-plane.

variables is identified and a failure mode and associated limit state function are defined, Monte Carlo simulation involves three major steps:

Step 1: Sampling on random input variables X.

Step 2: Evaluating the limit state function $g(x)$.

Step 3: Statistical analysis of the outcome of the limit state function.

We assume that the CDFs of these respective random variables are known. To simplify our discussion, we further assume that these random variables are independent. Note that Monte Carlo simulation is not limited to problems of independent random variables.

The purpose of sampling the input random variables is to generate samples that represent distributions of the input random variables from their respective CDFs $F_{X_i}(x_i)$, $i = 1, n$. For each random variable, a set of random variable values $z = [z_1, z_2, ..., z_m]^T$ between 0 and 1 is generated first. Note that m is the prescribed number of sample points. These samples $z \in [0,1]$ are then transformed into sample values of random variable X_i following a given CDF $F_{X_i}(x_i)$ by

$$x_{ij} = F_{X_i}^{-1}(z_j), \ j = 1, \ m \tag{5.45}$$

where $F_{X_i}^{-1}(z_j)$ is the inverse of the CDF of the ith random variable X_i. For example, if the random variable X_i is normally distributed with $N(\mu_{x_i}, \sigma_{x_i})$, then

$$z_j = F_{X_i}(x_{ij}) = \Phi\left(\frac{x_{ij} - \mu_{Xi}}{\sigma_{Xi}}\right), \ j = 1, \ m \tag{5.46a}$$

Thus,

$$x_{ij} = \mu_{X_i} + \sigma_{X_i}\Phi^{-1}(z_j), \ j = 1, \ m \tag{5.46b}$$

Note that some software, such as MATLAB, generates sample points for x_{ij} directly, in which case the steps just described may not be necessary. The MATLAB function `normrnd(MU, SIGMA, n, m)` returns a set of sample points in an $n \times m$ matrix chosen from the normal distribution with mean value μ and standard deviation σ.

Once the sample values of all random variables are generated, the limit state function $y_j = g(x_j)$ is solved for each sample point $x_j = [x_{1j}, x_{2j}, ..., x_{nj}]^T$, $j = 1, \ m$ in step 2. Note that $y = [y_1, y_2, ..., y_m]^T = [g(x_1), g(x_2), ..., g(x_m)]^T$ is the vector of m limit state function values for the respective m sample points.

After m samples of output y are obtained, a statistical analysis can be carried out to estimate the failure probability (in addition to characteristics of the output such as mean value and standard deviation) using

$$p_f = \frac{1}{m}\sum_{j=1}^{m} I(g(x_j)) = \frac{m_f}{m} \tag{5.47}$$

where $I(\bullet)$ is an indicator function defined as

$$I(g(x)) = \begin{cases} 1, & \text{if } g(x) \leq 0 \\ 0, & \text{otherwise} \end{cases} \tag{5.48}$$

and m_f is the number of sample points that yield a nonpositive limit state function. We are essentially counting the number of failures among the total sample points generated.

EXAMPLE 5.4

We employ Monte Carlo simulation to estimate the failure probability of the cantilever beam discussed in Section 5.2. Recall that the stress s and yield strength S_y are two random variables of normal distribution with $s \sim N(\mu_s, \sigma_s) = (40, 8)$ ksi and $S_y \sim N(\mu_{S_y}, \sigma_{S_y}) = (50, 6)$ ksi. The limit state function for the stress failure mode of the beam was defined as

$$g(\pmb{x}) = S_y - s$$

The failure probability calculated was $P_f = 0.159$ in Section 5.2, which is analytical.

Solution

If we simply use 10 sample points, $m = 10$, we can use the MATLAB function `normrnd` (`MU,SIGMA,1,10`) to generate sample points for s and S_y directly, as shown in Table 5.2 (left two columns). Note that with only 10 sample points, three values of the g function are less than 0; hence, the failure probability P_f is 0.3. It is obvious that the failure probability calculated using only 10 sample points is inaccurate compared with the analytical solution. We need more sample points.

Table 5.2a: Sample points for s, S_y and g function

s	S_y	$g = S_y - s$
27.9961	60.0879	32.0918
35.6724	42.0859	6.4135
43.7378	48.5422	4.8044
36.8566	45.0047	8.1481
41.2556	51.3974	10.1418
50.2375	37.4567	−12.7808
6.8483	45.6715	−1.1769
50.5365	42.6835	−7.8530
36.1466	60.8851	24.7385
32.7855	48.5281	15.7425

Table 5.2b: Increasing sample points to improve accuracy of probability estimate

m	P_f
10	0.3000
100	0.1300
1000	0.1510
10,000	0.1546
100,000	0.1598
1,000,000	0.1588

If we increase the number of sample points, for example from 10 to 1,000,000 as shown in Table 5.2b, the failure probability becomes 0.1588, which is very close to the analytical solution of 0.159. The table clearly shows that when we increase the number of sample points the failure probability calculated using Monte Carlo simulation approaches the analytical solution.

Calculating failure probability using Monte Carlo simulation can be implemented using the following MATLAB script.

```
%Matlab script

m=10

s=normrnd(40,8,1,m)

Sy=normrnd(50,6,1,m)

n=0

for i=1:1:m

  g=Sy(i)-s(i)

  if g < 0

    n=n+1

  end

end

Pf=n/m
```

5.4.3 The First-Order Reliability Method

Monte Carlo simulation is straightforward and reliable. However, as demonstrated in Example 5.4, it requires a large number of sample data to ensure an accurate estimate of failure probability. Engineering applications; for example, complex structural problems that require finite element analysis, often requires nontrivial computational time. For such applications, Monte Carlo simulation is infeasible because of the substantial computational effort it requires.

A number of methods have been developed to reduce the computational effort involved in Monte Carlo simulation while still offering sufficiently accurate failure probability estimates. These methods aim to provide acceptable estimates for the integral form of failure probability P_f defined in Eq. 5.44, which can be achieved in two important steps:

> *Step 1*: Simplify the joint probability density function $f_X(x)$ by transforming a given joint probability density function, which may be multidimensional, into a standard normal distribution function of independent random variables of the same dimensions.
> *Step 2*: Approximate the limit state function $g(x) = 0$ by the Taylor series expansion and by keeping the first few terms for approximation. This is mainly to ease calculation of the failure probability; it has nothing to do with analysis of product performance.

If only linear terms are included in the calculation, the method is referred to as first-order reliability method. When the second-order terms are also considered, the method is referred to as second-order reliability method. In this subsection we introduce FORM. SORM is discussed briefly in Section 5.4.4. For both methods, we assume that the random variables are independent and are normally distributed. Transforming dependent random variables of non-normal distribution is discussed in detail in Section 5.4.5.

FORM

The space that contains the given set of random variables $X = [X_1, X_2, ..., X_n]^T$ is called the X-space. These random variables are transformed into a standard normal space, called U-space, where the transformed random variables $U = [U_1, U_2, ..., U_n]^T$ follow the standard normal distribution (that is, with mean value 0 and standard deviation 1). Such a transformation is carried out based on the condition that the CDFs of the random variables remain the same before and after transformation:

$$F_{X_i}(x_i) = \Phi(u_i) \tag{5.49}$$

where $\Phi(\bullet)$ is the CDF of the standard normal distribution. The transformation can be written as

$$u_i = \Phi^{-1}(F_{X_i}(x_i)) \tag{5.50a}$$

Thus, the transformed random variable U_i can be written as

$$U_i = \Phi^{-1}(F_{X_i}(X_i)) \tag{5.50b}$$

If the random variables X_i are independent and normally distributed—that is,

$F_{X_i}(x_i) = \Phi\left(\dfrac{x_i - \mu_i}{\sigma_i}\right)$—the transformation, as illustrated in Figure 5.12, can be obtained as

$$u_i = \Phi^{-1}\left(\Phi\left(\frac{x_i - \mu_i}{\sigma_i}\right)\right) = \frac{x_i - \mu_i}{\sigma_i} \tag{5.51}$$

Thus,

$$x_i = \mu_i + \sigma_i \mu_i \tag{5.52}$$

Note that, as shown in Figure 5.12c, the projected contours of the transformed PDF on the u_1-u_2-plane are circles centered at the origin.

The limit state function is also transformed into U-space as

$$g(x) = g_u(u) = 0 \tag{5.53}$$

The transformed limit state function separates the U-space into safe regions and failure regions, as illustrated in Figure 5.12b and 5.12c. After the transformation, the probability integration of Eq. 5.44 becomes

$$P_f = P(g_u(u) \leq 0) \int_{g_u(u) \leq 0} \phi(u) du \tag{5.54}$$

Since all random variables U are independent, the joint PDF is the product of the individual PDFs of standard normal distribution:

$$\phi(u) = \prod_{i=1}^{n} \frac{1}{\sqrt{2\pi}} e^{-\frac{1}{2}u_i^2} \tag{5.55}$$

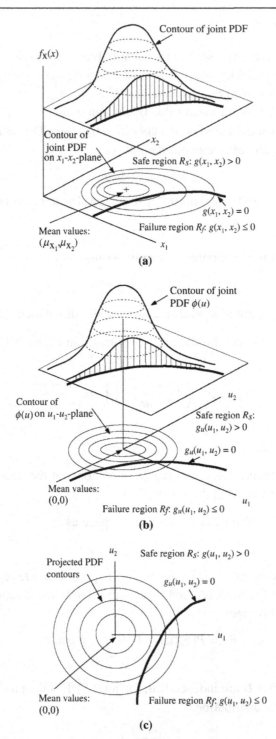

FIGURE 5.12: Transformation of random variables from **X** to **U** using a two-dimensional example: (a) PDF and projected contours in X-space, (b) standard normal distribution in U-space, and (c) projected view on the u_1-u_2-plane.

Therefore, the probability integration becomes

$$P_f = \underbrace{\iint \cdots \int}_{g_u(u) \leq 0} \prod_{i=1}^{n} \frac{1}{\sqrt{2\pi}} e^{-\frac{1}{2}u_i^2} \, du_1 \, du_2 \dots du_n \qquad (5.56)$$

Although it is obvious that Eq. 5.56 is relatively easier to calculate than Eq. 5.44, it is still difficult, if not entirely impossible, to do so since the limit state function $g_u(u)$ is in general a nonlinear function of variable u.

If the nonlinearity of the limit state function is not too severe, the integration of Eq. 5.56 can be approximated by

$$P_f \approx \int_{-\infty}^{-\beta} \frac{1}{\sqrt{2\pi}} e^{-\frac{1}{2}u^2} \, du = \Phi(-\beta) \qquad (5.57)$$

where

    Φ is the standard normal distribution function of single dimension.
    β is the shortest distance between the origin and the point on the limit state function $g_u(u) = 0$.

The point on the limit state function is depicted as u^* in Figure 5.13 and is called the β-point, design point, or most probable point (MPP). The value of β is referred to as the reliability index.

As illustrated in Figure 5.13a, the line that connects the origin and the MPP must be perpendicular to the tangent line $L_u(u_1, u_2) = 0$ at the MPP. Since the joint probability density function $\phi(u)$ of multidimension is axisymmetric, its projection on the plane that is normal to the tangent line is nothing but a probability density function of the standard normal distribution $\phi(u)$ of a single dimension. The probability integration of Eq. 5.56 can then be approximated by $\Phi(-\beta)$, as stated in Eq. 5.57.

Another way to understand the approximation of Eq. 5.57 is to linearize the limit state function at the MPP—that is, create a tangent line to it, which can be written as

$$g_u(u) \approx g_u(u^*) + \nabla g_u(u^*)(u - u^*)^T = L(u) \qquad (5.58)$$

where $\nabla g_u(u^*)$ is the gradient of function $g_u(u^*)$ defined as

$$\nabla g_u(u^*) = \left[\frac{\partial g_u(u)}{\partial u_1}, \frac{\partial g_u(u)}{\partial u_2}, \dots, \frac{\partial g_u(u)}{\partial u_n} \right]\Bigg|_{u=u^*} \qquad (5.59)$$

Since at the MPP u^*, $g_u(u) = 0$, Eq. 5.58 becomes

$$L(u) = \sum_{i=1}^{n} \frac{\partial g_u(u)}{\partial u_i}\Bigg|_{u_i=u_i^*} (u_i - u_i^*) = a_0 + \sum_{i=1}^{n} a_i u_i \qquad (5.60)$$

where

$$a_0 = -\sum_{i=1}^{n} \frac{\partial g_u(u)}{\partial u_i}\bigg|_{u_i=u_i^*} u_i^* \qquad (5.61a)$$

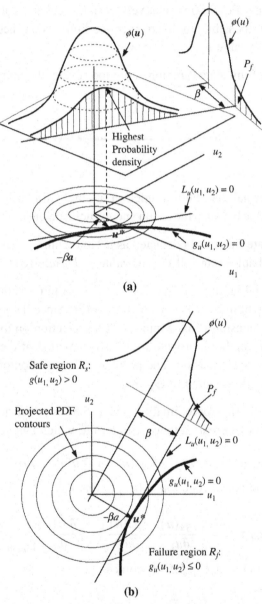

FIGURE 5.13: Approximation of failure probability integration using a two-dimensional example: (a) PDF and projected contours in *U*-space and (b) projected view on the u_1-u_2-plane.

and

$$a_i = \left. \frac{\partial g_u(\boldsymbol{u})}{\partial u_i} \right|_{u_i=u_i^*} \tag{5.61b}$$

Because U_i are random variables of standard normal distribution and $L(\boldsymbol{u})$ is a linear function of the random variables $\boldsymbol{U}$, $L(\boldsymbol{u})$ also has standard normal distribution. The mean value and standard deviation of $L(\boldsymbol{u})$ are, respectively,

$$\mu_L = a_0 = -\sum_{i=1}^{n} \left. \frac{\partial g_u(\boldsymbol{u})}{\partial u_i} \right|_{u_i=u_i^*} u_i^* \tag{5.62a}$$

and

$$\sigma_L = \sqrt{\sum_{i=1}^{n} a_i^2} = \sqrt{\sum_{i=1}^{n} \left(\left. \frac{\partial g_u(\boldsymbol{u})}{\partial u_i} \right|_{u_i=u_i^*} \right)^2} \tag{5.62b}$$

Therefore, the probability of failure can be approximated as

$$P_f \approx P(L(\boldsymbol{u}) \leq 0) = \Phi\left(\frac{-\mu_L}{\sigma_L}\right) = \Phi\left(\frac{\sum_{i=1}^{n} \left. \frac{\partial g_u(\boldsymbol{u})}{\partial u_i} \right|_{u_i=u_i^*} u_i^*}{\sqrt{\sum_{i=1}^{n} \left(\left. \frac{\partial g_u(\boldsymbol{u})}{\partial u_i} \right|_{u_i=u_i^*} \right)^2}} \right) = \Phi\left(\sum_{i=1}^{n} \alpha_i u_i^* \right) \tag{5.63}$$

where

$$\alpha_i = \frac{\left. \frac{\partial g_u(\boldsymbol{u})}{\partial u_i} \right|_{u_i=u_i^*}}{\sqrt{\sum_{i=1}^{n} \left(\left. \frac{\partial g_u(\boldsymbol{u})}{\partial u_i} \right|_{u_i=u_i^*} \right)^2}} = \frac{\nabla g_u(\boldsymbol{u}^*)}{||\nabla g_u(\boldsymbol{u}^*)||} \tag{5.64}$$

Then the probability of failure can be written as

$$P_f \approx \Phi\left(\boldsymbol{u}^* \cdot \boldsymbol{a}^T\right) \tag{5.65a}$$

where $\boldsymbol{a} = [\alpha_1, \alpha_2, \dots \alpha_n]$. As shown in Figure 5.13b, the position vector of the MPP is $\boldsymbol{u}^* = -\beta \boldsymbol{a}$; hence,

$$P_f \approx \Phi\left(-\beta \boldsymbol{a} \cdot \boldsymbol{a}^T\right) = \Phi(-\beta) \tag{5.65b}$$

since $\boldsymbol{a}$ is a normalized unit vector and $\boldsymbol{a} \cdot \boldsymbol{a}^T = 1$.

Note that in Eq. 5.63, $P(L(u) \leq 0)$ can also be found as

$$P_f \approx P(L(u) \leq 0) = P(\beta - a \cdot u^T \leq 0) = \Phi(-\beta) \qquad (5.66)$$

in which the failure zone is approximated as $L(u) \leq 0$, and a, as defined above, is the normalized gradient of the limit state function at the MPP, which is also the gradient of the tangent line that passes the MPP (i.e., $L(u^*) = 0$). Note that in Eq. 5.66 the approximated failure region $L(u) \leq 0$ can be represented by any point in the U-space that satisfies $\beta - a \cdot u^T \leq 0$, where $a \cdot u^T$ is the projection of the position vector u on vector a, which is the normalized gradient at the MPP.

EXAMPLE 5.5

Recall that in the cantilever beam example discussed in Section 5.2, we assumed that both the length and the diameter of the beam cross-section are uncorrelated and are normally distributed—that is, $X_\ell \sim N(\mu_\ell, \sigma_\ell) = (10, 1)$ in. and $X_d \sim N(\mu_d, \sigma_d) = (1, 0.1)$ in. The external force $P = 393$ lb is assumed deterministic. We want to define the maximum bending stress as the failure mode, and calculate the failure probability of the beam using FORM.

Solution
The limit state function for the stress failure mode of the beam can be defined as

$$g(x) = g(d, \ell) = S_y - s = S_y - \frac{32P\ell}{\pi d^3} = 50 - \frac{4\ell}{d^3} = 0 \text{ ksi}$$

Rewrite the limit state function (to simplify the math) as

$$g(d, \ell) = 12.5d^3 - \ell = 0$$

Function $g(d, \ell) = 0$ separates the d-ℓ-plane into safe and failure regions, as shown below.

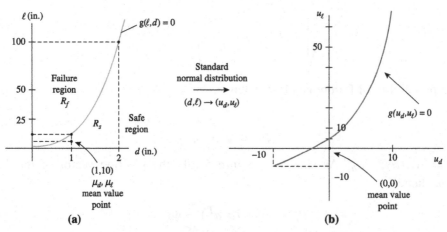

Separation of the response plane into safe and failure regions using the limit state function, (a) d-ℓ-plane by $g(d, \ell) = 0$, and (b) u_d-u_ℓ-plane by $g(u_d, u_\ell) = 0$.

The first step is to transform the random variables $X = [d, \ell]^T$ into $U = [u_d, u_\ell]^T$, in which u_d and u_ℓ are independent random variables of standard normal distribution, using Eq. 5.51:

$$u_d = \frac{d - \mu_d}{\sigma_d} = \frac{d - 1}{0.1} = 10d - 10$$

and

$$u_\ell = \frac{\ell - \mu_\ell}{\sigma\ell} = \frac{\ell - 10}{1} = \ell - 10$$

Thus,

$$g_u(u) = g(u_d, u_\ell) = 12.5(0.1u_d + 1)^3 - (u_\ell + 10) = 0$$

From Eq. 5.59, the gradient of function $g_u(u^*)$ can be calculated as

$$\nabla g_u(u^*) = \left[\frac{\partial g_u(u)}{\partial u_d}, \frac{\partial g_u(u)}{\partial u_\ell} \right]\Bigg|_{u=u^*} = \left[3.75(0.1u_d^* + 1)^2, -1 \right]$$

The MPP is located at $u^* = [u_d^*, u_\ell^*]^T$, with a distance β from the origin, as shown below.

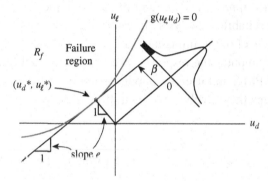

Location of the MPP at $u^* = [u_d^*, u_\ell^*]^T$ at distance β from the origin.

The straight line that is perpendicular to the tangent line at the MPP is thus

$$u^* = [u_d^*, u_\ell^*]^T = -\beta a = -\beta[\alpha_1, \alpha_2] = -\beta \frac{\nabla g_u(u^*)}{||\nabla g_u(u^*)||} = -\beta \frac{\left[3.75(0.1u_d^* + 1)^2, -1 \right]}{\sqrt{\left(3.75(0.1u_d^* + 1)^2 \right)^2 + (-1)^2}}$$

from which we obtain the following relationship between u_d^* and u_ℓ^* :

$$\frac{u_\ell^*}{u_d^*} = \frac{-1}{3.75(0.1u_d^* + 1)^2}$$

u_d^* and u_ℓ^* can be solved by bringing this relation back to the limit state function:

$$g(u_d^*, u_\ell^*) = 12.5(0.1u_d^* + 1)^3 - \left(\frac{-u_d^*}{3.75(0.1u_d^* + 1)^2} + 10 \right) = 0$$

Solving for this (using MATLAB for example), we obtain $u_d^* = -0.6553$ and $u_\ell^* = 0.2001$. Hence, the β value can be obtained as

$$\beta = \sqrt{u_d^{*2} + u_\ell^{*2}} = \sqrt{(-0.6553)^2 + (0.2001)^2} = 0.6852$$

and so the failure probability can be approximated, using Eq. 5.66, as

$$P_f \approx \Phi(-\beta) = \Phi(-0.6852) = 0.2466$$

As can be seen, the method for finding the MPP shown in the example is not general. Moreover, this ad hoc approach is not suitable for implementation in software. A more general and systematic approach for the MPP search is discussed next.

The key to calculating the failure probability is to locate the MPP in U-space. Many numerical methods have been developed for the MPP search. These can be categorized as two types: the reliability index approach (RIA) and the performance measure approach (PMA). RIA employs a forward reliability analysis algorithm that computes failure probability for a specified performance level in the limit state function. PMA employs an inverse reliability analysis algorithm that computes the response level for a specified failure probability. We introduce both RIA and PMA and use one popular numerical algorithm to illustrate the detailed computation steps they involve. We then discuss the pros and cons of each.

The Reliability Index Approach

The problem for an MPP search using RIA can be formulated as follows:

$$\begin{cases} \text{minimize:} & \|u\| \\ \text{subject to:} & g_u(u) = 0 \end{cases} \tag{5.67}$$

in which the MPP is identified by searching a point on the limit state function $g_u(u) = 0$, where the distance between the point to the origin of the U-space is minimum. Again, the distance β is called the reliability index—hence the name of this approach. The RIA was, in fact, illustrated in Example 5.5. Note that in Eq. 5.67, the performance level of the limit state function is specified. As seen in Example 5.5, the yield strength in the limit state function of the stress failure mode is specified as 50 ksi. In this case, the MPP can be searched only for the specified performance level.

One of the most popular algorithms uses a recursive formula and is based on linearization of the limit state function. The MPP search procedure is illustrated in Figure 5.14 using a two-dimensional example, in which the limit state function has been transformed into the U-space.

The basic idea of the algorithm is that it constructs a linear approximation to the limit state function at a search point, calculates the normalized gradients of the limit state function at

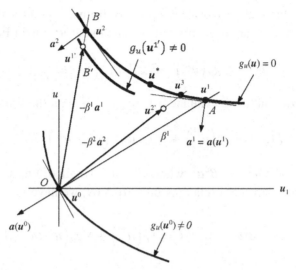

FIGURE 5.14: MPP search using the reliability index approach.

that search point using Eq. 5.64, and locates the next search point on the limit state function using a vector originating at the origin and pointing in a reverse direction from that of the gradient of the limit state function obtained at the current search point. The distance between the origin and the next search point identified is the β value at the search point. As illustrated in Figure 5.14, an initial search point u^0 is usually given as the mean value of the random variables—that is, at the origin O. It is apparent that u^0 is most likely not on the limit state function; therefore, $g_u(u^0) \neq 0$.

At this point, the β value is $\beta^0 = 0$. We calculate the normalized gradient of the limit state function at u^0 (i.e., $a^0 = a(u^0)$) using Eq. 5.64. We reverse the gradient vector to intersect it with the limit state function; the intersecting point A is assigned as the next search point u^1. Note that the distance between the origin and the search point u^1 (line OA) is β^1. We calculate the normalized gradient vector of the limit state function at u^1 (i.e., $a^1 = a(u^1)$), as shown in Figure 5.14, and move the gradient vector a^1 to the origin O, reversing its direction to intersect it with the limit state function; the intersecting point B is assigned as the next search point u^2. The process repeats until the search point approaches the MPP u^*.

One key step involved in the this process is calculating the search point u^{k+1} in the $(k + 1)$th iteration by intersecting vector a^k (directed at the origin) with the limit state function. Such a calculation can be extensive when the number of random variables becomes large. One possible way to avoid such calculations is outlined next.

Let us go back to Figure 5.14 and look more closely at iterations 1 and 2—that is, from points A to B. Note that u^1 (point A) is on the limit state function (i.e., $g_u(u^1) = 0$) and the length of the line segment OA is β^1. For the current iteration, it is obvious that $u^{1'} = -\beta^1 a^1$. If we locate a point $u^{1'} = -\beta^1 a^1$ from the origin O (as shown in Figure 5.14, the point is labeled as

B'), then $g_u(u^{1'}) \neq 0$. Our goal is to locate u^2 (point B) from $u^{1'}$ (point B') without calculating the intersection. First we linearize the limit state function at $u^{1'}$ (as in Eq. 5.58) as

$$L(u) = g_u\left(u^{1'}\right) + \nabla g_u\left(u^{1'}\right)\left(u - u^{1'}\right)^T \tag{5.68}$$

Equation 5.68 is not zero until we evaluate it at u^2, which is on the limit function:

$$L(u^2) = g_u\left(u^{1'}\right) + \nabla g_u\left(u^{1'}\right)\left(u^2 - u^{1'}\right)^T = 0 \tag{5.69}$$

Note that $u^{1'} = -\beta^1 a^1$ and $u^2 = -\beta^2 a^1$, where β^2 is the length of line segment OB, which is to be calculated. If we insert these relations into Eq. 5.69, we have

$$g_u\left(u^{1'}\right) + \nabla g_u\left(u^{1'}\right)\left(-\beta^2 a^1 + \beta^1 a^1\right)^T = g_u\left(u^{1'}\right) + \nabla g_u\left(u^{1'}\right)a^{1^T}\left(\beta^1 - \beta^2\right) = 0 \quad (5.70)$$

Thus,

$$\beta^2 = \beta^1 + \frac{g_u\left(u^{1'}\right)}{\nabla g_u(u^{1'})a^{1^T}} \tag{5.71}$$

Note that since everything on the right of Eq. 5.71 is known from the first iteration, β^2 can be readily calculated. We may generalize the iteration steps from k to $k + 1$, and Eq. 5.71 can be rewritten as

$$\beta^{k+1} = \beta^k + \frac{g_u\left(u^{k'}\right)}{\nabla g_u(u^{k'})a^{k^T}} \tag{5.72}$$

Therefore,

$$u^{k+1} = -a^k \beta^{k+1} = -a^k\left\{\beta^k + \frac{g_u\left(u^{k'}\right)}{\nabla g_u(u^{k'})a^{k^T}}\right\} \tag{5.73}$$

Several other algorithms, such as HL-RF (Hasofer and Lind 1974; Rackwitz and Fiessler 1978) and sequential quadratic programming (SQP) (see, for example, Boggs and Tolle 1995), are also popular for an MPP search.

EXAMPLE 5.6

We use the MPP search algorithm for the cantilever beam from Example 5.5. Recall that the normalized gradient vector of the limit state function is

$$a = \frac{\nabla g_u(u^*)}{||\nabla g_u(u^*)||} = \frac{\left[3.75\left(0.1u_d^* + 1\right)^2, -1\right]}{\sqrt{\left(3.75\left(0.1u_d^* + 1\right)^2\right)^2 + (-1)^2}}$$

Solution

We start from the mean value point; i.e., $u^{0'} = [0, 0]$, $\nabla g_u(u^{0'}) = [3.75, -1]$, $a^0 = [0.9662, -0.2577]$, and $\beta^0 = 0$. From Eqs. 5.72 and 5.73, we have

$$\beta^1 = \beta^0 + \frac{g_u(u^{0'})}{\nabla g_u(u^{0'})a^{0^T}} = \frac{2.500}{[3.75, -1][0.9662, -0.2577]^T} = 0.6442$$

Therefore,

$$u^1 = -\alpha^0 \beta^1 = -[0.9662, -0.2577](0.6442) = [-0.6224, 0.1660]$$

and

$$\alpha^1 = [0.9570, -0.2902]$$

$$u^{1'} = -\beta^1 \alpha^1 = -(0.6442)[0.9570, -0.2902] = [-0.6164, 0.1869]$$

The steps are repeated, and the data obtained are listed in Table 5.3. Note, between iterations 2 and 3, the normalized gradient vectors a and β values are identical (up to four digits) so the solution converges. The MPP found is identical to that found in Example 5.5. A MATLAB script employed for the computation is shown for reference.

Table 5.3: Numerical data obtained for the MPP search.

Iteration	u	a	β	u'
0	0,0	0.9662, −0.2577	0	
1	−0.6224, 0.1660	0.9570, −0.2902	0.6442	−0.6164, 0.1869
2	−0.6556, 0.1986	0.9564, −0.2921	0.6852	−0.6552, 0.2001
3	−0.6553, 0.2001	0.9564, −0.2921	0.6852	

```
%Matlab script

% initial iteration (from mean value point)

u=[0,0]

beta = 0

gu=12.5*(0.1*u(1)+1)^3-(u(2)+10)

vectora=[3.75*(0.1*u(1)+1)^2,-1]

a=normr(vectora)

beta=beta+gu/(vectora*a')

u=-a*beta

%Follow-on iterations

vectora=[3.75*(0.1*u(1)+1)^2,-1]

a=normr(vectora)

up=-a*beta
```

```
gu=12.5*(0.1*up(1)+1)^3-(up(2)+10)

vectora=[3.75*(0.1*up(1)+1)^2,-1]

a=normr(vectora)

beta=beta+gu/(vectora*a')

u=-a*beta
```

The Performance Measure Approach

As discussed previously, in the reliability index approach we are searching the MPP along the limit state function where a target performance is prescribed. The point on the limit state function that is the shortest distance to the origin is the MPP. Once the MPP is found, the distance β (reliability index) can be used to approximate the failure probability as $P_f = \Phi(-\beta)$.

The performance measure approach (PMA), on the other hand, is given a target reliability index β (or failure probability P_f and then $\beta = (\Phi^{-1}(P_f))$, and searches for the MPP by bringing the function $g_u(u)$ closer to $g_u(u) = 0$, in which the target performance level is achieved. The concept is illustrated in Figure 5.15a using a two-dimensional example. The required reliability index β is shown as a circle centered at the origin of the u_1-u_2-plane with

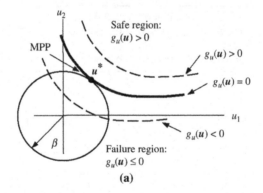

(a)

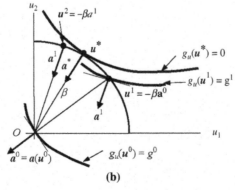

(b)

FIGURE 5.15: MPP search using PMA: (a) concept illustration and (b) MPP search scheme.

radius β. Depending on the u value entered for the MPP search, the limit state function $g_u(u)$ is usually nonzero. If the u value entered is in the safe region, then $g_u(u) > 0$.

On the other hand, if u is in the failure region, $g_u(u) < 0$. The MPP search becomes one that searches the u to brings the limit state function $g_u(u)$ to $g_u(u) = 0$. The problem can be formulated as following

$$\begin{cases} \text{minimize:} & |g_u(u)| \\ \text{subject to:} & \beta = ||u|| \end{cases} \tag{5.74}$$

in which the reliability index $\beta = \Phi^{-1}(P_f)$, and P_f is the required failure probability. Also, in Eq. 5.74 the limit state function $g_u(u)$ can be rewritten as

$$g_u(u) = g'(u) - g^t \tag{5.75}$$

where

    $g'(u)$ is the performance measure corresponding to the failure mode.
    g^t is the target performance level.

As seen in Example 5.5, the yield strength in the limit state function of the stress failure mode is specified as 50 ksi, which is the target performance level g^t. Note that during the MPP search the search point u^i at the ith iteration yields a nonzero limit state function:

$$e(u^i) = |g'(u^i) - g^t| > 0 \tag{5.76}$$

where $e(u^i)$ represents the quantity to be minimized to bring the curves shown in Figure 5.15a closer to $g_u(u) = 0$, in which the MPP is determined.

We discuss one search algorithm similar to that discussed earlier, using a recursive formula. The algorithm is illustrated in Figure 5.15b with a 2D example in which the limit state function has been transformed into the U-space.

Similar to our earlier discussion, an initial search point u^0 is usually given as the mean value of the random variables—that is, at the origin O. It is apparent that u^0 is most likely not on the limit state function; therefore, $g_u(u^0) = g^0 \neq 0$. At this point, the β value is $\beta^0 = 0$. We calculate the normalized gradient at u^0 (i.e., $a^0 = a(u^0)$) using Eq. 5.64. We locate the next search point u^1 by simply setting $u^1 = -\beta a^0$ as shown in Figure 5.15b. Next we calculate the normalized gradient vector of the limit state function at u^1 (i.e., $a^1 = a(u^1)$) and then locate the next search point u^2, again by setting $u^2 = -\beta a^1$. The process can be generalized for the iterations k to $(k+1)$:

$$u^{k+1} = -\beta \alpha^k \tag{5.77a}$$

and

$$\alpha^k = \frac{\nabla g_u(u^k)}{||\nabla g_u(u^k)||} \tag{5.77b}$$

The process is repeated until the search point approaches the MPP u^*.

EXAMPLE 5.7

We illustrate the MPP search using PMA, as discussed earlier, with the same cantilever beam of Example 5.6. Recall that the required reliability index value at the MPP is $\beta = 0.6852$.

Solution

As before, we start from the mean value point, where $u^0 = [0, 0]$, $g(u^0) = 2.5$, $\nabla g_u(u^0) = [3.75, -1]$, and $a^0 = [0.9662, -0.2577]$. From Eq. 5.77a, we have
$$u^1 = -\beta a^0 = -(0.6852)[0.9662, -0.2577] = [-0.6621, 0.1766]$$
and

$$g(u^1) = 0.0015$$

which is nonzero so the process continues. At the search point u^1, we calculate the normalized gradient vector $a^1 = [0.9563, -0.2925]$; then

$$u^2 = -\beta a^1 = -(0.6852)[0.9563, -0.2925] = [-0.6552, 0.2004]$$

and

$$g(u^2) = 0.00006278$$

which is very close to zero so the process terminates. The MPP is found at $u^* = u^2 = [-0.6552, 0.2004]$, which is very close to the results obtained in Example 5.6.

As illustrated, PMA seems to offer a simpler way to search for the MPP than RIA. It takes fewer calculations in each search iteration since, unlike RIA, it does not need to calculate $u^{i'}$. The advantage of PMA is more significant for reliability-based design optimization (RBDO) in which an objective function (such as structure weight) is to be minimized subject to probabilistic constraints. These probabilistic constraints are often written in terms of reliability index β if FORM is employed for reliability analysis. All probabilistic constraints must be satisfied when an optimal design is found. In each design iteration, they must be evaluated, and those violated must be corrected in successive design iterations.

If RIA is used, Eq. 5.67 must be solved for all probabilistic constraints to see if they are violated by comparing their respective β values obtained at the MPP with the respective target reliability indices. In this case, the MPP search must be carried out for all probabilistic constraints. However, if PMA is used, this search is required only for those that have been violated. This is because a negative value of the limit state function indicates that a failure has occurred, so the sign of the limit state function can be used as a measure to determine whether a probabilistic constraint is satisfied—without actually obtaining a reliability index value by going through the MPP search. For this reason, PMA is more attractive computationally for support of reliability-based design, as reported by Lee et al. (2002). More discussion on this subject can be found in *Design Theory and Methods using CAD/CAE*, a book in The Computer Aided Engineering Design Series.

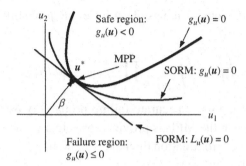

FIGURE 5.16: Comparison of FORM and SORM using a two-dimensional example.

5.4.4 The Second-Order Reliability Method

For a limit state function of high curvature at the MPP, first-order approximation may not provide adequate accuracy in failure probability calculation. The reason is that the probability integration of Eq. 5.44 (or Eq. 5.56) using FORM ignores the volume underneath the surface of the probability density function and the area between the true limit state function $g_u(u) = 0$ and the linearized function $L_u(u) = 0$. This is illustrated in Figure 5.16.

For limit state functions with a large curvature at the MPP, the second-order reliability method offers a more accurate approximation of failure probability. Instead of discussing the mathematics of SORM, we briefly go over a few key equations to understand how the method works.

SORM uses the second-order Taylor series expansion to approximate the limit state function at the MPP u^*. The approximation is given as

$$g_u(u) \approx q(u) = g_u(u^*) + \nabla g_u(u^*)(u - u^*)^T + \frac{1}{2}(u - u^*)H(u^*)(u - u^*)^T \qquad (5.78)$$

where $H(u^*)$ is the Hessian matrix of $n \times n$ at the MPP:

$$H(u^*) = \begin{bmatrix} \dfrac{\partial^2 g_u}{\partial u_1^2} & \dfrac{\partial^2 g_u}{\partial u_1 \partial u_2} & \cdots & \dfrac{\partial^2 g_u}{\partial u_1 \partial u_n} \\[2mm] \dfrac{\partial^2 g_u}{\partial u_2 \partial u_1} & \dfrac{\partial^2 g_u}{\partial u_2^2} & \cdots & \dfrac{\partial^2 g_u}{\partial u_2 \partial u_n} \\[2mm] & & \cdots & \\[2mm] \dfrac{\partial^2 g_u}{\partial u_n \partial u_1} & \dfrac{\partial^2 g_u}{\partial u_n \partial u_2} & \cdots & \dfrac{\partial^2 g_u}{\partial u_n^2} \end{bmatrix}_{u^*} \qquad (5.79)$$

where n is the number of random variables. Therefore, the probability integration becomes, according to Eq. 5.56,

$$P_f = \int\int \cdots \int\int_{q_u(u) \leq 0} \prod_{i=1}^{n} \frac{1}{\sqrt{2\pi}} e^{-\frac{1}{2}u_i^2} du_1 du_2 \ldots du_n \tag{5.80}$$

As illustrated in Figure 5.16, a more accurate failure probability estimate, especially for a function of large curvature at the MPP, is provided by integration using the boundary curve $q_u(u) = 0$, which is a better approximation than the straight line $L_u(u) = 0$ used in FORM for the true limit state function $g_u(u) = 0$.

Although the limit state function is simplified using a quadratic function $q_u(u)$, Eq. 5.80 is still difficult to calculate, so an approximation has been derived (e.g., Wei 2006). When β is large enough, an asymptotic solution of the failure probability can be derived as

$$P_f = \Phi(-\beta) \prod_{i=1}^{n-1} \left(\frac{1}{\sqrt{1 + \beta k_i}} \right) \tag{5.81}$$

where k_i is the ith main curvature of the limit state function $g_u(u)$ at the MPP.

EXAMPLE 5.8

We calculate the failure probability of the cantilever beam of Examples 5.6 and 5.7 using SORM. Recall that the limit state function of the cantilever beam is

$$g_u(u) = g_u(u_d, u_\ell) = 12.5(0.1u_d + 1)^3 - (u_\ell + 10) = 0$$

The MPP is found at $u^* = (-0.6552, 0.2004)$, and β is 0.6852.

Solutions
The Hessian matrix can be found using Eq. 5.79 as

$$H(u^*) = \begin{bmatrix} \dfrac{\partial^2 g_u}{\partial u_d^2} & \dfrac{\partial^2 g_u}{\partial u_d \partial u_\ell} \\ \dfrac{\partial^2 g_u}{\partial u_\ell \partial u_d} & \dfrac{\partial^2 g_u}{\partial u_\ell^2} \end{bmatrix}_{u^*} = \begin{bmatrix} 0.75(0.1u_d + 1) & 0 \\ 0 & 0 \end{bmatrix}_{u^*=[-0.6552, 0.2004]} = \begin{bmatrix} 0.7009 & 0 \\ 0 & 0 \end{bmatrix}$$

The quadratic approximation given in Eq. 5.78 can be obtained as

$$q(u) = g_u(u^*) + \nabla g_u(u^*)(u - u^*)^T + \frac{1}{2}(u - u^*)H(u^*)(u - u^*)^T$$

$$= 0 + [3.275, -1] \begin{bmatrix} u_d + 0.6553 \\ u_\ell - 0.2001 \end{bmatrix}$$

$$+ \frac{1}{2}[u_d + 0.6553, u_\ell - 0.2001] \begin{bmatrix} 0.7009 & 0 \\ 0 & 0 \end{bmatrix} \begin{bmatrix} u_d + 0.6553 \\ u_\ell - 0.2001 \end{bmatrix}$$

$$= 3.275(u_d + 0.6553) - (u_\ell - 0.2001) + \frac{1}{2}(0.7009)(u_d + 0.6553)^2$$

$$= 0.3505u_d^2 + 3.734u_d - u_\ell + 2.490$$

Again, bringing this function into Eq. 5.80 for failure probability calculation is very difficult. We use Eq. 5.81 for the calculation instead. For this two-dimensional problem, Eq. 5.81 can be reduced as

$$P_f = \Phi(-\beta) \prod_{i=1}^{n-1} \left(\frac{1}{\sqrt{1 + \beta \kappa_i}} \right) = \Phi(-\beta) \left(\frac{1}{\sqrt{1 + \beta \kappa_1}} \right)$$

The curvature for a two-dimensional curve given implicitly by $g_u(u_d, u_\ell) = 0$ is given by

$$\kappa_1 = \frac{\dfrac{\partial^2 g_u}{\partial u_d^2} \left(\dfrac{\partial g_u}{\partial u_\ell} \right)^2 - 2 \dfrac{\partial^2 g_u}{\partial u_\ell \partial u_d} \dfrac{\partial g_u}{\partial u_d} \dfrac{\partial g_u}{\partial u_\ell} + \dfrac{\partial^2 g_u}{\partial u_\ell^2} \left(\dfrac{\partial g_u}{\partial u_d} \right)^2}{\left(\left(\dfrac{\partial g_u}{\partial u_d} \right)^2 + \left(\dfrac{\partial g_u}{\partial u_\ell} \right)^2 \right)^{\frac{3}{2}}} \Bigg|_{u^*} = \frac{0.7009(1)^2 - 2(0) + 0}{\left((3.275)^2 + (-1)^2 \right)^{\frac{3}{2}}} = 0.0175$$

Thus, the failure probability can be obtained as

$$P_f^{SORM} = \Phi(-\beta) \left(\frac{1}{\sqrt{1 + \beta \kappa_1}} \right) = \Phi(-0.6852) \left(\frac{1}{\sqrt{1 + (0.6852)(0.0175)}} \right) = 0.2451$$

Recall that the failure probability calculated using FORM in Example 5.5 is $P_f^{FORM} = 0.2466$. If we use Monte Carlo simulation for 100,000 sample points, the failure probability $P_f^{MCS} = 0.2450$. The failure probability calculated using SORM is $P_f^{SORM} = 0.2451$, which is closer to that of Monte Carlo simulation, so it is considered more accurate than FORM. However, the difference in results obtained from FORM and SORM is not significant because for this problem the curvature of the limit state function at the MPP is small. For cases like this, as illustrated in Figure 5.16, FORM provides an excellent failure probability estimate.

5.4.5 Transformation of Random Variables

So far in our discussion, we have assumed that all random variables are independent and are normally distributed. They can be easily transformed into those of standard normal distribution using Eqs. 5.51 and 5.52. Methods such as FORM and SORM assume random variables that are independent with standard normal distribution.

In engineering applications, random variables are often correlated and reveal non-normal distributions. To employ FORM or SORM for failure probability calculation, these variables must be transformed into those of independent and standard normal distribution. As discussed before, such transformations can be written mathematically (see Eq. 5.50a) as

$$u_i = \Phi^{-1}(F_{X_i}(x_i)) \tag{5.82}$$

where $\Phi(\bullet)$ is the CDF of the standard normal distribution.

To apply FORM or SORM for failure probability calculation, transformation of random variables is indispensable. It is an important topic that must be addressed carefully. In this subsection, we discuss transformation for three types of random variable: those that are correlated with normal distribution, those that are independent with non-normal distribution, and those that are correlated with non-normal distribution.

Correlated Random Variables of Normal Distribution

Let X be a vector of correlated random variables $X = [X_1, X_2, ..., X_n]^T$ with joint probability density function $f_X(x)$ that are of normal distribution. The elements in the vectors of expected values and the covariance matrix are, respectively, $\mu_i = E[X_i]$, $i = 1, n$, and $C_{ij} = \text{Cov}[X_i, X_j]$, $i, j = 1, n$, which can be written in a matrix form as

$$C_X = \begin{bmatrix} \text{Var}[X_1] & ... & & ... & \text{Symmetric} \\ \text{Cov}[X_1, X_2] & \text{Var}[X_2] & & ... & \\ \text{Cov}[X_1, X_n] & \text{Cov}[X_2, X_n] & & ... & \text{Var}[X_n] \end{bmatrix}_{n \times n} \tag{5.83}$$

where

$$\begin{aligned} \text{Cov}(X_i, X_j) &= E\big[(X_i - \mu_i)(X_j - \mu_j)\big] \\ &= \int\limits_{-\infty}^{\infty} \int\limits_{-\infty}^{\infty} (x_i - \mu_i)(x_j - \mu_j) f_{X_i X_j}(x_i, x_j) dx_i dx_j \end{aligned} \tag{5.84a}$$

and, for $i = j$,

$$\text{Cov}(X_i, X_i) = E\Big[(X_i - \mu_i)^2\Big] = \text{Var}[X_i] = \sigma_i^2 \tag{5.84b}$$

in which σ_i is the standard deviation of the random variable X_i. The correlation coefficient between X_i and X_j, as discussed in Section 5.3 (Eq. 5.34), is

$$\rho_{ij} = \frac{c_{ij}}{\sigma_i \sigma_j}; \quad i, j = 1, n; \quad -1 \le \rho_{ij} \le 1 \tag{5.85}$$

which can be written in matrix form as

$$\rho_X = \begin{bmatrix} 1 & ... & \text{Symmetric} & \\ \rho_{21} & 1 & ... & \\ & & ... & \\ \rho_{n1} & \rho_{n2} & ... & 1 \end{bmatrix}_{n \times n} \tag{5.86}$$

Transformation of correlated random variables involves two steps:

Step 1: Transform random variables X into Y, in which $Y = [Y_1, Y_2, ..., Y_n]^T$ is a vector of random variables of standard normal distribution (i.e., $Y_i \sim N(1, 0)$ for $i = 1, n$).
Step 2: Transform random variables Y of correlated standard normal distribution into $U = [U_1, U_2, ..., U_n]^T$, which are random variables of uncorrelated (independent) standard normal distribution.

The step 1 transformation can be carried out as before simply using

$$y_i = \frac{x_i - \mu_{x_i}}{\sigma_{x_i}}, i = 1, n \tag{5.87}$$

It is important to point out that the covariance matrix of Y (that is, C_Y, as defined in Eq. 5.83) is equal to the correlation coefficient matrix of X (that is, ρ_X, as defined in Eq. 5.88):

$$C_Y = \rho_X \tag{5.88}$$

Proof of Eq. 5.88 is straightforward and therefore left as an exercise.

Step 2 transforms random variables Y of correlated standard normal distribution into $U = [U_1, U_2, ..., U_n]^T$, which are random variables of uncorrelated (independent) standard normal distribution. This step can be written as

$$Y = TU \tag{5.89}$$

where T is a lower triangular matrix (i.e., $T_{ij} = 0$ for $j > i$). Using Eq. 5.88, the covariance matrix C_Y for Y can be written as

$$C_Y = E[YY^T] = E[TUU^TT^T] = TE[UU^T]T^T = TT^T = \rho_X \tag{5.90}$$

The elements in T can be determined from $TT^T = \rho_X$ as

$$T_{jj} = \sqrt{1 - \sum_{k=1}^{j-1} T_{jk}^2}, j = 1, n \tag{5.91a}$$

and

$$T_{ij} = \frac{\rho_{ji} - \sum_{k=1}^{j-1} T_{jk}T_{ik}}{T_{jj}}, i > j \tag{5.91b}$$

Thus, the transformation from X to U can be obtained in matrix form as

$$X = \mu_X + DTU \tag{5.92}$$

where D is a diagonal matrix with standard deviations σ_{X_i} in the diagonal.

Using Eq. 5.92, the limit state function can be written as $g(X) = g_u(\mu_X + DTU) = 0$. Methods such as FORM or FORM can then be employed for failure probability calculation. Derivation of Eqs. 5.91a and 5.91b is straightforward and is left as an exercise. A simple example illustrates the transformation.

EXAMPLE 5.9

A limit state function is defined as

$$g(x) = x_1 - x_2 + x_3^2$$

which consists of three random variables X_1, X_2, and X_3 of normal distribution with respective mean values and standard deviations $X_1 \sim N(25, 0.25)$, $X_2 \sim N(4, 0.2)$, and $X_3 \sim N(2, 0.1)$. The correlated coefficient of the random variables $X = [X_1, X_2, X_3]^T$ is given as

$$\rho_X = \begin{bmatrix} 1 & 0.5 & 0.2 \\ 0.5 & 1 & 0.4 \\ 0.2 & 0.4 & 1 \end{bmatrix}_{3 \times 3}$$

We want to write the limit state function in terms of independent random variables U of standard normal distribution.

Solution
We start by transforming the random variables X into U as discussed previously. Step 1 involves transforming X into Y, which are random variables of standard normal distribution; that is,

$$Y_1 = \frac{X_1 - 25}{0.25}, \; Y_2 = \frac{X_2 - 4}{0.2}, \text{ and } Y_3 = \frac{X_3 - 2}{0.1}$$

As we saw earlier, random variables Y are transformed into U in step 2, following Eqs. 5.91a and 5.91b. From the given correlated coefficient matrix ρ_X, we have

$$T_{11} = 1, \; T_{21} = \rho_{12} = 0.5$$

and

$$T_{31} = \rho_{13} = 0.2$$

Then

$$T_{22} = \sqrt{1 - T_{21}^2} = \sqrt{1 - 0.5^2} = 0.866$$

$$T_{32} = \frac{\rho_{23} - T_{21}T_{31}}{T_{22}} = \frac{0.4 - (0.5)(0.2)}{0.866} = 0.346$$

$$T_{32} = \sqrt{1 - T_{31}^2 - T_{32}^2} = \sqrt{1 - 0.2^2 - 0.346^2} = 0.919$$

Thus, the transformation matrix T is

$$T = \begin{bmatrix} 1 & 0 & 0 \\ 0.5 & 0.866 & 0 \\ 0.2 & 0.346 & 0.919 \end{bmatrix}_{3 \times 3}$$

Now, since $Y = TU$, we can find U by

$$U = T^{-1}Y = \begin{bmatrix} 1 & 0 & 0 \\ 0.5 & 0.866 & 0 \\ 0.2 & 0.346 & 0.919 \end{bmatrix}^{-1} \begin{bmatrix} Y_1 \\ Y_2 \\ Y_3 \end{bmatrix} = \begin{bmatrix} Y_1 \\ -0.577Y_1 + 1.15Y_2 \\ -0.000251Y_1 - 0.435Y_2 + 1.09Y_3 \end{bmatrix}$$

And we can express X in terms of U, using Eq. 5.92:

$$X = \mu_X + DTU = \begin{bmatrix} 25 \\ 4 \\ 2 \end{bmatrix} + \begin{bmatrix} 0.25 & 0 & 0 \\ 0 & 0.2 & 0 \\ 0 & 0 & 0.1 \end{bmatrix} \begin{bmatrix} 1 & 0 & 0 \\ 0.5 & 0.866 & 0 \\ 0.2 & 0.346 & 0.919 \end{bmatrix} \begin{bmatrix} U_1 \\ U_2 \\ U_3 \end{bmatrix}$$

$$= \begin{bmatrix} 25 + 0.25U_1 \\ 4 + 0.1U_1 + 0.173U_2 \\ 2 + 0.02U_1 + 0.0346U_2 + 0.0919U_3 \end{bmatrix}$$

Thus, the limit state function can be written in U as

$$g_u(u) = (25 + 0.25U_1)$$
$$- (4 + 0.1U_1 + 0.173U_2) + (2 + 0.02U_1 + 0.0346U_2 + 0.0919U_3)^2$$

Independent Random Variables of Non-Normal Distribution

Transformation of independent random variables X of non-normal distribution to random variables U of standard normal distribution is straightforward. Since there is no correlation between X_i and X_j, the two can be transformed individually using Eq. 5.50a:

$$x_i = F_{X_i}^{-1}(\Phi(u_i)), i = 1, n$$

Then the limit state function can be written in terms of U as

$$g(x) = g(x_1, x_2, \ldots, x_n) = g_u\left(F_{X_1}^{-1}(\Phi(u_1)), F_{X_2}^{-1}(\Phi(u_2)), \ldots, F_{X_n}^{-1}(\Phi(u_n))\right) = 0$$

For example, if X is a random variable of lognormal distribution (i.e., $X \sim LN(\mu_X, \sigma_X)$), X can be transformed into a random variable U of standard normal distribution following the definition of lognormal distribution shown in Eqs. 5.63b and 5.63c. In other words, the cumulative distribution function of a lognormal distribution is found to be

$$F_X(x) = \Phi\left(\frac{\ln(x) - \mu_L}{\sigma_L}\right) \int_{-\infty}^{\ln(x)} \frac{1}{\sqrt{2\pi\sigma_y}} e^{-\frac{1}{2}\left(\frac{s-\mu_L}{\sigma_L}\right)^2} ds \qquad (5.93a)$$

in which

$$\sigma_L = \sqrt{\ln\left(\left(\frac{\sigma_X}{\mu_X}\right)^2 + 1\right)}, \quad \mu_L = \ln(\mu_X) - \frac{1}{2}\sigma_L^2 \qquad (5.93b)$$

Thus, the transformation is simply

$$x = F_X^{-1}(\Phi(u)) \qquad (5.93c)$$

Therefore,

$$u = \frac{\ln(x) - \mu_L}{\sigma_L} \qquad (5.94)$$

and

$$x = e^{\mu_L + u\sigma_L} \qquad (5.95)$$

For a random variable X with a Gumbel distribution (also called a type 1 extreme value distribution)—$X \sim EXI(\mu_X, \sigma_X)$—the CDF as defined in Section 5.3 (Eq. 5.38b) is

$$F_X(x) = e^{-e^{-a(x-b)}}, \quad -\infty < x < \infty, a > 0 \qquad (5.96)$$

where a and b are scale and location parameters, respectively. For

$$a = \frac{\pi}{\sqrt{6}\sigma_X}, b = \mu_X - \frac{0.5772}{a}$$

the random variable X can be transformed into U as follows. First, we equate the CDF of X with a standard normal CDF of U by

$$F_X(x) = e^{-e^{-a(x-b)}} = \Phi(u) \qquad (5.97)$$

from which we obtain

$$x = b - \frac{1}{a}\ln(-\ln(\Phi(u))) \tag{5.98}$$

For random variables with a CDF other than lognormal or Gumbel, we follow the same approach discussed earlier for such transformations.

EXAMPLE 5.10

A limit state function is defined as

$$g(x) = x_1 + x_2^2$$

which consists of two random variables X_1 and X_2. X_1 has a lognormal distribution with mean value 20 and standard deviation 5 (i.e., $X_1 \sim LN(20, 5)$). X_2 has a Gumbel distribution with mean value 1 and standard deviation 0.1 (i.e., $X_2 \sim EXI(1, 0.1)$). These two random variables are independent. We want to write the limit state function in terms of independent random variables U of standard normal distribution.

Solution

Using Eqs. 5.95 and 5.98, X_1 and X_2 can be transformed, respectively, into U_1 and U_2:

$$x_1 = e^{\mu_L + u_1 \sigma_L} = e^{2.97 + 0.2462 u_1}$$

where

$$\sigma_L = \sqrt{\ln\left(\left(\frac{\sigma_X}{\mu_X}\right)^2 + 1\right)} = \sqrt{\ln\left(\left(\frac{5}{20}\right)^2 + 1\right)} = 0.2462$$

and

$$\mu_L = \ln(\mu_X) - \frac{1}{2}\sigma_L^2 = \ln(20) - \frac{1}{2}(0.2462)^2 = 2.965$$

$$x_2 = b - \frac{1}{a}\ln(-\ln(\Phi(u_2))) = 0.9550 - 0.07794\ln(-\ln(\Phi(u_2)))$$

where

$$a = \frac{\pi}{\sqrt{6}\sigma_X} = \frac{\pi}{\sqrt{6} \times 0.1} = 12.83, b = \mu_X - \frac{0.5772}{a} = 1 - \frac{0.5772}{12.83} = 0.9550$$

Hence, the limit state function can be written in U:

$$g_u(u) = \left(e^{2.97 + 0.2462 u_1}\right) + (0.9550 - 0.07794\ln(-\ln(\Phi(u_2))))^2$$

Correlated Random Variables of Non-Normal Distribution

Transformation of correlated random variables of non-normal distribution is more involved than the transformations just discussed. There are two widely used techniques for this: the Rosenblatt transformation (Rosenblatt 1952) and the Nataf transformation (Nataf 1962, Liu and Der Kiureghian 1986). We briefly discuss only the Rosenblatt transformation with an example to provide some basic understanding. Interested readers may refer, for example, to Hurtado (2004) for a more in-depth discussion.

Transformation of correlated random variables X of non-normal distribution to the U-space of uncorrelated random variables U of normalized normal distribution can be defined as

$$\begin{cases} x_1 = F_{X_1}^{-1}(\Phi(u_1)) \\ x_2 = F_{X_2|X_1}^{-1}(\Phi(u_2)|X_1 = x_1) \\ \quad \cdots \\ x_n = F_{X_n|X_1\ldots X_{n-1}}^{-1}(\Phi(u_n)|X_1 = x_1, \ldots X_{n-1} = x_{n-1}) \end{cases} \tag{5.99}$$

where $F_{X_i|X_1,\ldots,X_{i-1}}^{-1}(\Phi(u_i)|X_1 = x_1, \ldots X_{i-1} = x_{i-1})$ is the distribution of X_i given $X_1 = x_1, \ldots, X_{i-1} = x_{i-1}$. The conditional distribution function $F_{X_i|X_1,\ldots X_{i-1}}(x_1, \ldots, x_{i-1})$ can be written as

$$F_{X_i|X_1,\ldots X_{i-1}}(x_i|x_1, \ldots, x_{i-1}) = \frac{\displaystyle\int_{-\infty}^{x_i} f_{X_1\ldots X_{i-1}X_i}(x_1, \ldots, x_{i-1}, s)ds}{f_{X_1\ldots X_{i-1}}(x_1, \ldots, x_{i-1})} \tag{5.100}$$

where $f_{X_1\ldots X_i}(x_1, \ldots, x_i)$ is the joint probability density function of random variables $X_1, \ldots, X_i$.

Note that the probability density functions and cumulative distribution functions $f_{X_1,\ldots X_i}(x_1, \ldots, x_i)$ and $F_{X_i|X_1,\ldots X_{i-1}}(x_1, \ldots, x_{i-1})$ must be calculated for each ith random variable before Eq. 5.99 can be employed for the transformation. The transformation starts by calculating $f_{X_1}(x_1)$ and $F_{X_1}(x_1)$; then x_1 can be obtained as $x_1 = F_{X_1}^{-1}(\Phi(u_1))$. Once x_1 in terms of u_1 is obtained, $F_{X_2|X_1}(x_2|x_1)$ can be calculated using $f_{X_1}(x_1)$ and $f_{X_1,\ldots X_n}(x_1, \ldots, x_n)$ following Eq. 5.100, and then x_2 can be obtained as $x_2 = F_{X_2|X_1}^{-1}(\Phi(u_2)|X_1 = x_1)$. The following example illustrates the computation process.

EXAMPLE 5.11

This example has been slightly modified from that given in Ditlevsen and Madsen (1996). A limit state function is defined as

$$g(x) = 2x_1^2 + \ln(1 + x_2) - (1 + x_1)x_2$$

which consists of two random variables X_1 and X_2. Also, X_1 and X_2 are correlated with a joint distribution function

$$F_{X_1X_2}(x_1,x_2) = 1 - e^{-x_1} + e^{-(x_1+x_2+x_1x_2)}, x_1 > 0, x_2 > 0$$

The joint probability density function $f_{X_1X_2}(x_1,x_2)$ can be obtained by

$$f_{X_1X_2}(x_1,x_2) = \frac{\partial^2 F_{X_1X_2}(x_1,x_2)}{\partial x_1 \partial x_2} = (x_1 + x_2 + x_1x_2)e^{-(x_1+x_2+x_1x_2)}, \ x_1 > 0, x_2 > 0$$

Solution

We first calculate the probability and distribution functions for these random variables. Those for X_1 can be calculated, respectively, as

$$f_{X_1}(x_1) = \int_0^\infty f_{X_1X_2}(x_1,x_2)dx_2 = \int_0^\infty (x_1 + x_2 + x_1x_2)e^{-(x_1+x_2+x_1x_2)}dx_2$$

$$= e^{-x_1}\left[x_1 \int_0^\infty e^{-(1+x_1)x_2}dx_2 + (1+x_1)\int_0^\infty x_2 e^{-(1+x_1)x_2}dx_2 \right]$$

$$= e^{-x_1}\left[\frac{x_1}{1+x_1} + \frac{1+x_1}{(1+x_1)^2} \right] = e^{-x_1}$$

and

$$F_{X_1}(x_1) = \int_0^{x_1} f_{X_1}(s)ds = \int_0^{x_1} e^{-s}ds = e^{-s}\big|_0^{x_1} = 1 - e^{-x_1}$$

Those for X_2 given X_1 can be calculated, respectively, as

$$f_{X_2}(x_2|x_1) = \frac{f_{X_1X_2}(x_1,x_2)}{f_{X_1}(x_1)} = \frac{(x_1 + x_2 + x_1x_2)e^{-(x_1+x_2+x_1x_2)}}{e^{-x_1}}$$

$$= (x_1 + x_2 + x_1x_2)e^{-(1+x_1)x_2}$$

with the corresponding CDF

$$F_{X_2|X_1}(x_2|x_1) = \int_0^{x_2} (x_1 + s + x_1s)e^{-(1+x_1)s}ds$$

$$= \int_0^{x_2} \left(x_1 e^{-(1+x_1)s} + (1+x_1)se^{-(1+x_1)s} \right)ds = 1 - (1+x_2)e^{-(1+x_1)x_2}$$

Following Eq. 5.99, we have

$$F_{X_1}(x_1) = 1 - e^{-x_1} = \Phi(u_1)$$

Thus,

$$x_1 = -\ln(1 - \Phi(u_1))$$

Now, for X_2, we have

$$F_{X_2|X_1}(x_2|x_1) = 1 - (1 + x_2)e^{-(1+x_1)x_2} = \Phi(u_2)$$

Hence,

$$\ln(1 + x_2) - (1 + x_1)x_2 = -\ln(1 - \Phi(u_2))$$

Therefore, the limit state function can be written in U as

$$g(\boldsymbol{x}) = 2x_1^2 + \ln(1 + x_2) - (1 - x_1)x_2 = 2(\ln(1 - \Phi(u_1)))^2 - \ln(1 - \Phi(u_2))$$

5.4.6 Importance Sampling

As shown previously, the computation cost of Monte Carlo simulation is very high, especially for applications of small failure probability, which are common in design engineering. The reason is that only samples that fall into the failure region contribute to the failure probability estimate. If most sample points are taken with a distribution that concentrates in an area in the safe zone and is away from the limit state function, the number of sample points used for Monte Carlo simulation must be very large to obtain an accurate failure probability estimate.

To improve the computational efficiency of Monte Carlo simulation, importance sampling methods are commonly employed. The idea of important sampling is to sample the random variables according to an alternative set of distributions such that more sample points fall into the failure region. As a result, more sample points contribute to the probability estimation and so fewer overall sample points are required for a desired accuracy. This concept is illustrated in Figure 5.17, in which the sampling is based on the MPP. We introduce the MPP-based importance sampling method in this subsection.

As illustrated in Figure 5.17, sample points generated from the original distribution of random variables X_1 and X_2 fall into the safe region, in which the sampling is often centered on the variables' mean value point. If a new set of distributions of X_1 and X_2 is selected, for example with the mean value point moving to the MPP, many more sample points fall into the failure region. Since more sample points in a given simulation

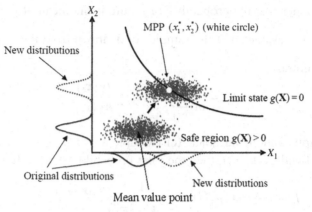

FIGURE 5.17: Importance sampling.

contribute to the failure probability estimate, a more accurate estimate can be expected with a smaller sample set.

After the MPP is obtained, samples are selected around the MPP to evaluate the probability of failure through importance sampling. To do so, an importance sampling density, $h_X(x)$, is introduced into the Monte Carlo estimation to obtain the probability of failure:

$$P_f = \int_{g(x) \leq 0} f_X(x)dx = \int I[g(x)]f_X(x)dx = \int \left\{ I[g(x)] \frac{f_X(x)}{h_X(x)} \right\} h_X(x)dx \qquad (5.101)$$

where the importance sampling density $h_X(x)$ is the same as $f_X(x)$ except that the mean values of X are replaced by the MPP $x^* = [x_1^*, x_2^*, \ldots, x_n^*]^T$. For example, if random variable X_1 is normally distributed with $N(\mu_1, \sigma_1)$ and

$$f_{X_1}(x_1) = \frac{1}{\sqrt{2\pi}\sigma_1} e^{-\frac{1}{2}\left(\frac{x_1 - \mu_1}{\sigma_1}\right)^2}$$

the corresponding importance sampling distribution is $N(x_1^*, \sigma_1)$ and

$$h_{X_1}(x) = \frac{1}{\sqrt{2\pi}\sigma_1} e^{-\frac{1}{2}\left(\frac{x_1 - x_1^*}{\sigma_1}\right)^2}$$

It is noted that the importance sampling density $h_{X_1}(x)$ is centered at x_1^*.

Equation 5.101 indicates that the probability of failure is the mean of the integrand $\left\{ I[g(x)] \dfrac{f_X(x)}{h_X(x)} \right\}$ that is evaluated at the samples of X drawn from the importance sampling density $h_X(x)$. Therefore,

$$P_f = \frac{1}{m} \sum_{j=1}^{m} I[g(x_j)] \frac{f_X(x_j)}{h_X(x_j)} \tag{5.102}$$

In the case of multiple random variables, for example $X = [X_1, X_2, \ldots, X_n]^T$, if they are all normally distributed with $N(\mu_i, \sigma_i)$, then the joint probability density function becomes

$$f_X(x) = \phi(x, \mu, C) = \frac{1}{\sqrt{|C|(2\pi)^n}} e^{-\frac{1}{2}(x-\mu)^T C^{-1}(x-\mu)} \tag{5.103}$$

where μ is an $n \times 1$ vector of the respective random variables' mean values, and C is the covariance matrix of $n \times n$, which is symmetric and positive definite, as discussed before. In this case, the corresponding importance sampling distribution is $N(x^*, \sigma)$ and

$$h_X(x) = \phi(x, x^*, C) = \frac{1}{\sqrt{|C|(2\pi)^n}} e^{-\frac{1}{2}(x-x^*)^T C^{-1}(x-x^*)} \tag{5.104}$$

We use the same cantilever beam to illustrate the method in Example 5.12.

EXAMPLE 5.12

We use MPP-based importance sampling on the cantilever beam discussed in this subsection. Recall from Example 5.8 that the MPP is found at $u^* = [u_d^*, u_\ell^*]^T = [-0.6552, 0.2004]^T$ and that the diameter and length random variables are $X_\ell \sim N(\mu_\ell, \sigma_\ell) = (10, 1)$ in. and $X_d \sim N(\mu_d, \sigma_d) = (1, 0.1)$ in. respectively. Also, the failure probability estimated using FORM and SORM are, respectively, $P_f^{FORM} = 0.2466$ and $P_f^{SORM} = 0.2451$. Note that the MPP is transformed back into the X-space as

$$x^* = [d^*, \ell^*]^T = [0.1u_d^* + 1, u_\ell^* + 10]^T = [0.9345, 10.2004]^T$$

The limit state function in the X-space is

$$g(X) = g(d, \ell) = 12.5d^3 - \ell$$

It is not necessary to transform the MPP and limit state function back into the X-space for importance sampling. The procedure shown next is applicable to the simulation in U-space as well.

Solution
From Eq. 5.103, the joint probability density function becomes

$$f_X(x) = \frac{1}{\sqrt{|C|(2\pi)^n}} e^{-\frac{1}{2}(x-\mu)^T C^{-1}(x-\mu)} = \frac{1}{\sqrt{(0.01)(2\pi)^2}} e^{-\frac{1}{2}\left(100(d-1)^2 + (\ell-10)^2\right)}$$

where the covariance matrix C is

$$C = \begin{bmatrix} \sigma_d^2 & 0 \\ 0 & \sigma_\ell^2 \end{bmatrix} = \begin{bmatrix} 0.1^2 & 0 \\ 0 & 1^2 \end{bmatrix} = \begin{bmatrix} 0.01 & 0 \\ 0 & 1 \end{bmatrix}$$

and

$$h_X(x) = \frac{1}{\sqrt{|C|(2\pi)^n}} e^{-\frac{1}{2}(x-x^*)^T C^{-1}(x-x^*)} = \frac{1}{\sqrt{(0.01)(2\pi)^2}} e^{-\frac{1}{2}\left(100(d-0.9345)^2 + (\ell-10.2004)^2\right)}$$

Thus, using MPP-based sampling, the failure probability can be estimated as

$$P_f = \frac{1}{m} \sum_{j=1}^m I[g(x_j)] \frac{f_X(x_j)}{h_X(x_j)} = \frac{1}{m} \sum_{j=1}^m I[g(x_j)] \frac{e^{-\frac{1}{2}\left(100(d_j-1)^2 + (\ell_j-10)^2\right)}}{e^{-\frac{1}{2}\left(100(d_j-0.9345)^2 + (\ell_j-10.2004)^2\right)}}$$

Table 5.4 shows the simulation results. The left column contains the number of sample points. The middle and right columns contain the failure probability obtained from Monte Carlo simulation and MPP-based importance sampling, respectively. As shown, 1000 sample points gives a reasonably accurate estimate for the failure probability using the MPP-based sampling method. The MATLAB scripts that perform the calculation are shown for reference.

Table 5.4: Numerical data to illustrate the importance sampling method.

m	P_f (Monte Carlo)	P_f (MPP-based sampling)
100	0.2200	0.2165
1000	0.2640	0.2368
10,000	0.2422	0.2429
100,000	0.2444	0.2443
1,000,000	0.2437	

```
% Monte Carlo simulation

m=10000

d=normrnd(1,0.1,1,m)

l=normrnd(10,1,1,m)

n=0

for i=1:1:m

 if (12.5*d(i)^3-l(i)) < 0
```

```
% MPP-based sampling

m=100000

d=normrnd(0.9345,0.1,1,m)
```

```
l=normrnd(10.2004,1,1,m)

n=0

for i=1:1:m

g=12.5*d(i)^3-l(i)

 if g < 0

 f= exp(-0.5*(100*(d(i)-1)^2+(l(i)-10)^2))

 h= exp(-0.5*(100*(d(i)-0.9345)^2+(l(i)-10.2004)^2))

 n=n+ f/h

end
```

5.4.7 The Response Surface Method

The practicality of reliability methods for a specific limit state depends on how complex the formulation of the limit state function is. Often the limit state function is not available in explicit form, but rather defined implicitly through a complicated numerical procedure, given, for example, by finite element analysis. For such limit state formulations, the needed calculations may require prohibitive computational effort.

One way to alleviate such expensive computations is to approximate the limit state surface in a numerical-experimental way by using a surface in explicitly mathematical form and then performing a reliability analysis. This procedure is referred to as the response surface method. In this subsection, we discuss a basic form of this method and use a simple example to illustrate it.

Let $X = [X_1, X_2, ..., X_n]^T$ be the vector of n random variables. The central idea of the response surface method is to approximate the exact limit state function $g(x)$, which is usually known through an algorithmic procedure, by a polynomial function $\hat{g}(x)$. In practice, quadratic functions are commonly used in the form

$$g(x) \approx \hat{g}(x) = a_0 + \sum_{i=1}^{n} a_i x_i + \sum_{i=1}^{n} a_{ii} x_i^2 + \sum_{i=1}^{n} \sum_{j=1, j\neq i}^{n} a_{ij} x_i x_j \qquad (5.105)$$

where the set of coefficients $a = \{a_0, a_i, a_{ii}, a_{ij}\}$ that correspond to the constant, linear, square, and cross terms, respectively, are to be determined.

A limited number of evaluations of the limit state function are required to build the surface. A reliability analysis can then be performed by means of the analytical expression $\hat{g}(x)$ in Eq. 5.105 instead of the true limit state function $g(x)$. This approach is particularly attractive when Monte Carlo simulation is used to obtain reliability results.

The unknown coefficient a is often determined using the least-squares method. After choosing a set of fitting points, $x^k, k = 1, m$ (k is the index instead of the power order) for which the exact function values $y^k = g(x^k)$ are calculated using, for example, finite element analysis. An error measure $e(a)$, defined by

$$e(a) = \sum_{k=1}^{m} (y^k - \hat{g}(x^k))^2 \tag{5.106}$$

is minimized with respect to a. Reformulating Eq. 5.105 in the form

$$\hat{g}(x) = [1, x_i, x_{ii}, x_{ij}] [a_0, a_i, a_{ii}, a_{ij}]^T \equiv V(x) \cdot a^T \tag{5.107}$$

where $i, j = 1, n$, and $j \neq i$, the least-square problem becomes

$$\text{Minimize} \left\{ \sum_{k=1}^{m} (y^k - V(x^k) \cdot a^T)^2 \right\} \tag{5.108}$$

After some basic algebra (left as exercise), the solution to the problem yields

$$a = (v^T v)^{-1} v^T y \tag{5.109}$$

where v is the matrix whose rows are the vectors $V(x^k)$ and y is the vector whose components are $y^k = g(x^k)$.

The various response surface methods proposed in the literature differ only in the terms retained in the polynomial expression (Eq. 5.105) and in the selection of the coordinates of the fitting points—that is, the experimental design used in the regression analysis. It is emphasized that $m \geq n'$, where n' is number of coefficients in a, is required to solve Eq. 5.106. Furthermore, the fitting points have to be chosen in a consistent way to obtain independent equations.

EXAMPLE 5.13

Once again we use the cantilever beam discussed in this subsection to illustrate the response surface method. The limit state function in X-space is rewritten as follows (to avoid division by zero):

$$g(x) = g(d, \ell) = 12.5 * d^3 - \ell = 0 \tag{5.110}$$

Note that if we choose a cubic function as the response surface, we obtain the exact solution. This most likely will not happen since in real applications the limit state function is unknown and is highly non-linear. In this example, we use a quadratic function to create a response surface that approximates the cubic limit state function.

Solution

The response surface using quadratic function can be stated as:

$$\hat{g}(x) = a_0 + a_1 x_1 + a_2 x_2 + a_{11} x_1^2 + a_{22} x_2^2 + a_{12} x_1 x_2$$
$$= [1\ x_1\ x_2\ x_1^2\ x_2^2\ x_1 x_2][a_0\ a_1\ a_2\ a_{11}\ a_{22}\ a_{12}]^T = V(x) \cdot a^T$$

We choose nine arbitrary points to create the response surface:

$$[-1,1],\ [0,1],\ [1,1],\ [-1,0],\ [0,0],\ [1,0],\ [-1,-1],\ [0,-1],\ [1,-1]$$

For the first data point $x^1 = [-1, 1]$, the vector $V(x^1)$ can be found as

$$V(x^1) = [1, -1, 1, 1, 1, -1]_{1\times 6}$$

We repeat the same for all 9 points to form the matrix V, which is 9×6. In the meantime, we calculate vector $y^k = g(x^k)$ for them, using Eq. 5.110, as

$$y_{9\times 1} = [-13.5, -1, 11.5, -12.5, 0, 12.5, -11.5, 1, 13.5]^T$$

In practical applications, y^k can be obtained only from numerical solutions—for example using finite element analysis.

Solve a using Eq. 5.109 as

$$a_{6\times 1} = [0, 12.5, -1, 0, 0, 0]^T$$

The response surface found is thus

$$\hat{g}(x) = 12.5 x_1 - x_2$$

which is nothing but a straight line. A straight line is certainly not able to exactly represent a cubic function of the true limit state function. However, the function $\hat{g}(x)$ is obtained in a closed form that can be evaluated very quickly. If we use a different set of data points, x^k, we will most likely obtain a different straight line, which may or may not provide a better approximation. Therefore, selecting data points is an important step in the response surface method.

5.4.8 Short Summary

Among the methods discussed, Monte Carlo simulation is certainly the simplest to use, especially for those who have only limited working knowledge of probability and statistics. It can be applied to virtually any performance function and distribution. In addition, it is computationally robust; with a sufficient number of simulations, it can always converge. For reliability analysis, Monte Carlo simulation is generally computationally expensive. The higher the reliability, the larger the simulation size needed. Because of its accuracy, Monte Carlo simulation is widely used in two areas: engineering applications where model evaluations (deterministic analyses) are not computationally expensive, and validation of other methods. However, its computational inefficiency prevents its regular use for problems where deterministic analyses are expensive.

Importance sampling is essentially a form of Monte Carlo simulation in which sampling uses a new set of distributions for the random variables—for example, moving the mean value point to the MPP so that many more sample points fall into the failure region. As a result, many fewer sample points are needed for an accurate failure probability estimate. However, this method requires that the MPP first be identified.

MPP search is also an essential step in FORM and SORM. Of the two, FORM is more popular and widely used for engineering applications, especially those requiring finite element analysis for calculating limit state functions. This is because once the MPP is found, the estimation of failure probability is extremely straightforward. However, if the limit state function has high curvature at the MPP, the estimate may not be accurate. SORM improves the accuracy in probability estimate, but it requires that the curvature of the limit state function be calculated. For problems with large numbers of random variables, curvature calculation may be computationally expensive.

The response surface method is also relatively simple to implement. One of its key issues is the selection of sample data for response surface construction. Once a response surface is created, Monte Carlo simulation can be employed in estimating failure probability since a closed-form equation of the response surface is available. If the response surface closely resembles the true limit state function, this method can be very attractive for general applications, in which evaluating the limit state function is computationally expensive.

One possible combination is the use of FORM to provide a first estimate for failure probability and then use importance sampling at the MPP to improve the accuracy of the FORM approximation, if necessary.

5.5 Multiple Failure Modes

So far in our discussion, we have assumed, for the failure probability calculation, one single failure mode in a component. In this section, we address the failure probability calculation for a mechanical system or a component with multiple failure modes. The questions to be answered are how the individual limit states interact with each other and how overall reliability can be estimated.

Series and parallel systems are discussed in Sections 5.5.1 and 5.5.2, respectively. FORM approximation for a series system is introduced in Section 5.5.3, which is brief, providing only basic concepts and methods.

5.5.1 Series System

We first define a failure element to be a mathematic model of a specific failure mode at a specific location in the structure or a specific failure mode for a component in a mechanical system. A failure element is presented as a rectangular block in Figure 5.18a.

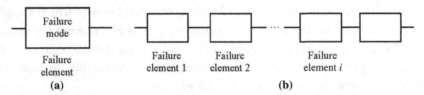

FIGURE 5.18: Block diagram for system reliability: (a) failure element and (b) series system.

One thing must be clarified first, and that is that "system" reliability does not necessarily mean that we are dealing with a mechanical system. This term is also used when there are multiple failure elements for an individual component.

A series system, as shown in Figure 5.18b, is one that fails when any failure element fails. For example, a frame structure with three members, shown in Figure 5.19, is considered. Each member is assumed to have two failure modes: stress yielding and buckling. As seen in this structure, load-carrying capacity is lost after any failure occurs in any individual component. A series system is also called a weakest-link system. Note that in the system shown, there are a total of six failure elements: two (yielding and buckling) for each member.

If the reliability of an individual failure element R_i is calculated, the reliability of a series system with m failure elements can be modeled as

$$R_{ss} = \prod_{i=1}^{m} R_i \qquad (5.111)$$

For example, if a product has 20 failure elements ($m = 20$), each with a reliability of $R_i = 0.99$, the system reliability of the product is $R_{ss} = \prod_{i=1}^{20} R_i = (0.99)^{20} = 0.818$.

Note that, since reliability is less than or equal to 1, series system reliability is always less than that of the individual failure elements. Most consumer products exhibit series reliability.

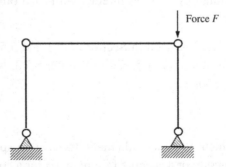

FIGURE 5.19: Frame structure with three truss members.

5.5.2 Parallel System

A much better arrangement is a parallel system, in which it is necessary for all failure elements in the system to fail for the system to fail. Such a system is represented in the block diagram shown in Figure 5.20. A system in which the components are arranged to give parallel reliability is said to be redundant; that is, more than one mechanism is necessary for the system functions to be carried out. In case one fails, a backup mechanism takes over to perform an identical function. In a system with full active redundancy, all but one component may fail before the system fails. For example, the frame structure with four members, shown in Figure 5.21, is holding a downward vertical load. Each member is assumed to have one failure mode, stress yielding. There are a total of four failure elements in this system. As seen in the structure, the system does not lose its load-carrying capacity unless all members fail.

If the reliability of individual failure element R_i is available, the reliability of a parallel system with m failure elements can be modeled as

$$R_{ps} = 1 - \prod_{i=1}^{m} (1 - R_i) \tag{5.112}$$

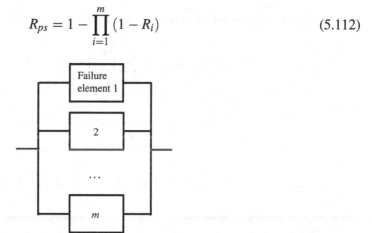

FIGURE 5.20: Reliability diagram for a parallel system.

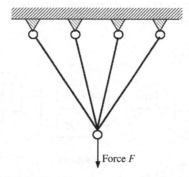

FIGURE 5.21: Frame structure with four members.

For example, if a product has four failure elements ($m = 4$), each with a reliability of $R_i = 0.90$, then the system reliability of the product is

$$R_{ps} = 1 - \prod_{i=1}^{m}(1 - R_i) = 1 - (1 - 0.90)^4 = 0.9999$$

Note that reliability of a parallel system is always greater than that of individual failure elements. Although a parallel system offers better reliability, it is much more expensive to build because of its redundancy.

Some systems have partial active redundancy, in which certain components can fail without causing system failure but more than one component must remain functioning to keep the system operating. One example is a four-engine airplane, which can fly on two engines, but loses stability and control if only one engine is operating. This type of system is known as an n-out-of-m unit network. At least n units must function normally for the system to succeed rather than only one unit in the parallel case and all units in the series case.

The reliability of an n-out-of-m unit system is given by a binomial distribution:

$$R_{n|m} = \sum_{i=n}^{m} \binom{m}{i} R^i (1 - R)^{m-i} \tag{5.113}$$

where

$$\binom{m}{i} = \frac{m!}{i!(m-i)!} \tag{5.114}$$

and the reliability of each failure element is identical and equal to R.

EXAMPLE 5.14

A complex engineering design is described by the reliability block diagram shown in below. In subsystem A three components must operate for the subsystem to function successfully. In subsystem C, two components must operate for the subsystem to function. Subsystem D has

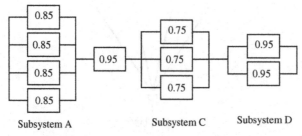

The block diagram for system reliability analysis.

true parallel reliability. We want to calculate the reliability of each subsystem and the overall system reliability.

Solution

For subsystem A, the reliability R_A is

$$R_A = R_{3|4} = \sum_{i=3}^{4} \binom{4}{i} 0.85^i (1 - 0.85)^{4-i}$$

$$= \binom{4}{3} 0.85^3 (1 - 0.85)^{4-3} + \binom{4}{4} 0.85^4 (1 - 0.85)^{4-4}$$

$$= 4 \times 0.85^3 (1 - 0.85)^1 + 1 \times 0.85^4 \times 1 = 0.8905$$

For subsystem C, the reliability R_C is

$$R_C = R_{2|3} = \sum_{i=2}^{3} \binom{3}{i} 0.75^i (1 - 0.75)^{3-i}$$

$$= \binom{3}{2} 0.75^2 (1 - 0.75)^{3-2} + \binom{3}{3} 0.75^3 (1 - 0.75)^{3-3}$$

$$= 3 \times 0.75^2 (1 - 0.75)^1 + 1 \times 0.75^3 \times 1 = 0.8438$$

For subsystem D, the reliability R_D is

$$R_D = 1 - \prod_{i=1}^{2} (1 - R_i) = 1 - (1 - 0.95)^2 = 0.9975$$

The overall system reliability is then

$$R = R_{ss} = \prod_{i=1}^{4} R_i = R_A R_B R_C R_D = 0.8905 \times 0.95 \times 0.8438 \times 0.9975 = 0.7120$$

5.5.3 FORM Approximation for a Series System

We assume that the reliability of individual failure elements in a system is known. In many applications, the reliability of individual failure elements is estimated using, for example, FORM. This subsection introduces FORM failure probability estimation for a series system. Instead of a thorough discussion here, we present important concepts and basic equations sufficient for an idea about this subject. For a detailed discussion with analytical examples, the reader may refer to Lee (2012) and Enevoldsen and Sørensen (1994).

As was shown earlier (Eq. 5.66), failure probability for a single failure element can be estimated using FORM as

$$P_f \approx P(L(\boldsymbol{u}) \le 0) = P(\beta - \boldsymbol{a} \cdot \boldsymbol{u}^T \le 0) \tag{5.115}$$

in which the failure zone is approximated as $L(u) \leq 0$, and a is the normalized gradient of the limit state function at the MPP, which is also the gradient of the tangent line that passes the MPP (i.e., $L(u^*) = 0$).

For a series system with m failure elements, the failure probability for the ith element can be written as

$$P_f = P\big(g_{u_i}(u) \leq 0\big) \approx P(L_i(u) \leq 0) = P\big(\beta_i - a_i \cdot u^T \leq 0\big) = \Phi(-\beta_i) \qquad (5.116)$$

where $g_{u_i}(u)$ and $L_i(u)$ are, respectively, the limit state function and its linearized Taylor expansion at the MPP for the ith failure element of the series system. a_i is the normalized gradient of the limit state function at the MPP for the ith failure element.

As discussed in Section 5.5.1, a series system fails if any one of the failure elements fails. Therefore, the failure region for a series system is the union of those regions of the individual failure elements, as illustrated in Figure 5.22a, in which three limit state functions representing the respective failure elements are included in a two-dimensional problem. Using FORM, the failure region is first approximated by the union of linearized limit state functions at their respective MPPs, as shown in Figure 5.22b.

The failure probability of the series system can be estimated as

$$
\begin{aligned}
P_f^s &= P\big(\cup_{i=1}^m \big(g_{u_i}(u) \leq 0\big)\big) \approx P\big(\cup_{i=1}^m (L_i(u) \leq 0)\big) \\
&= P\big(\cup_{i=1}^m \big(\beta_i - a_i \cdot u^T \leq 0\big)\big) = 1 - P\big(\cap_{i=1}^m \big(\beta_i - a_i \cdot u^T \geq 0\big)\big)
\end{aligned}
\qquad (5.117)
$$

Note that the last term on the right of Eq. 5.117 has been obtained by applying the well-known De Morgan's law to the set $\cup_{i=1}^m (L_i(u) \leq 0) = \cup_{i=1}^m (\beta_i - a_i \cdot u^T \leq 0)$. Equation 5.117 can be further reduced to

$$
\begin{aligned}
P_f^s &\approx 1 - P\left(\bigcap_{i=1}^m \big(\beta_i - a_i \cdot u^T \geq 0\big)\right) = 1 - P\left(\bigcap_{i=1}^m \big(a_i \cdot u^T \leq -\beta_i\big)\right) \\
&= 1 - \Phi_m(\beta, \rho)
\end{aligned}
\qquad (5.118)
$$

in which Φ_m is the m-dimensional normal distribution function defined as

$$\Phi_m(\beta, \rho) \equiv \int\limits_{-\infty}^{\beta_1} \int\limits_{-\infty}^{\beta_2} \cdots \int\limits_{-\infty}^{\beta_m} \phi_m(x, \rho) dx_1 dx_2 \ldots dx_n \qquad (5.119)$$

where $\phi_m(x, \rho)$ is the m-dimensional normal probability density function defined as

$$\phi_m(x, \rho) = \frac{1}{(2\pi)^{m/2} \sqrt{|\rho|}} e^{-\frac{1}{2} x^T \rho^{-1} x} \qquad (5.120)$$

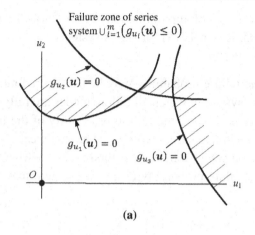

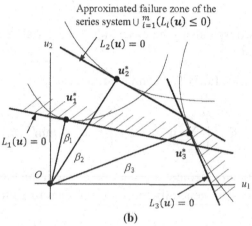

FIGURE 5.22: Failure probability estimate using FORM for the series system: (a) safe zone of the series system formed by $\bigcup_{i=1}^{m}(g_{u_i}(\boldsymbol{u}) \leq 0)$; (b) safe zone of the series system formed by $\bigcup_{i=1}^{m}(L_i(\boldsymbol{u}) \leq 0)$.

In Eqs. 5.118, 5.119, and 5.120, $\boldsymbol{\rho}$ is an $m \times m$ matrix, defined as

$$\rho_{ij} = \boldsymbol{a}_i^T \cdot \boldsymbol{a}_j \tag{5.121}$$

It can be shown that ρ_{ij} is the correlated coefficient of random variables X_i and X_j (Sørensen 2004). $|\boldsymbol{\rho}|$ in Eq. 5.120 is the determinant of the Jacobian of the matrix $\boldsymbol{\rho}$.

The descriptions just given show that it is very important to be able to calculate the multidimensional standard normal distribution function $\boldsymbol{\phi}_m(\boldsymbol{x}, \boldsymbol{\rho})$, which, except for special cases, must be estimated by approximation. We simply mention a few popular methods, in decreasing order of general precision:

- Gollwitzer and Rackwitz's approximation (Gollwitzer and Rackwitz 1983)
- Hohenbichler's approximation (Hohenbichler 1984)

- Average correlation coefficient approximation (Thoft-Christensen and Sørensen 1982)
- Ditlevsen's bounds (Ditlevsen 1979) for series systems
- Simple bounds (Cornell 1967)

The simple bounds on the failure probability may be established merely on the basis of consideration. For a series system in which all failure modes are fully correlated, it is realized that the failure probability is equal to the failure probability of the failure element with the largest failure probability—in this case, a system where the weakest link may be clearly identified. As the correlation between the failure modes is somewhere between zero and one, the simple bounds on the failure probability of a series system may thus be given as

$$\max_{i=1}^{m} \left\{ P\big(g_{u_i}(\boldsymbol{u}) \leq 0\big) \right\} \leq P_f^s \leq \sum_{i=1}^{m} \left\{ P\big(g_{u_i}(\boldsymbol{u}) \leq 0\big) \right\} \tag{5.122}$$

where the lower bound corresponds to the exact value of P_f^s if all the elements in the series system are fully correlated.

In terms of reliability indices, Eq. 5.122 can be written as

$$-\Phi^{-1}\left(\sum_{i=1}^{m} \Phi^{-1}(-\beta_i) \right) \leq \beta^s \leq \min_{i=1}^{m} \beta_i \tag{5.123}$$

When the failure of one failure element does not dominate in relation to the other failure elements, the simple bounds are generally too wide and therefore often of minor interest for practical use.

5.6 General-Purpose Reliability Analysis Tools

As discussed in Section 5.4.1, a limit state function associated with the respective failure mode must be defined before failure probability can be calculated. Although in this chapter we have included limit state functions of structural problems, the methods and theory presented have not been restricted to structural problems. For example, if we are dealing with a rigid-body dynamic problem (as discussed in Chapter 3) in the design of a mechanical system, there may be a need to calculate the failure probability of a critical performance measure identified in the system because of uncertainties in parameters, such as mass properties or size dimensions of individual moving components, external forces, and the like.

One instance might be the design of a single-piston engine (see Chapter 3) in which the maximum dynamic load applied to the connecting rod resulting from the firing load and the inertia of the individual components must not be greater than an upper limit because of concern about the rod's structural integrity. Uncertainties in physical parameters may have to be considered, such as magnitude of the firing force, geometric dimension variation in various components including the rod, and mass density of the rod, which can be characterized by certain distributions determined by physical measurement. The failure probability of the limit state function associated with the load failure mode can be estimated using reliability analysis methods, such as Monte Carlo simulation or FORM, as discussed.

Thus far, limit state functions have been represented by a simple algebraic function of random variables representing the aforementioned uncertainties. In these examples, we can easily perform reliability analysis. In dealing with many complex structural systems or failure mechanisms, however, evaluation of limit state functions require sophisticated structural analyses or mechanism simulations employing analysis tools.

Although significant progress has been made in reliability analysis methods, the development of corresponding software is still lagging behind (Pellissetti and Schuëller 2006), especially in the commercial sector. There are very few options in commercial reliability analysis tools. Those available have been developed mainly to support reliability analysis for structural problems that require finite element analysis for evaluating limit state functions—often termed finite element-reliability analysis (FE-RA) (e.g., Sudret and Der Kiureghian 2000, Lee et al. 2008). In this section, we briefly review some FE-RA software tools that are readily (i.e., relatively) available for use. Note that many of the existing FE-RA software packages are primarily for component reliability analysis, in which the FE reliability analysis is performed for individual failure modes of a structural member or location that are represented by single limit state function.

One of the most reputable reliability analysis software programs is NESSUS®, a modular computer software program for performing probabilistic analysis of structural/mechanical components and systems. NESSUS was initially developed by the Southwest Research Institute (SwRI) for NASA to perform probabilistic analysis of space shuttle main engine components. SwRI continues to develop and apply NESSUS to a diverse range of problems, among them aerospace structures, automotive structures, and biomechanics. Instead of including its own FE code, NESSUS has been interfaced with many well-known third-party and commercial deterministic analysis programs, including ABAQUS®, ANSYS®, MSC/Nastran®, and MATLAB.

Integrating with commercial third-party FE software tools seems to be the most logical and economic approach for structural reliability analysis. The only drawback is that the type of limit state functions are restricted to those offered by the FE tools. However, these tools are powerful and support a broad range of structural analysis problems. Even if certain types of structural problem or finite element are not found in one code, integrating with multiple FE software tools greatly diminishes this concern. Moreover, the same approach supports other types of engineering problems when proper analysis software is integrated to provide results for the limit state functions. Details can be found in Schuëller and Pellissetti (2008).

There are several other commercial tools that follow the same approach, such as COSSAN, Pro-FES®, STRUREL , Proban, FReET , and UNIPASS®.

Many research or semi-research codes are worthy of mention here. These codes, developed by research groups in universities, are occasionally upgraded; they are offered as free downloads and are customizable for various purposes, such as graduate research. One code in this category is FERUM (Finite Element Reliability Using MATLAB), which is primarily intended for pedagogical purposes although it is also useful for research and engineering production. FERUM is a set of functions within MATLAB that carry out finite element-reliability analysis as well as reliability analysis for prescribed analytical limit state functions. Another code in this category is CalREL, a general-purpose structural reliability analysis program that is available for purchase from UC Berkeley in both object and source codes.

All of the noncommercial software codes mentioned offer excellent reliability analysis capabilities, including FORM, SORM, Monte Carlo simulation, and response surface. However, the ANSYS Probabilistic Design System (PDS) is the only commercial FE tool with such capability. ANSYS PDS automates the reliability analysis process. Using simple menu picks (or commands), users can specify many input variables and their variations in statistical terms (Gaussian, Weibull, etc.). ANSYS then manages the many runs that are necessary for an accurate reliability estimate. Instead of more advanced FORM or SORM, ANSYS PDS offers Monte Carlo simulation and the response surface method.

5.7 Case Study

A high-mobility multipurpose wheeled vehicle (HMMWV) roadarm is presented here as a case study to demonstrate reliability analysis in practical failure probability calculation. A deterministic fatigue life prediction of the roadarm was discussed in Section 4.8.1 together with a rigid-body dynamic simulation of the HMMWV on the Aberdeen Proving Ground 4 (APG4). Here random variables are defined and probabilistic fatigue life predictions are made for the roadarm.

As discussed in Section 4.8.1, a rigid-body dynamic simulation on the APG4 was carried out first to obtain loads applied to the roadarm for a total of 20 seconds in simulation. A finite element analysis was performed to obtain the roadarm's stress influence coefficients (SICs), using ANSYS to apply 18 quasistatic loads. The dynamic stresses at finite element nodes were then calculated by superposing the SICs with their corresponding external forces and accelerations and velocities in the time domain obtained from the dynamic simulation. To compute the multiaxial crack initiation life of the roadarm, the equivalent von Mises strain approach (see Chapter 4) was employed. The fatigue life contour is given in Figure 5.23. The shortest fatigue life of the roadarm was identified at node 1216 (Chang et al. 1997).

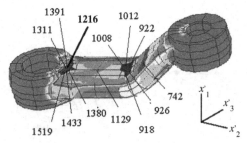

FIGURE 5.23: Crack initiation life contour for the HMMWV roadarm.

The random variables and their statistical values for the crack initiation life prediction are listed in Table 5.5, including those for material and tolerance. The eight tolerance random variables were defined to characterize the four cross-sectional shapes of the roadarm, as shown in Figure 5.24a (Yu, et al. 1997). The profile of the cross-sectional shape was composed of four straight lines and four cubic curves (Figure 5.24b). Side variations (x'_2-direction) of the cross-sectional shapes are defined as random variables b1, b3, b5, and b7 for sections 1 through 4, respectively (Figure 5.24b). The vertical variations (x'_3-direction) of the cross-sectional shapes are defined as the remaining four random variables (Figure 5.24b).

The first-order reliability method (FORM) was used to calculate the failure probability of the crack initiation life. The deterministic fatigue life at node 1216 was the shortest, with 9.63E + 06 blocks (20 sec per block). The cumulative distribution function of the crack initiation life (number of blocks to failure) at node 1216 is shown in Figure 5.25. The horizontal axis in the figure is the required number of service blocks, and the vertical axis is the failure probability. The CDF in the figure was obtained by reliability analysis at the seven required numbers of service blocks, ranging from 1×10^5 to 5×10^6 blocks, which are marked in the figure.

Table 5.5: Random variables for crack initiation life prediction.

Random Variables	Mean Value	Standard Deviation	Distribution
Young's modulus, E	30.0E+6	0.75E+6	Lognormal
Fatigue strength coefficient, σ'_f	1.77E+5	0.885E+4	Lognormal
Fatigue ductility coefficient, ε'_f	0.41	0.0205	Lognormal
Fatigue strength exponent, b	−0.07300	0.00365	Normal
Fatigue ductility exponent, c	−0.6	0.003	Normal
Tolerance b1	3.2496	0.032450	Normal
Tolerance b2	1.9675	0.019675	Normal
Tolerance b3	3.1703	0.031703	Normal
Tolerance b4	1.9675	0.019675	Normal
Tolerance b5	3.1703	0.031703	Normal
Tolerance b6	2.6352	0.026352	Normal
Tolerance b7	3.2496	0.032496	Normal
Tolerance b8	5.0568	0.050568	Normal

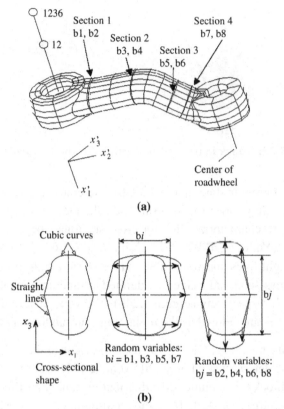

(a)

(b)

FIGURE 5.24: HMMWV roadarm: (a) FE model with four sections and (b) geometric parameters included for modeling uncertainty in manufacturing tolerance.

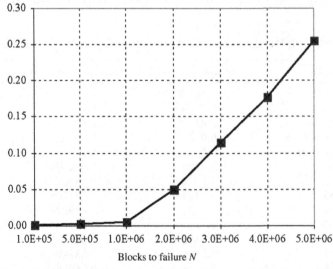

Blocks to failure N

FIGURE 5.25: CDF graph of failure probability for the HMMWV crack initiation life.

As discussed previously, one FORM analysis is equivalent to solving a deterministic optimization. For the roadarm example, seven FORM analyses were performed to create the graph shown in Figure 5.25. In actual design applications, the CDF curve can be used to obtain the failure probability for the required number of service blocks before crack initiation, or the required number of service blocks before crack initiation with a required failure probability. For example, it can be seen in the figure that, if the required number of service blocks before crack initiation is 3.0E+06, the failure probability is about 11%.

5.8 Summary

In this chapter, we demonstrated that a deterministic approach using the safety factor method or the worst-case scenario is not sufficient to provide a full picture of the safety or reliability of a product design. These methods can lead to a design that is either not as reliable as desired or to a product or component that is overdesigned. A safe and reliable product can be ensured only by bringing probabilistic or statistical considerations into the design process.

Mechanical engineers must understand the importance of the probabilistic aspect of product design, and they must be able to perform adequate reliability analysis in solving various engineering problems. For this reason, we introduced several basic but widely used methods for reliability analysis: FORM, SORM, Monte Carlo simulation, importance sampling, and response surface. None of these is a clear-cut choice above the others. All have strengths and weaknesses. The key is to understand how each works and to choose the one appropriate for solving the problem at hand—that is, the one that requires the least computational effort to achieve an accurate enough estimate of failure probability. For general applications in engineering design, evaluation of limit state functions can require substantial computation, such as analyzing large-scale structures using finite element method.

As mentioned in this chapter, one way to address such applications is to search the MPP and use FORM to estimate failure probability, and then use MPP-based importance sampling to further improve the accuracy of the FORM estimate if necessary. In using FEA for evaluating the limit state function, the designer may develop an interface program to integrate an FEA code into reliability analysis code. Details are provided in Schuëller and Pellissetti (2008). Alternatively, the FEA input data file may be manually modified in accordance with the random variable values.

In either case, it is important for design engineers to understand the importance of reliability in product design from a probabilistic perspective and to be able to use adequate methods to obtain failure probability estimates. With this in mind, design engineers should incorporate failure probability into their formulation of design problems and, for accurate estimates of product failure probability, strive to acquire information and statistical data to characterize physical parameters involved in the limit state functions.

In this chapter, we focused on reliability analysis and only touched on design. More in-depth discussion of engineering design for reliability, including reliability-based design, robust design, and probabilistic design, is provided in *Design Theory and Methods using CAD/CAE*, a book in The Computer Aided Engineering Design Series..

Questions and Exercises

5.1. Let $X = [X_1, X_2, ..., X_n]^T$ be a vector of random variables with normal distribution, and let $Y = [Y_1, Y_2, ..., Y_n]^T$ be the corresponding vector of random variables of standard normal distribution, which is obtained by transforming random vector X_i, for $i = 1, n$, by

$$y_i = \frac{x_i - \mu_{x_i}}{\sigma_{x_i}}$$

Show that the covariance matrix of Y (that is, C_Y, as defined in Eq. 5.83) is equal to the correlation coefficient matrix of X (that is, ρ_X, as defined in Eq. 5.88):

$$C_Y = \rho_X$$

5.2. Let $Y = [Y_1, Y_2, ..., Y_n]^T$ be a vector of random variables of the standard normal distribution transforms with covariance matrix C_Y. Show that a matrix T defined as shown transforms random variables Y of correlated standard normal distribution into $U = [U_1, U_2, ..., U_n]^T$ that are random variables of uncorrelated (independent) standard normal distribution:

$$Y = TU$$

where

$$T_{jj} = \sqrt{1 - \sum_{k=1}^{j-1} T_{jk}^2}, j = 1, n$$

and

$$T_{ij} = \frac{\rho_{ji} - \sum_{k=1}^{j-1} T_{jk} T_{ik}}{T_{jj}}, i > j$$

5.3 Derive Eq. 5.109 from Eq. 5.108.

5.4 A beam clamped at both ends, shown in the following figure, is subject to reliability assessment for its current design.

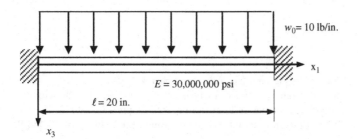
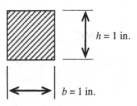

Assume a displacement failure mode that states that the maximum displacement of the beam cannot exceed 0.02 in. If two random variables, the width and height of the cross-sectional area, are considered, calculate the failure probability using Monte Carlo simulation. Note that both random variables are uncorrelated and of normal distribution, with their respective mean values and standard deviations defined as $X_b \sim N(1, 0.1)$ in. and $X_h \sim N(1, 0.15)$ in.

5.5 Continue with Exercise 5.4, but use FORM to estimate failure probability. Calculate the MPP and find the reliability index β. Compare the failure probability found using FORM with that found by Monte Carlo simulation. Is the FORM result accurate compared with the Monte Carlo result?

5.6 Use the MPP obtained from Exercise 5.5 to estimate the failure probability of the same beam problem using MPP-based importance sampling. How many sample points are needed to achieve an accurate failure probability?

5.7 If the width and height random variables of the beam are correlated with correlated coefficient $\rho_{bh} = 0.25$, transform them into a U-space of standard normal distribution and express the limit state function in terms of random variables u_b and u_d.

5.8 Use the response surface method to solve the same beam problem. Find a quadratic function $\hat{g}(x)$ to approximate the true limit state function $g(x)$. Use the following two sets of sample points:

Set A: $[-1,1]$, $[0,1]$, $[1,1]$, $[-1,0]$, $[0,0]$, $[1,0]$, $[-1,-1]$, $[0,-1]$, $[1,-1]$

Set B: $[-2,1]$, $[-1,1]$, $[0,1]$, $[1,1]$, $[2,1]$, $[-2,0]$, $[-1, 0]$, $[0,0]$, $[1,0]$, $[2,0]$, $[-2,-1]$, $[-1,-1]$, $[0,-1]$, $[1,-1]$, $[2,-1]$

Compare the functions $\hat{g}_A(x)$ and $\hat{g}_B(x)$ obtained from these respective sample data sets. Use Monte Carlo simulation to calculate failure probabilities using both functions $\hat{g}_A(x)$ and $\hat{g}_B(x)$. Which one gives a more accurate failure probability estimate? Can you explain why one function is better than the other?

References

Boggs, P.T., Tolle, J.W., 1995. Sequential quadratic programming. Acta Numerica 4, 1–51.

Chang, K.H., Yu, X., Choi, K.K., 1997. "Shape Design Sensitivity Analysis and Optimization For Structural Durability". International Journal of Numerical Methods in Engineering Vol. 40, pp. 1719–1743.

Choi, S.K., Grandhi, R.V., Canfield, R.A., 2006. Reliability-based Structural Design. ISBN-10: 1-84628-444-9. Springer.

Cornell, C.A., 1967. Bounds on the reliability of structural systems. ASCE Structural Division Journal 93 (1), 171–200.

Ditlevsen, O., 1979. Narrow reliability bounds for structural systems. J. Struct. Mechanics. 7 (4), 453–472.

Ditlevsen, O., Madsen, H.O., 1996. Structural Reliability Methods. John Wiley & Sons.

Enevoldsen, I., Sørensen, J.D., 1994. Reliability-based optimization in structural engineering. Structural Safety 15, 169–196.

Gollwitzer, S., Rackwitz, R., 1983. Equivalent components in first-order systems reliability. Reliability Engineering 5, 99–115.

Haldar, A., Mahadevan, S., 2000. Probability, Reliability, and Statistical Methods in Engineering Design. John Wiley & Sons.

Hasofer, A.M., Lind, N.C., 1997. An Exact and Invariant First Order Reliability Format. Journal of the Engineering Mechanics Division Vol. 100, No. EM1, pp. 111–121.

Hohenbichler, M., 1984. An asymptotic formula for the probability of intersections, Berichte zur Zuverltissigkeitstheorie der Bauwerke. Heft 69, LKI. Technische Universitlt München.

Hurtado, J.E., 2004. Structural Reliability: Statistical Learning Perspectives. Springer-Verlag.

Lee, Y.J., 2012. Finite-element-based system reliability analysis and updating of fatigue-induced sequential failures. PhD dissertation. University of Illinois at Urbana-Champaign.

Lee, Y.-J., Song, J., Tuegel, E.J., 2008. Finite element system reliability analysis of a wing torque box. Proceedings 10th AIAA Nondeterministic Approaches Conference April 7–10.

Lee, J.-O., Yang, Y.-S., Ruy W, -S., 2002. A comparative study on reliability-index and target-performance-based probabilistic structural design optimization. Computers and Structures 80, 257–269.

Liu, P.L., Der Kiureghian, A., 1986. Multivariate distribution models with prescribed marginals and covariances. Probabilistic Engineering Mechanics 1 (2), 105–112.

Nataf, A., 1962. Détermination des distribution dont les marges sont données. Comptes rendus de l'academie des sciences 225.

Pellissetti, M.F., Schuëller, G.I., 2006. On general purpose software in structural reliability. Structural Safety 28 (1–2), 3–16.

Rackwitz, R., Fiessler, B., 1998. Structural Reliability Under Combined Random Load Sequences. Journal of Composite Structures Vol. 9, pp. 489–494.

Rosenblatt, M., 1952. Remarks on a multivariate transformation. Annals of Mathematical Statistic 23, 470–472.

Schuëller, G.I., Pellissetti, M.F., 2008. Computational tools for structural analysis with uncertainties: Software technology and large-scale applications. In: Topping, B.H.V., Papadrakakis, M. (Eds.), Trends in Computational Structures Technology. Saxe-Coburg Publications.

Sørensen, J.D., 2004. Notes in Structural Reliability Theory and Risk Analysis. Aalborg University, Aalborg.

Sudret, B., Der Kiureghian, A., 2000. Stochastic finite element methods and reliability: A state-of-the-art report. Department of Civil and Environmental Engineering, University of California, Berkeley.

Thoft-Christensen, P., Sørensen, J.D., 1982. Reliability of structural systems with correlated elements. Applied Mathematical Modelling 6, 171–178.

Wei, D., 2006. A univariate decomposition method for higher order reliability analysis and design optimization. PhD dissertation. University of Iowa.

Yu, X., Choi, K.K., Chang, K.H., 1997. A Mixed Design Approach for Probabilistic Structural Durability. Journal of Structural Optimization 14, pp. 81–90.

Sources

ABAQUS: www.mscsoftware.com
ANSYS PDS: www.ansys.com
COSSAN: www.cossan.co.uk
CalREL: www.ce.berkeley.edu
FReET: www.cervenka.cz
MATLAB: www.mathworks.com
MSC/Nastran: www.mscsoftware.com
NESSUS: www.nessus.swri.org
Proban: www.dnvusa.com
Pro-FES®: www.ara.com
STRUREL: www.strurel.de/feat.htm
UNIPASS: www.predictionprobe.com

Project P2 Motion Analysis Using Pro/ENGINEER Mechanism Design

Chapter Outline

Motion analysis was reviewed in Chapter 3, in which both theoretical and practical aspects of the subject were discussed. There is no need to emphasize the importance of understanding theory and using analytical methods to solve engineering problems. However, as discussed in Chapter 3, analytical methods can only go so far. In many design applications, such as mechanical system analysis that involves multiple bodies, analytical methods can support only a small number of rigid bodies; that is, one or two at best. In many cases, engineers must rely on tools such as motion analysis software to evaluate product performance in mechanisms with high accuracy in support of design decisions. It is critical for design engineers to be able to use software tools for solving problems that are beyond hand calculations relying on analytical methods.

In Project P2, we introduce Pro/ENGINEER Mechanism Design for kinematic and dynamic analysis of mechanical systems. We include two simple examples to help you get started on learning and using the software. These two examples are a sliding block and a single-piston engine, which should provide you with a good overview of the motion analysis capabilities offered by Pro/ENGINEER Mechanism Design. You may find the example files on the book's companion website (http://booksite.elsevier.com/9780123984609).

Overall, the objective of this project is to enable readers to use Pro/ENGINEER Mechanism Design for basic applications. If you are interested in learning more, you may want to run through the examples provided by Mechanism Design or review other tutorial books on the subject—for example, *Mechanism Design and Analysis with Pro/ENGINEER Wildfire 5.0* (*www.sdcpublications.com*)

There are three lessons included in this project. Lesson 1 (Section P2.1) offers a brief overview on Mechanism Design. Lesson 2 provides detailed steps in carrying out motion analysis for the sliding block example. Lesson 3 does the same for the single-piston engine example. All three lessons included in this project are developed using Pro/ENGINEER Wildfire 5.0 (Date Code M040). If you are using a different version, you may see slightly different menu options or dialog boxes. Since Pro/ENGINEER is fairly intuitive to use, these differences should not be too difficult to figure out.

P2.1 Introduction to Pro/ENGINEER Mechanism Design

Pro/ENGINEER Mechanism Design (or Mechanism Design) is a computer software tool that supports users in analysis of kinematic and dynamic performance of mechanical systems, as well as mechanism design. The main objective of this tutorial is to help you, as a new user, to become familiar with Mechanism Design and to provide you with a brief overview of the concept and the detailed steps in its use.

Mechanism Design is fully integrated and embedded in Pro/ENGINEER. It can be accessed through Pro/ENGINEER's menus and windows. The transition is seamless. The same parts and assemblies created in Pro/ENGINEER are directly employed for creating motion models in Mechanism Design. In general, part geometry is essential for mass property computations. In Mechanism Design, all mass properties calculated in Pro/ENGINEER are ready for use. The detailed part geometry is critical for interference checking in Mechanism Design is also available. Menus and dialog boxes for Mechanism

Design are just like those in Pro/ENGINEER. Those who are familiar with Pro/ENGINEER will find Mechanism Design relatively easy to learn.

P2.1.1 Overall Process

The overall process of using Mechanism Design consists of three main steps: model generation (or preprocessing), analysis, and result visualization (or postprocessing), as illustrated in Figure P2.1. Basic entities that constitute a motion model include bodies, joints, initial conditions, and force or driver. A body can be stationary (a ground body) or movable. Revolute joints, cylindrical joints, and the like are defined between bodies to constrain the relative movement between them. In addition to joints, you may use placement constraints to bring the components into the assembly to create a motion model that captures essential characteristics and closely resembles the behavior of the physical mechanism. Mechanism Design accepts both placement constraints and joints for a motion model. In motion models, at least one degree of freedom (dof) must be free to allow the mechanism to move. The free degrees of freedom are usually driven either by a servomotor for a kinematic analysis or by an external load (force and torque) for a dynamic simulation.

The analysis capabilities in Mechanism Design include position (initial assembly), static (equilibrium configuration), motion (kinematic and dynamic), and force balance (to maintain the system in a prescribed configuration). For example, the position analysis brings bodies closer within a prescribed tolerance at each joint to create an initial assembled configuration for the mechanism.

The analysis results can be visualized in various forms. You may animate motion of the mechanism, or generate graphs for more specific information, such as the reaction force of

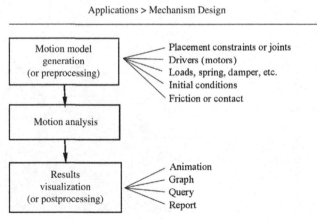

Figure P2.1: Using Pro/ENGINEER Mechanism Design.

a joint in the time domain. You may also query results at specific locations for a given time. Furthermore, you may ask for a report on results that you specified, such as the acceleration of a moving body in the time domain.

P2.1.2 User Interface

The user interface for Mechanism Design is identical to that for Pro/ENGINEER, as shown in Figure P2.2. Pro/ENGINEER users should find it straightforward to maneuver in Mechanism Design. As shown in the figure, the user interface window consists of pull-down menus, shortcut buttons, scroll-down menus, and Graphics, Prompt/Message, and Model Tree windows.

The Graphics window displays the motion model with which you are working. The pull-down menus and the shortcut buttons at the top of the screen provide typical Pro/ENGINEER functions. The Mechanism Design shortcut buttons to the right provide all the functions required to create and modify motion models, create and run analyses, and visualize results. When you click the menu options, the Prompt/Message window shows brief messages describing the menu commands. It also shows system messages following command execution. The shortcut buttons in Mechanism Design and their functions are summarized in Table P2.1.

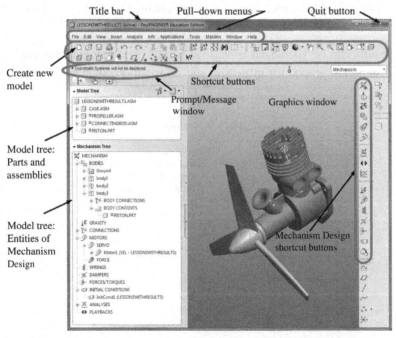

Figure P2.2: Pro/ENGINEER Mechanism Design user interface.

Table P2.1: Shortcut buttons in Mechanism Design.

Button	Name	Function
	Display Entities	Turn icon visibility on or off in assembly
	Cam-Follower Connection Definition	Create new cam-follower
	Gear Pairs	Create new gear pair
	Servo Motors	Define servomotor (driver)
	Analysis Definition	Define and run analysis
	Playbacks	Play back results of analysis run
	Measure Results	Create measures and select measures and result sets to display; also graph results or save them to a table
	Gravity	Define gravity
	Force Motors	Define new force motor
	Springs	Define new spring
	Dampers	Define new damper
	Forces/Torques	Define force or torque
	Initial Condition Definition	Specify initial position snapshots and define velocity initial conditions for point, motion axis, or body
	Mass Properties	Specify mass properties for part or specify density for assembly

P2.1.3 Defining Motion Entities

As mentioned before, the basic entities of a valid Mechanism Design simulation model consist of a ground part (or ground body), moving parts (or moving bodies), joints (defined directly or imposed implicitly by equivalent placement constraints created in Pro/ENGINEER assembly), initial conditions (usually position and velocity of a moving body), and forces or drivers respectively for dynamic and kinematic analyses.

A ground part, or a ground body, represents a fixed reference in space. The first component brought into the assembly is usually stationary, therefore becoming a ground part. Parts (or subassemblies) assembled to the stationary components without any possibility of moving become part of the ground body. A moving part or body is an entity that represents a single rigid component moving relatively to other parts (or bodies). A moving part may consist of a single Pro/ENGINEER part or a subassembly composed of multiple parts. When a subassembly is designated as a moving part, none of its composing parts is allowed to move relative to one another within the subassembly.

An unconstrained rigid body in space has six degrees of freedom: three translational and three rotational. When you add a joint (or a placement constraint) between two rigid bodies, you remove degrees of freedom between them. Each independent movement permitted by a joint (or placement constraint) is a free degree of freedom. The free degree(s) of freedom that a joint allows can be translational or rotational along the three perpendicular axes. It is extremely important to understand the definition and characteristics of joints and placement constraints in support of generating successful motion models.

P2.1.4 Motion Simulation

Mechanism Design is capable of solving typical engineering problems, including position, static, motion (kinematic and dynamic), and force balance. The position (or assembly analysis) that brings the mechanism together, as illustrated in Figure P2.3a, is often performed before any other type of analysis. It determines an initial configuration of the mechanism based on the body geometry, joints, and initial conditions of bodies. The points, axes, or planes chosen for defining joints are brought within a small prescribed tolerance.

Static analysis is used to find the rest position (equilibrium condition) of a mechanism in which none of the bodies are moving. A simple example of static analysis is illustrated in Figure P2.3b, in which an equilibrium position of a block is to be determined according to its own mass m, the two spring constants k_1 and k_2, and the gravity g. Kinematics is the study of motion without regard for forces or torque. A mechanism can be driven by a motion driver for a kinematic analysis, where the position, velocity, and acceleration of individual bodies of the mechanism can be analyzed at any given time. Figure P2.3c

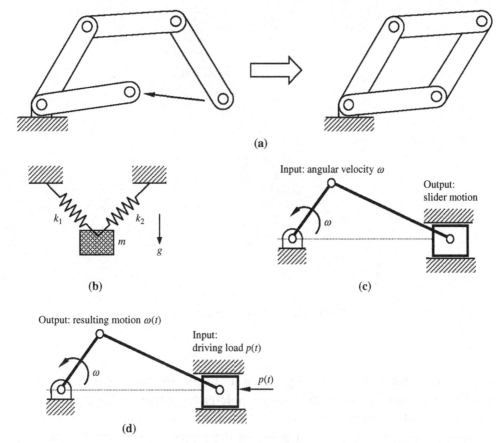

Figure P2.3: Pro/ENGINEER Mechanism Design motion analysis capabilities: (a) position (assembly) analysis, (b) static analysis, (c) kinematic analysis, and (d) dynamic analysis.

shows a servomotor driving a mechanism at a constant angular velocity. Dynamic analysis is employed for studying the mechanism motion in response to loads, as illustrated in Figure P2.3d. This is the most complicated and common, and usually a more time-consuming analysis.

P2.1.5 Viewing Results

In Mechanism Design, results of the motion analysis can be realized using animations, graphs, reports, and queries. Animations show the configuration of the mechanism in consecutive time frames. Animations also give you a global view of how the mechanism behaves. An example of such animation is shown in Figure P2.4, in which a single-piston engine is in motion.

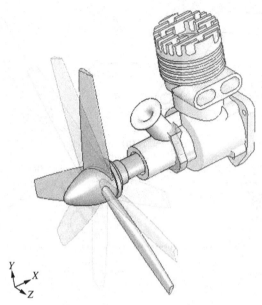

Figure P2.4: Motion animation.

You may choose a joint or a point to generate result graphs; for example, the graph in Figure P2.5 shows the piston position of a single-piston engine (please see Section P2.3 for more details). Graphs give you a quantitative understanding of the behavior of the mechanism. You may also pick a point on the graph to query the results of your interest at a specific time frame. In addition, you may ask Mechanism Design for a report that includes a complete set of results output in the form of numerical data.

In addition to the capabilities discussed above, Mechanism Design allows you to check interference between bodies during motion. Furthermore, the reaction forces calculated can be used to support structural analysis using, for example, Pro/MECHANICA Structure, a p-version finite element analysis module of Pro/ENGINEER.

P2.1.6 Examples

Two simple examples are included in this tutorial project, which illustrate step-by-step details of modeling, analysis, and result visualization capabilities in Mechanism Design. We will start with a very simple sliding block in Section P2.2. This example provides you a quick start and a quick run-through for solving a simple particle dynamic problem. The second example (Section P2.3) illustrates the steps for carrying out kinematic analysis for a single-piston engine, in which the propeller is driven by a rotary motor at

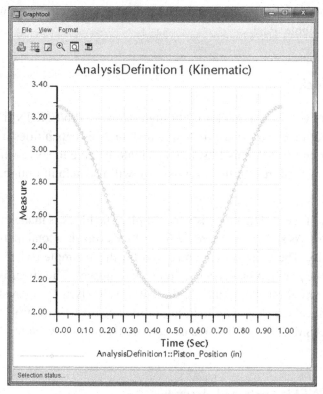

Figure P2.5: Piston position versus time.

Table P2.2: Project examples.

Example	FEA Model	Problem Type	Things to Learn
Sliding block		Particle dynamics	Become familiar with basic Mechanism Design operation Create motion model from Pro/ENGINEER assembly, define and run motion analysis, visualize simulation results Verify simulation results using analytical equations of motion
Single-piston engine		Kinematic analysis	Kinematic analysis of four-bar linkage mechanism Create joints and placement constraints between parts and subassemblies for four-bar linkage Create rotary motor to drive mechanism, carry out motion simulation, visualize simulation results Verify simulation results using analytical equations of motion

constant angular speed (Sections P2.4 and P2.5). Topics to be discussed in both examples are summarized in Table P2.2.

P2.2 Sliding Block

This lesson provides you with a quick run-through on using Pro/ENGINEER Mechanism Design. You will learn pre- and post-processing and analysis capabilities using a simple sliding block. After completing this lesson, you should be able to solve similar problems following the same procedures. In this exercise, we will use default options for most of the selections.

The physical model of the sliding block is very simple. The block is made of steel with a size of 1 in. × 1 in. × 1 in. As shown in Figure P2.6, the block travels a total of 10 in. on the slope surface due to gravity. The units system employed for this example is 1 lb_m in./sec^2 (Pro/E Default), in which the gravitational acceleration is 386 in./sec^2. The slider is released from a rest position at the top of the slope (that is, the initial velocity is 0), then slides down along the slope surface due to gravity pointing in the negative Y-direction of the assembly coordinate system ASM_DEF_CSYS. We assume no friction between the block and the ground.

P2.2.1 Pro/ENGINEER Parts and Assembly

For this lesson, parts and assembly have been created in Pro/ENGINEER. There are four model files prepared for you: block.prt, ground.prt, Lesson2.asm, and Lesson2withresults.asm. We will start with Lesson2.asm, in which the block is assembled to the ground and no motion entities have been added. The assembly file

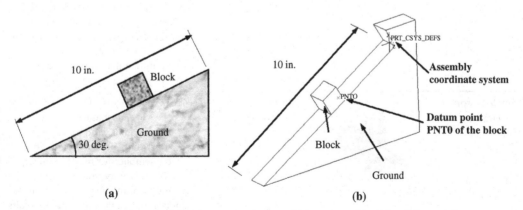

Figure P2.6: Sliding block: (a) schematic view and (b) assembly of solid models in CAD (Pro/ENGINEER).

Lesson2withresults.asm contains a complete motion model with simulation results. You may open this file and review the results before going over this lesson.

The block part has an extruded solid feature, three datum planes (FRONT, TOP, and RIGHT), a coordinate system (PRT_CSYS_DEF), and a datum point PNT0 at the mid-point of its rear bottom edge (see Figure P2.6b), which is used to measure the position of the block. Similarly, the ground part has an extruded solid feature, three datum planes (FRONT, TOP, and RIGHT), and a coordinate system (PRT_CSYS_DEF).

The ground part was brought into the assembly by aligning the respective coordinate systems PRT_CSYS_DEF and ASM_DEF_CSYS. The assembly coordinate system is assigned as the World Coordinate System (WCS) that the physical measures of the motion simulation refer to.

There are two placement constraints created for the block—surface mate and surface align—as shown in Figure P2.7. Note that these two constraints leave one translational degree of freedom for the block; that is, the block is only allowed to move on the slope surface.

P2.2.2 Motion Model

The block and ground parts (or bodies) are assumed rigid. The block is the only movable body in this example. Gravity is added in the negative Y-direction referring to the assembly coordinate system ASM_DEF_CSYS (or WCS), as shown in Figure P2.6b. The block is sitting on top of the slope surface and will travel for 10 in. before passing the bottom edge of the slope surface in about 0.32 seconds, as shown in Figure P2.8a. The graph of Figure P2.8b shows the Y-position of the block, measured

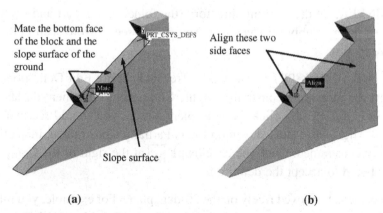

(a)   **(b)**

Figure P2.7: Placement constraints defined in Lesson2.asm: (a) surface mate and (b) surface alignment.

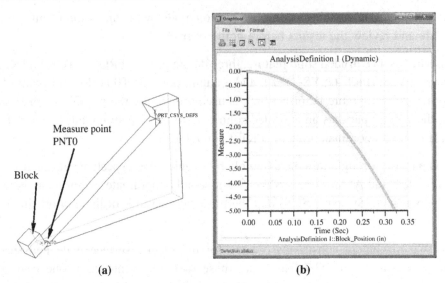

(a)   (b)

Figure P2.8: Sliding block in Pro/ENGINEER Mechanism Design: (a) passing the end of the slope surface and (b) Y-position of the measure point PNT0.

at the datum point PNT0, which coincides with the WCS initially. The *Y*-position of PNT0 should range between 0 and −5 in. This is because the distance the block travels is 10 in. and the slope is 30 deg. Therefore, the *Y*-distance the block travels should be $10 \times \sin 30 \ \deg = 5$ in.

P2.3 Using Pro/ENGINEER Mechanism Design for the Sliding Block

Start Pro/ENGINEER, set the working directory (the folder where part and assembly files reside), and open the assembly Lesson2.asm. You should see Lesson2.asm in the Graphics window (Figure P2.6b).

You may right click BLOCK.PRT in the Model Tree and choose Edit Definition (see Figure P2.9a) to review the placement constraints. In the Component Placement dashboard (upper left, as shown in Figure P2.9b), click the Placement button; you should see two constraints listed: Mate and Align. The status shown on top is Partially Constrained, indicating that the block is not fully constrained. Click the checkmark ✔ at the right of the Component Placement dashboard to accept the definition.

Note that the block can be moved freely on the 30-deg. plane. For example, you may click the Drag Packed Components button 👆, pick the block in the Graphics window, and move the block by moving the mouse.

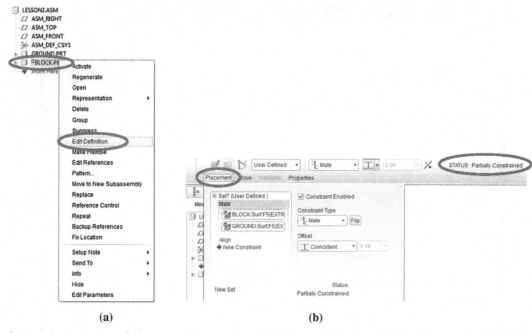

Figure P2.9: Review of placement constraints: (a) right-click the part in Model Tree, and (b) click Placement to see constraints.

Before creating any motion entities, it is always a good idea to check the units system. Choose from the pull-down menu File > Properties. In the Model Properties dialog box (Figure P2.10), choose the default units system, 1 lb_m in./sec^2. Close the dialog box.

From the pull-down menu, choose Applications > Mechanism to enter Mechanism Design. Note that a Mechanism Tree appears right below the model tree, as shown in Figure P2.11. The Mechanism Tree lists existing motion entities. Expand Bodies to see Ground and body1 (block.prt), which are the only motion entities carried over from Pro/ENGINEER (in addition to placement constraints).

P2.3.1 Defining Gravity

Click the Define Gravity shortcut button ⟨8⟩ on the right to bring out the Gravity dialog box (Figure P2.12a). Again, the default values of acceleration (386.088) and direction (0, −1, 0) are what we need. No change is necessary. An arrow appears in the Graphics window indicating that the gravity is acting downward, as shown in Figure P2.12b. Simply click OK in the Gravity dialog box. You need to activate the gravity when you define a dynamic analysis.

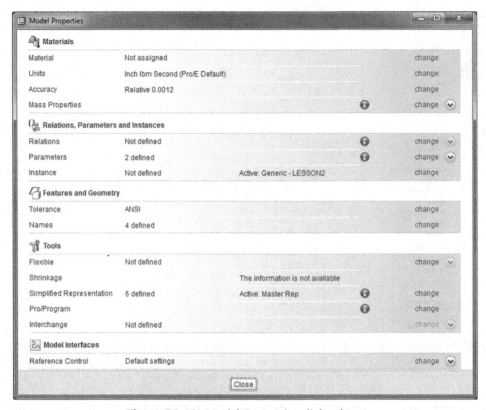

Figure P2.10: Model Properties dialog box.

P2.3.2 Defining Initial Position

We want to bring the block to the upper end of the slope surface as its initial position. The block is then released from this position to simulate sliding without friction. We use the Drag Packed Components to mate the front face at the top of the ground and the rear face of the block, and use it for the initial condition.

Click the Drag Packed Components button [image] at the top of the Graphics window, to see the Drag dialog box (Figure P2.13a). Click the Constraints tab and choose the Mate Two Entities button (second on the left). Pick the front face at the top of the slope surface and the rear side face of the block from the Graphics window (see Figure P2.13b). The block should now be sitting on top of the slope surface. Click the Current Snapshots button [image] to create a snap shot Snapshot1, as shown in Figure P2.13a. Close the Drag dialog box.

Click the Define Initial Conditions button [image] (second from the bottom on the right) to bring out the Initial Condition Definition dialog box (Figure P2.14). Pull down the snapshot field and choose Snapshot1. Click OK to accept the initial condition. Note that since the initial velocity is zero, velocity is left undefined.

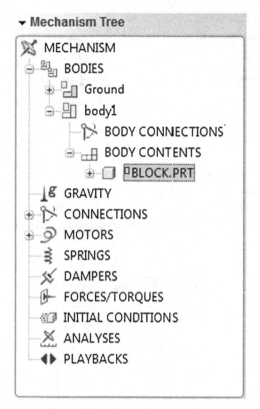

Figure P2.11: Mechanism Tree.

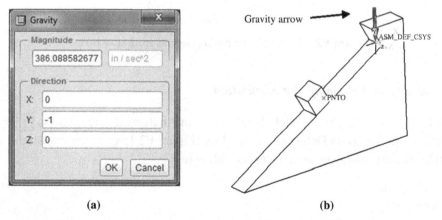

(a)   (b)

Figure P2.12: Define gravity: (a) Gravity dialog box and (b) gravity arrow.

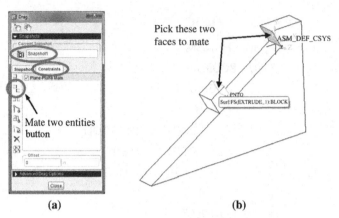

Figure P2.13: Define packed component: (a) Drag dialog box and (b) pick two faces to mate.

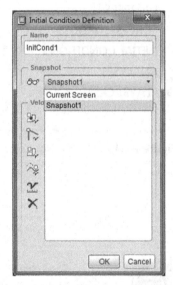

Figure P2.14: Initial Condition Definition dialog box.

P2.3.3 Defining and Running the Simulation

From the button list on the right, click the Mechanism Analysis shortcut button ⊠ to create an analysis. In the Analysis Definition dialog box (Figure P2.15a), leave the default name, AnalysisDefinition1, and choose or enter the following:

Type: Dynamic
Duration: 0.32
Frame Rate: 1000

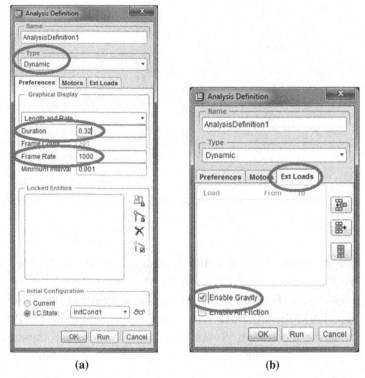

(a)   **(b)**

Figure P2.15: Analysis Definition dialog box: (a) Preferences tab and (b) Ext Loads tab.

Minimum Interval: 0.001
Initial Configuration: I.C. State: InitCond1

Then choose the Ext Loads tab, select Enable Gravity, and click Run (see Figure P2.15b). The progress of the analysis is shown at the top of the Graphics window, and the block starts sliding down on the slope surface. Click OK to save the analysis definition.

P2.3.4 Saving and Reviewing Results

Click the Replay button ◀▶ on the right to bring out the Playbacks dialog box. In the Playbacks dialog box, click the Save button 🖫 to save the results as a *.pbk* file and click the Play button ◀▶ (at the top of the Playback dialog box shown in Figure P2.16a). Play the motion animation by clicking the Play button in the Animation dialog box (Figure P2.16b). Adjust speed by moving the slider if necessary.

Next, we create a measure to monitor the position of the block. We choose the datum point PNT0 of the block and *Y*-component for the measure. To do so, click the Generate

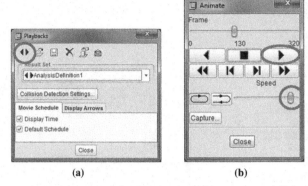

Figure P2.16: Play motion animation: (a) Playbacks dialog box and (b) Animate dialog box.

Measure Results of Analyses shortcut button ⊠ on the right. In the Measure Results dialog box (Figure P2.17), click the Create New Measure button ⬚. The Measure Definition dialog box opens (Figure P2.18). Enter Block_Position for Name. Under Type, select Position. Pick PNT0 in the block from the Graphics window for Point or Motion Axis. Leave WCS as the Coordinate System (default). Choose *Y*-component as the Component. Under Evaluation Method, leave Each Time Step (default). Click OK to accept the definition.

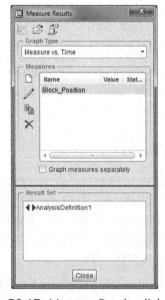

Figure P2.17: Measure Results dialog box.

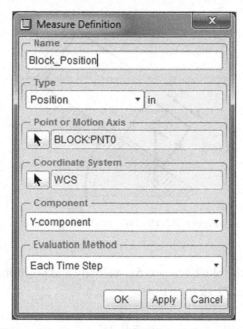

Figure P2.18: Measure Definition dialog box.

In the Measure Results dialog box (Figure P2.17) choose AnalysisDefinition1 in the Result Set and click the Graph button ⊠ on the top left corner to graph the measure. The graph should be similar to that in Figure P2.8b. Go back to Pro/ENGINEER by choosing from the pull-down menu Applications > Standard. Next you can save your model before exiting from Pro/ENGINEER.

P2.3.5 Results Verification

In this section, we verify analysis results obtained from Mechanism Design. We assume that the block is of a concentrated mass so that the particle dynamics theory is applicable. In addition, we assume that there is no air friction and no friction between the sliding faces.

It is well known that the equations of motion can be derived from Newton's second law for the block. By sketching a free-body diagram, shown in Figure P2.19, we have the following force equilibrium equation:

$$F = ma = mg \sin 30 = 0.5mg \qquad \text{(P2.1a)}$$

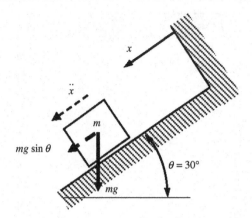

Figure P2.19: Free-body diagram.

Hence, the acceleration of the block is $a = 0.5g$. Velocity and position of the block can then be obtained by integrating Eq. P2.1a over time:

$$v = at = 0.5gt \qquad (P2.1b)$$

$$s = \frac{1}{2}at^2 = 0.25gt^2 \qquad (P2.1c)$$

The Y-position of the block can be obtained as

$$P_y = -\frac{1}{2}at^2 \sin 30 = -\frac{1}{8}gt^2 \qquad (P2.2)$$

These equations can be implemented using, for example, Microsoft Excel for numerical solutions. The Y-position of the block from 0 to 3.22 seconds is shown in Figure P2.20. Comparing the graph with that of Pro/ENGINEER Mechanism Design (Figure P2.8b), they agree very well.

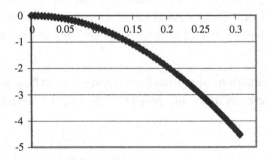

Figure P2.20: Y-position of the block obtained from spreadsheet calculations.

P2.4 Single-Piston Engine

In this lesson, we learn how to create a simulation model for the single-piston engine shown in Figure P2.21a. Instead of using placement constraints to assemble these components and parts, as in Section P2.2, we use joints, such as pin, sliders, and so on, for assembling parts and components. Pro/ENGINEER provides two approaches—placement constraints and joints—for assembly. In general, a joint is kinematically equivalent to, and can be represented by, one placement constraint, or multiple constraints combined. For example, a pin joint (also called revolute joint or hinge joint) that allows only rotation between two components is equivalent to a combination of axis alignment and surface mate placement constraints.

You will learn how to select joints to connect components for an adequate motion model. We drive the mechanism by rotating the crankshaft with a constant angular velocity, basically, conducting a kinematic analysis. We start this example with a brief overview of the parts and subassemblies created in Pro/ENGINEER. At the end, we verify the kinematic simulation results using theory and computational methods commonly found in a mechanism design textbook.

Kinematically, the single-piston engine is a four-bar linkage, as shown in Figure P2.21b. For an internal combustion engine, the linkage is driven by a firing load that pushes the piston, converting the reciprocal motion into rotational motion at the crank. However, in this example, since our goal is to conduct a kinematic analysis, we create a rotary motor at the crank to drive the four-bar linkage. The rotational motion is then converted into a reciprocal motion at the piston.

Note that for a four-bar linkage, the length of the crank must be smaller than that of the rod to allow the mechanism to operate. This is commonly referred to as Grashof's law. In this example, the lengths of the crank and rod are 0.58 and 2.25 in., respectively, which satisfies the requirement of Grashof's law. The units system chosen for this example is IPS (in.-lb$_f$-sec). No friction is present between any pair of components (parts or subassemblies).

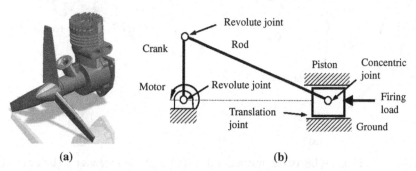

(a)   **(b)**

Figure P2.21: Single-piston engine: (a) CAD model and (b) schematic view of the kinematic model.

P2.4.1 Pro/ENGINEER Parts and Assembly

The engine assembly consists of three subassemblies (case, propeller, and connecting rod) and one part (piston), as shown in Figure P2.22a. There are 23 model files created, including four assemblies. Among the four assemblies, three are subassemblies mentioned above. The remaining assembly is Lesson3withresults.asm, which contains a complete simulation model with simulation results. You may open this assembly to review parts and assembly, as well as simulation results. Note that joints, instead of placement constraints, are employed to assemble the part and subassemblies. If you open Lesson3withresults.asm and choose pull-down menu Applications > Mechanism to enter Mechanism Design, you should see four joints, a pin joint (Pin1) between the propeller and case, another pin joint (Pin2) between the connecting rod and the crankshaft (propeller), a slider joint between the piston and the case, and a bearing joint between the piston and the piston pin (mounted on the connecting rod). Note that datum features, including axes and points, are created in parts and subassemblies to aid the joint definition and for defining measures and generating graphs.

P2.4.2 Motion Model

As mentioned before, four joints are created for the engine assembly in Pro/ENGINEER. These four joints are carried over to Mechanism Design as part of the motion model.

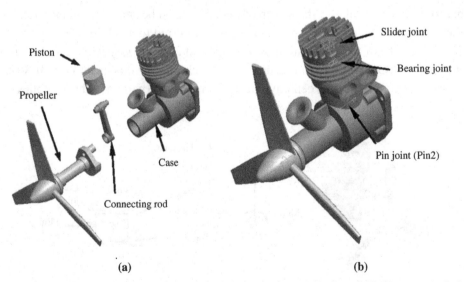

(a)   (b)

Figure P2.22: Pro/ENGINEER parts and assembly: (a) exploded view and (b) unexploded view with joints.

The pin joint (Pin1) allows one rotational motion between the propeller and the ground body (case), therefore removing 5 dof. The second pin joint (Pin2) allows rotation motion between the propeller and the connecting rod. After assembling the propeller and the connecting rod, the system should have 2 dof, allowing the propeller and the connecting rod to rotate along their respective joint axes independently.

Next, the piston is assembled to the piston pin (which is rigidly assembled to the connecting rod) by defining a bearing joint. The piston is free to translate along the axis of the piston pin and free to rotate in all three directions. That is, a bearing joint removes 2 dof. The total dof of the assembly increases to six.

Finally, the piston is assembled to the ground body (case) by defining a slider joint, which allows only one translational movement between the piston and the ground, removing 5 dof. The slider joint enforces the bearing joint between the piston and piston pin to behave more like a pin joint, allowing only rotation along the common axes. The engine becomes a four-bar linkage mechanism restricted to planar motion. Individually, the propeller and the connecting rod are allowed to rotate. The piston is allowed for a translational movement. However, all rotations and translational motion are related to form a closed-loop mechanism, leaving only 1 free dof, which can be the rotation of the propeller or the translation of the piston.

The total dof of the engine can also be calculated as follows:

$$3(\text{bodies}) \times 6(\text{dof}/\text{body}) - 2(\text{pins}) \times 5(\text{dof}/\text{pin}) - 1(\text{slider joint}) \times 5(\text{dof}/\text{slider})$$
$$- 1(\text{bearing joint}) \times 2(\text{dof}/\text{bearing}) = 18 - 17 = 1$$

It is important to point out that the way the joints are defined is not unique. One would probably create three pin joints (replacing the bearing joint with a pin joint between the piston and the piston pin) and one slider joint, which still generates a valid motion model. However, the total dof becomes -2. This is because there are 3 redundant dof created in the system. This is fine since Mechanism Design filters out the redundant dof in conducting motion analysis.

For this engine example, since there is only 1 free dof, one single driver or force will move the mechanism and uniquely determine the motion of individual bodies in the mechanism in the time domain.

We need to define a servomotor (driver) that rotates the propeller through the rotation axis of the joint Pin1 (between the propeller and the ground body). As a result, the motor drives the four-bar linkage mechanism. The driver rotates the propeller at a constant angular velocity of 360 deg./sec (60 rpm) for a kinematic analysis. The position and velocity of the piston obtained from the motion analysis are shown in Figure P2.23a and P2.23b, respectively. Note

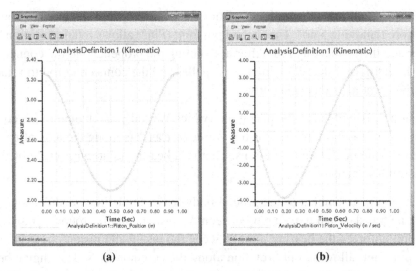

(a)   (b)

Figure P2.23: Result graphs obtained from Pro/ENGINEER Mechanism Design: (a) *Y*-position of the piston and (b) *Y*-velocity of the piston.

that from the position graph, the piston moves approximately between about 2.1 and 3.3 in. vertically. The total travel distance is about 1.2 in., which can be easily verified by the radius of the crank, which is 0.58 in. The piston travel distance is 2 times the radius of the crank, which is 1.16 in.

P2.5 Using Pro/ENGINEER Mechanism Design for the Single-Piston Engine

Start Pro/ENGINEER and open assembly file Lesson3withresults.asm. First take a look at the exploded view by right-clicking the root assembly (Lesson3), and choose Explode. You should see the assembly in an exploded view similar to that of Figure P2.22a. Right-click the root assembly and choose Collapse to collapse the assembly. If you enter Mechanism Design, joint symbols like those in Figure P2.22b appear.

Start Pro/ENGINEER, set the working directory, create a new assembly, and name it Lesson3. You should see a default assembly with three datum planes and one datum coordinate system. Before proceeding, you may want to make sure that the unit system is properly chosen for the part. The unit system for all four parts and the assembly should be IPS (in.-lb$_f$-sec).

Now you are ready to bring in components for assembly.

The *case* is brought in first. It is fixed to the assembly by aligning their respective coordinate system; hence, it becomes a ground body. As mentioned previously, there are four joints to be

Table P2.3: Joints defined in the motion model.

	Case (ground body)	Propeller	Connecting rod	Piston
Propeller	Pin1: AA_1 (propeller)/AA_1 (case), and end faces		Pin2: AA_2 (propeller)/ AA_2 (connecting rod), and end faces	
Connecting rod		Pin2: AA_2 (propeller)/ AA_2 (connecting rod), and end faces		Bearing: PNT2 (piston)/ AA_1 (connecting rod)
Piston	Slider: A_1 (piston)/AA_2 (case), and DTM1 (piston)/ASM_RIGHT (assembly)		Bearing: PNT2 (piston)/ AA_1 (connecting rod)	

defined. Joints to be created in this simulation model are summarized in Table P2.3. Please review the table before moving to the next steps.

Now we are ready to bring in the first component, case, which is attached to the assembly by aligning their respective coordinate systems, CASE_COORD_CSYS and ASM_DEF_CSYS, as shown in Figure P2.24a.

Click the Add component shortcut button ![] and choose case.asm. You may want to turn off the datum features, except for the datum coordinate system display to show only coordinate systems. Click the Placement button on top (see Figure P2.24b). In the Component Placement dashboard, select Coord Csys, and then select CASE_COORD_CSYS (case) and ASM_DEF_CSYS (assembly). The case is fully assembled. Click the checkmark button ![✓] to accept the definition.

Next, we bring in the propeller. Before we do that, please note that we define a pin joint by aligning axis AA_1 (propeller) with AA_1 (assembly) and choosing two mating faces between the propeller and the case to restrain the translational motion, as shown in Figures P2.25a and P2.25b. The pin joint allows 1 rotation dof along the X-direction of the assembly coordinate system, which is chosen as the WCS by Mechanism Design.

Click the Add component shortcut button ![] and choose propeller.asm. You may want to move and rotate the propeller to a position and orientation like those of Figure P2.25a. You may also want to turn on only the datum axis display to show datum axes.

In the Component Placement dashboard (see Figure P2.25c), choose the Pin joint from the User Defined list, and pick AA_1 (propeller) and AA_1 (assembly). The propeller is repositioned to a configuration where the two axes align. Next, click Translation node (see

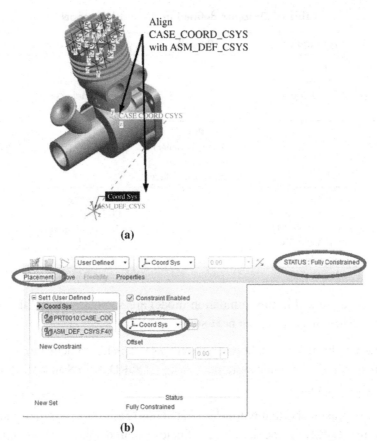

Figure P2.24: Assembling case.asm: (a) choosing two respective coordinate systems and (b) Component Placement dashboard.

Figure P2.25d) and pick the two end faces from the Graphics window, as shown in Figure P2.25d. The status at the top of the Graphics window should show Connection Definition Complete. This means that the pin joint is defined completely (not necessarily implying a fully constrained component). A pin joint symbol appears. Even though you are able to drag/rotate the propeller, it is recommended that you stay with the current configuration and not drag the component. Click the ✅ button to accept the definition.

Similarly, we assemble the connecting rod using a pin joint. Follow the steps described above to define a pin joint between the propeller (at the crankshaft) and the connecting rod. Note that you need to pick two axes, AA_2 (propeller) and AA_2 (connecting rod), and two faces, as shown in Figure P2.26 for the pin joint.

The final part we are bringing in is the piston. The piston is assembled to the piston pin (as part of connecting rod subassembly) and the ground body, using a bearing and a slider joint,

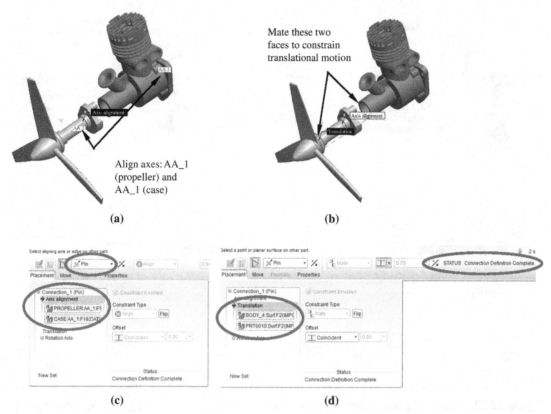

Figure P2.25: Assembling propeller.asm: (a) choosing two axes for rotation, (b) choosing two mate faces for constraining translational motion, (c) Component Placement dashboard, (d) constraining translation dofs and checking joint status.

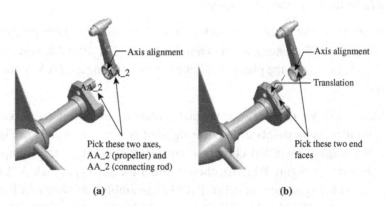

Figure P2.26: Assembling connectingrod.asm: (a) choosing two axes for rotation and (b) choosing two mate faces for constraining translational motion.

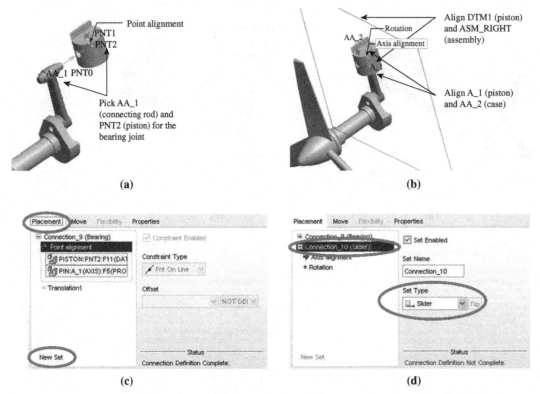

Figure P2.27: Assembling piston.prt: (a) choosing two axes for the bearing joint, (b) choosing two axes and two datum planes for the slider joint, (c) selecting New Set in the Placement dashboard, and (d) choosing the Slider joint.

respectively. You may want to translate and rotate the piston to a configuration similar to that of Figure P2.27a to facilitate the joint definition.

In the Component Placement dashboard, click Bearing from the User Defined list. Pick AA_1 (connecting rod) and PNT2 (piston), as shown in Figure P2.27a. PNT2 is now constrained to move on axis AA_1; therefore, the piston is allowed to move along axis AA_1 and rotate in all three directions.

Next, we will define a slider joint between the piston and the case. Click the Placement button in the Component Placement dashboard; a bearing joint is listed, as shown in Figure P2.27c. Click the New Set button (lower left corner) to create a new joint. In the Component Placement dashboard (see Figure P2.27d), choose Slider for Set Type. Pick A_1 (piston) and AA_2 (case), and DTM1 (piston) and ASM_RIGHT (assembly), as shown in Figure P2.27b, to define the slider joint. You should see that the piston is properly assembled and a slider joint symbol appears.

Now we have completely assembled the mechanism. We will create a snapshot for the current configuration of the assembly using the same steps learned in Section P2.2, and use the snapshot for defining the initial condition for motion analysis later.

Click the Drag Packed Components button ⬚ at the top of the Graphics window, and the Drag dialog box appears (see Figure P2.13a of Section P2.2). Click the Current Snapshots button ⬚ to create a snapshot, Snapshot1.

P2.5.1 Creating a Motion Study

Now we are ready to enter Mechanism Design. We create a kinematic analysis model by defining the initial condition, adding a rotary motor to the propeller, and defining a kinematic analysis to simulate the motion of the single-piston engine.

From the pull-down menu, choose Applications > Mechanism. We first define an initial condition for the simulation model. The initial condition can be created following steps learned in Section P2.2.

Click the Define Initial Conditions button ⬚ (second from the bottom on the right) to bring up the Initial Condition Definition dialog box (see Figure P2.14). Pull down the snapshot field and choose Snapshot1. Click OK to accept the initial condition.

Now we create a servomotor at the rotational axis of the pin joint between the propeller and the ground body (Pin1). The servomotor rotates the propeller at a constant angular velocity of 360 deg./sec.

From the shortcut buttons on the right, click Define Servo Motors ⬚ (sixth from the top), or choose from the pull-down menu Insert > Servo Motors. The Servo Motor Definition dialog box appears (Figure P2.28a). Enter Motor1 for Name, leave Motion Axis (default) for Driven Entity (under the Type tab), then pick Pin1 from the Graphics window. Note that you may want to turn off all datum features to see the pin joint clearly (see Figure P2.28b to locate the pin joint). After selecting the pin joint axis, a larger arrow appears to confirm your selection.

The next step is to specify the profile of the motor. From the Servo Motor Definition dialog box, pick the Profile tab (Figure P2.28c), choose Velocity in Specification, and leave Constant (default) in Magnitude. Enter 360 for the constant A, and click OK. A motor symbol should appear at the pin joint in the Graphics window (Figure P2.28d).

P2.5.2 Defining and Running the Simulation

From the button list on the right, click the Mechanism Analysis shortcut button ⬚ to create an analysis. In the Analysis Definition dialog box (see Figure P2.15a), leave the default name, AnalysisDefinition1, and choose or enter

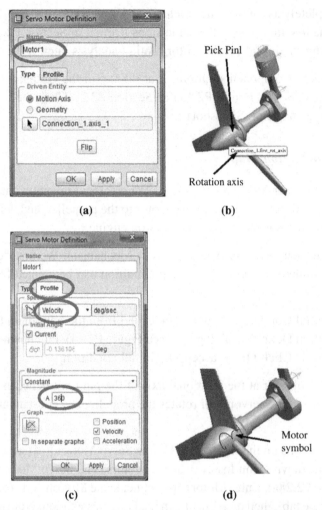

Figure P2.28: Defining a servomotor: (a) Servo Motor Definition dialog box, (b) choosing the pin joint Pin1, (c) Servo Motor Definition dialog box (Profile tab), and (d) motor created.

Type: Kinematic
Duration: 1
Frame Rate: 100
Minimum Interval: 0.01
Initial Configuration: I.C.State: InitCond1

Then choose the Motors tab, and click the Add New Row button (first on the right). The servomotor Motor1 should appear (see Figure P2.29).

Figure P2.29: Motors tab of the Analysis Definition dialog box.

Click Run. The progress of the analysis is shown at the top of the Graphics window, and the propeller starts turning. The propeller makes a complete turn in 1 second. Click OK to save the analysis definition.

P2.5.3 Saving and Reviewing Results

As in Section P2.2, click the Replay button [◀▶] on the right to bring up the Playbacks dialog box and repeat the motion animation. On the Playbacks dialog box, click the Save button [💾] to save the results as a *.pbk* file and click the Play button [◀▶] (on top of the Playback dialog box). Play the motion animation by clicking the Play button in the Animation dialog box.

Next, we create two measures to monitor the vertical position and velocity of the piston. We choose the datum point PNT0 of the piston and *Y*-component for the measure. You may follow the same steps as in Section P2.2 to create these two graphs. These two graphs should be the same as those shown in Figures P2.23a and P2.23b, respectively. Save your model.

P2.5.4 Results Verification

In this section, we verify the motion analysis results using the kinematic analysis theory often found in mechanism design textbooks. Note that in kinematic analysis, position, velocity, and acceleration of given bodies, points, or axes of joints in the mechanism are analyzed.

In kinematic analysis, forces and torques are not involved. All bodies (or links) are assumed massless. Hence, mass properties defined for bodies do not influence the analysis results.

The motion characteristics of the single-piston engine can be modeled as a slider-crank mechanism, which is a planar kinematic analysis problem of a four-bar linkage. A vector plot that represents the positions of joints of the planar mechanism is shown in Figure P2.30. The vector plot serves as the first step in computing position, velocity, and accelerations of the mechanism.

The position equations of the system can be described by the following vector summation:

$$\mathbf{Z}_1 + \mathbf{Z}_2 = \mathbf{Z}_3 \tag{P2.3}$$

where

$$\mathbf{Z}_1 = Z_1 \cos \theta_A + i Z_1 \sin \theta_A = Z_1 e^{i\theta_A}$$

$$\mathbf{Z}_2 = Z_2 \cos \theta_B + i Z_2 \sin \theta_B = Z_2 e^{i\theta_B}$$

$Z_3 = Z_3$, since θ_C is always 0.

The real and imaginary parts of Eq. P2.3, corresponding to the X- and Y-components of the vectors, can be written as

$$Z_1 \cos \theta_A + Z_2 \cos \theta_B = Z_3 \tag{P2.4a}$$

$$Z_1 \sin \theta_A + Z_2 \sin \theta_B = 0 \tag{P2.4b}$$

In Eqs. P2.4a and P2.4b, Z_1, Z_2, and θ_A are given. We are solving for Z_3 and θ_B. Equations P2.4a and P2.4b are nonlinear in terms of θ_B. Solving them directly for Z_3 and θ_B is not

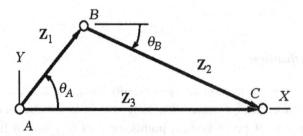

Figure P2.30: Vector plot of the slider-crank mechanism.

straightforward. Instead, we calculate Z_3 first, using trigonometric relations for the triangle ABC shown in Figure P2.30:

$$Z_2^2 = Z_1^2 + Z_3^2 - 2Z_1Z_3 \cos \theta_A$$

Hence,

$$Z_3^2 - 2Z_1 \cos \theta_A Z_3 + Z_1^2 - Z_2^2 = 0$$

Solving Z_3 from the above quadratic equation, we have

$$Z_3 = \frac{2Z_1 \cos \theta_A \pm \sqrt{(2Z_1 \cos \theta_A)^2 - 4(Z_1^2 - Z_2^2)}}{2} \qquad (P2.5)$$

where two solutions of Z_3 represent the two possible configurations of the mechanism shown in Figure P2.31. Note that point C can be either at C or C′ for a given Z_1 and θ_A.

From Eq. P2.4b, θ_B can be solved by

$$\theta_B = \sin^{-1}\left(\frac{-Z_1 \sin \theta_A}{Z_2}\right) \qquad (P2.6)$$

Similarly, θ_B has two possible solutions, corresponding to vector $\mathbf{Z}_3$.

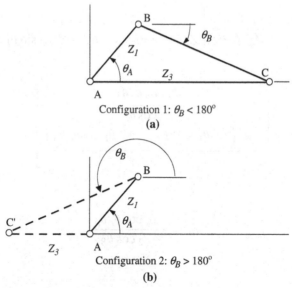

Configuration 1: $\theta_B < 180°$

(a)

Configuration 2: $\theta_B > 180°$

(b)

Figure P2.31: Two possible configurations of the mechanism.

Taking derivatives of Eqs. P2.4a and P2.4b with respect to time, we have

$$-Z_1 \sin \theta_A \dot{\theta}_A - Z_2 \sin \theta_B \dot{\theta}_B = \dot{Z}_3 \qquad \text{(P2.7a)}$$

$$Z_1 \cos \theta_A \dot{\theta}_A + Z_2 \cos \theta_B \dot{\theta}_B = 0 \qquad \text{(P2.7b)}$$

where $\dot{\theta}_A = \dfrac{d\theta_A}{dt} = \omega_A$ is the angular velocity of the rotation driver, which is a constant in this example.

Note that Eqs. P2.7a and P2.7b are linear functions of $\dot{Z}_3$ and $\dot{\theta}_B$. Rewrite the equations in a matrix form:

$$\begin{bmatrix} Z_2 \sin \theta_B & 1 \\ Z_2 \cos \theta_B & 0 \end{bmatrix} \begin{bmatrix} \dot{\theta}_B \\ \dot{Z}_3 \end{bmatrix} = \begin{bmatrix} -Z_1 \sin \theta_A \dot{\theta}_A \\ -Z_1 \cos \theta_A \dot{\theta}_A \end{bmatrix} \qquad \text{(P2.8)}$$

Equation P2.8 can be solved by

$$\begin{bmatrix} \dot{\theta}_B \\ \dot{Z}_3 \end{bmatrix} = \begin{bmatrix} Z_2 \sin \theta_B & 1 \\ Z_2 \cos \theta_B & 0 \end{bmatrix}^{-1} \begin{bmatrix} -Z_1 \sin \theta_A \dot{\theta}_A \\ -Z_1 \cos \theta_A \dot{\theta}_A \end{bmatrix}$$

$$= \frac{1}{-Z_2 \cos \theta_B} \begin{bmatrix} 0 & -1 \\ -Z_2 \cos \theta_B & Z_2 \sin \theta_B \end{bmatrix} \begin{bmatrix} -Z_1 \sin \theta_A \dot{\theta}_A \\ -Z_1 \cos \theta_A \dot{\theta}_A \end{bmatrix}$$

$$= \frac{1}{-Z_2 \cos \theta_B} \begin{bmatrix} Z_1 \cos \theta_A \dot{\theta}_A \\ Z_1 Z_2 \cos \theta_B \sin \theta_A \dot{\theta}_A - Z_1 Z_2 \sin \theta_B \cos \theta_A \dot{\theta}_A \end{bmatrix}$$

$$= \begin{bmatrix} -\dfrac{Z_1 \cos \theta_A \dot{\theta}_A}{Z_2 \cos \theta_B} \\ -\dfrac{Z_1 \left(\cos \theta_B \sin \theta_A \dot{\theta}_A - \sin \theta_B \cos \theta_A \dot{\theta}_A \right)}{\cos \theta_B} \end{bmatrix} \qquad \text{(P2.9)}$$

Hence,

$$\dot{\theta}_B = -\frac{Z_1 \cos \theta_A \dot{\theta}_A}{Z_2 \cos \theta_B} \qquad \text{(P2.10)}$$

and

$$\dot{Z}_3 = Z_1 \left(\tan \theta_B \cos \theta_A \dot{\theta}_A - \sin \theta_A \dot{\theta}_A \right) \qquad \text{(P2.11)}$$

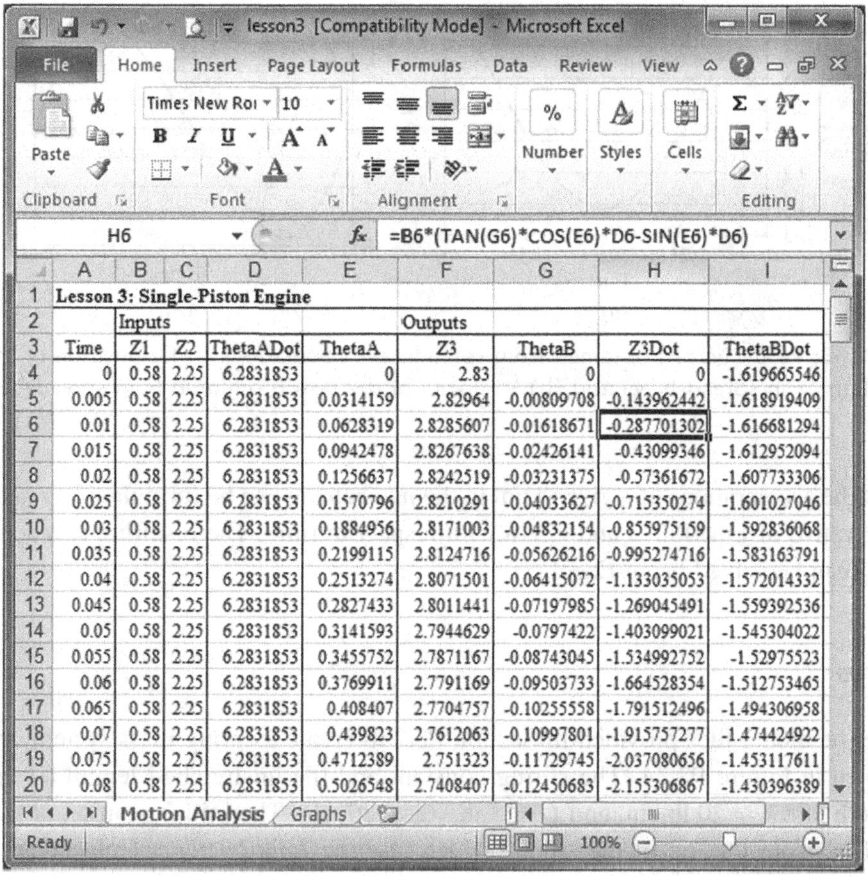

The Excel spreadsheet window shows:

Title bar: lesson3 [Compatibility Mode] - Microsoft Excel

Cell reference box: H6, formula bar: f_x =B6*(TAN(G6)*COS(E6)*D6-SIN(E6)*D6)

	A	B	C	D	E	F	G	H	I
1	Lesson 3: Single-Piston Engine								
2		Inputs				Outputs			
3	Time	Z1	Z2	ThetaADot	ThetaA	Z3	ThetaB	Z3Dot	ThetaBDot
4	0	0.58	2.25	6.2831853	0	2.83	0	0	-1.619665546
5	0.005	0.58	2.25	6.2831853	0.0314159	2.82964	-0.00809708	-0.143962442	-1.618919409
6	0.01	0.58	2.25	6.2831853	0.0628319	2.8285607	-0.01618671	-0.287701302	-1.616681294
7	0.015	0.58	2.25	6.2831853	0.0942478	2.8267638	-0.02426141	-0.43099346	-1.612952094
8	0.02	0.58	2.25	6.2831853	0.1256637	2.8242519	-0.03231375	-0.57361672	-1.607733306
9	0.025	0.58	2.25	6.2831853	0.1570796	2.8210291	-0.04033627	-0.715350274	-1.601027046
10	0.03	0.58	2.25	6.2831853	0.1884956	2.8171003	-0.04832154	-0.855975159	-1.592836068
11	0.035	0.58	2.25	6.2831853	0.2199115	2.8124716	-0.05626216	-0.995274716	-1.583163791
12	0.04	0.58	2.25	6.2831853	0.2513274	2.8071501	-0.06415072	-1.133035053	-1.572014332
13	0.045	0.58	2.25	6.2831853	0.2827433	2.8011441	-0.07197985	-1.269045491	-1.559392536
14	0.05	0.58	2.25	6.2831853	0.3141593	2.7944629	-0.0797422	-1.403099021	-1.545304022
15	0.055	0.58	2.25	6.2831853	0.3455752	2.7871167	-0.08743045	-1.534992752	-1.52975523
16	0.06	0.58	2.25	6.2831853	0.3769911	2.7791169	-0.09503733	-1.664528354	-1.512753465
17	0.065	0.58	2.25	6.2831853	0.408407	2.7704757	-0.10255558	-1.791512496	-1.494306958
18	0.07	0.58	2.25	6.2831853	0.439823	2.7612063	-0.10997801	-1.915757277	-1.474424922
19	0.075	0.58	2.25	6.2831853	0.4712389	2.751323	-0.11729745	-2.037080656	-1.453117611
20	0.08	0.58	2.25	6.2831853	0.5026548	2.7408407	-0.12450683	-2.155306867	-1.430396389

Sheet tabs: Motion Analysis / Graphs

Figure P2.32: Excel spreadsheet.

In this example, $Z_1 = 0.58$, $Z_2 = 2.25$, and the initial conditions are $\theta_A(0) = 0$ and $\theta_B(0) = 0$.

The solutions can be implemented using a spreadsheet. The Excel spreadsheet file, *lesson3.xls*, can be found on the book's companion website (http://booksite.elsevier.com/ 9780123984609). As shown in Figure P2.32, columns A to I represent time, Z_1, Z_2, $\dot{\theta}_A$, θ_A, Z_3, θ_B, $\dot{Z}_3$, and $\dot{\theta}_B$, respectively. Note that in this calculation, $Z_3(0) > 0$ is assumed, hence $\theta_B(0) < 0$ (clockwise), as illustrated in Figure P2.31. This is consistent with the initial configuration we created in Mechanism Design.

Figures P2.33a and P2.33b show the graphs of data in columns F and H; that is, piston position and velocity, respectively. Comparing Figures P2.33a and P2.33b with Figures P2.23a and P2.23b, the simulation analysis results are verified. Note that in Figure P2.33a, the piston position varies between 1.7 and 2.8 in.; that is, the distance the piston travels is about 1.1 in., which is the same as that observed in Figure P2.23a.

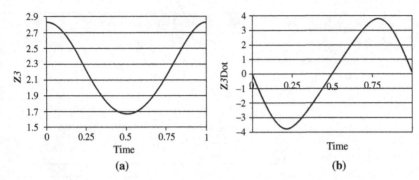

Figure P2.33: Result graphs obtained from spreadsheet calculations: (a) Y-position of the piston (column F in spreadsheet) and (b) Y-velocity of the piston (column H in spreadsheet).

Note that the accelerations of a given joint in the mechanism can be formulated by taking one more derivative of Eqs. P2.7a and P2.7b with respect to time. The resulting two coupled equations can be solved using Excel.

Exercises

P2.1 Use the model files provided in Section P2.2 to create a spring-mass system, as shown in Figure P2.34. The spring constant and the unstretched length (or free length) are $k = 20$ lb$_f$/in. and $U = 3$ in., respectively. The initial position of the block is 4 in. from the top of the slope surface (use the *LimitDistance1* placement constraint). As a result, we are simulating a free vibration problem, where the block is stretched 1 in. downward along the 30-deg. slope. No friction is assumed between

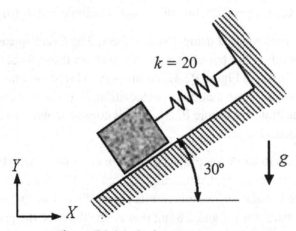

Figure P2.34: Spring-mass system.

the block and the slope surface. We assume a gravity of $g = 386$ in./sec^2 in the negative Y-direction.

(a) Carry out a motion analysis for the system, and graph the position, velocity, and acceleration of the block in the Y-direction.

(b) Derive equations of motion for the system, and implement the solutions of the equations of motions using a spreadsheet. Graph the results and compare them with those of (a). Are you getting accurate motion simulation results from Mechanism Design?

(c) Calculate the natural frequency of the system and compare your calculation with that of Mechanism Design. How do you figure out natural frequency of the system from results obtained from Mechanism Design?

P2.2 Derive the acceleration equations for the slider-crank mechanism, by taking derivatives of Eqs. P2.7a and P2.7b with respect to time. Solve these equations for the linear acceleration of the piston and the angular acceleration of the pin joint Pin2, using a spreadsheet. Compare your solutions with those obtained from Mechanism Design.

Project S2 Motion Analysis Using SolidWorks® Motion

Chapter Outline

Motion analysis was reviewed in Chapter 3, in which both theoretical and practical aspects of the subject were discussed. There is no need to emphasize the importance of understanding theory and using analytical methods to solve engineering problems. However, as discussed in Chapter 3, analytical methods can only go so far. For many design applications, such as mechanical system analysis, which involves multiple bodies, analytical methods can only support a small number of rigid bodies—that is, one or two, at best. In many cases, engineers must rely on tools such as motion analysis software to evaluate product performance in mechanisms with high accuracy in support of design decision making. It is critical for design engineers to learn and be able to use software tools for solving problems that are beyond hand calculations relying on analytical methods.

In this project, we introduce SolidWorks Motion for kinematic and dynamic analysis of mechanical systems. We include two simple examples: a sliding block and a single-piston engine, which should provide a good overview of the motion analysis capabilities offered by SolidWorks Motion. You may find the example files on the book's companion website (http://booksite.elsevier.com/9780123984609).

Overall, the objective of this project is to enable readers to use SolidWorks Motion for basic applications. Those who are interested in learning more may want to go over the examples provided by Motion or review other tutorials on the subject, such as *Motion Simulation and Mechanism Design with SolidWorks Motion* (available at http://www.sdcpublications.com).

Note that the lessons in this project are developed using SolidWorks 2011 SP2.0. A different version of SolidWorks may have slightly different menu options or dialog boxes. Since SolidWorks is fairly intuitive to use, these differences should not be too difficult to figure out.

S2.1 Introduction to SolidWorks Motion

SolidWorks Motion (or Motion) is a computer software tool for analyzing the kinematic and dynamic performance of mechanical systems, as well as for the support of mechanism design. The main objective of this tutorial is to help new users to become familiar with it and to provide a brief overview of the concept and detailed steps in using the software.

Motion is fully integrated and embedded in SolidWorks as an add-in module. The transition from SolidWorks to Motion is seamless. All solid parts, materials, assembly mates, and so forth defined in SolidWorks are automatically carried over into Motion. Also, assembly mates created in solid models are directly utilized for defining motion analysis models. Motion can be accessed through menus and windows in SolidWorks. The same parts and assemblies created in SolidWorks are directly employed for motion models in Motion, whose menus and dialog boxes are just like those in SolidWorks. For those who are familiar with SolidWorks, Motion is relatively easy to learn.

S2.1.1 Overall Process

The use of SolidWorks Motion for mechanism design and analysis consists of three main steps: model generation (or preprocessing), analysis (or simulation), and result visualization (or postprocessing), as illustrated in Figure S2.1. Basic entities that constitute a motion model include bodies, joints, initial conditions, and force or driver. A body can be stationary (ground body) or movable. Joints, such as revolute joints and cylindrical joints, are defined between bodies to constrain the relative movement between them. In Motion, joints are

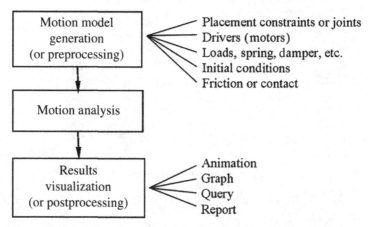

Figure S2.1: Overall Process for SolidWorks Motion.

implicitly defined by assembly mates (as in Motion 2011 and beyond). Users do not create joints to constrain body motion directly. Instead, body motion is governed by assembly mates, which must therefore be properly defined for the mechanism so that the motion model captures essential characteristics and closely resembles the behavior of the physical mechanism. In motion models, at least one degree of freedom must be free to allow the mechanism to move. The free degrees of freedom are usually either driven by a servomotor for kinematic analysis or by an external load (force and torque) for dynamic simulation.

The analysis capabilities in Motion employ a simulation engine, Adams/Solver, which solves the equations of motion for the mechanism. Adams/Solver calculates the position, velocity, and acceleration of individual moving bodies; as well as reaction forces acting on individual moving parts at joints. Typical simulation problems, including static (equilibrium configuration) and motion (kinematic and dynamic), are supported.

The analysis results can be visualized in various forms. You may animate motion of the mechanism or generate graphs for more specific results, such as the reaction force of a joint in the time domain. You may also query results at specific locations for a given time frame. Furthermore, you may ask for a report on results that you specify, such as the acceleration of a moving part in the time domain. You may also save the motion animation to an AVI for file portability.

S2.1.2 User Interface

The Motion user interface is identical to that of SolidWorks, as shown in Figure S2.2. SolidWorks users should find it is straightforward to maneuver in Motion. The main interface is through MotionManager below the graphics area, as shown in Figure S2.2. MotionManager creates and plays animations as well as conducts motion analysis. When an

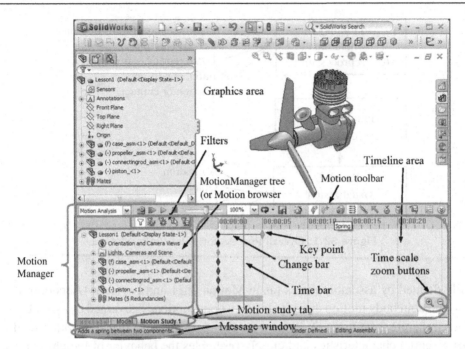

Figure S2.2: SolidWorks Motion user interface—MotionManager.

existing assembly (or part) in SolidWorks is opened, the Motion Study tab (with the default name Motion Study 1) appears at the bottom of the graphics area. Clicking it brings up the MotionManager window.

As shown in Figure S2.2, the user interface window of MotionManager consists of the MotionManager tree (or Motion browser), the Motion toolbar, filters, timeline area, and so forth. Components are mapped from the SolidWorks assembly into the MotionManager tree automatically, including root assembly, parts, subassemblies, and mates. Each part and subassembly entity can be expanded to show its components. Motion entities, such as spring and force, are added to the MotionManager tree once they are created. A result branch is added to the tree once a motion analysis is completed and result graphs are created. As with SolidWorks, right-clicking on a node in the MotionManager tree brings up command options that you can choose to modify or adjust the entity.

The graphics area displays the motion model you are working on. The Motion toolbar shown in Figure S2.3 (and in more detail in Table S2.1) provides the major functions required to create and modify motion models, including creating and running analyses and visualizing results. The toolbar includes type of study (animation, basic motion, or motion analysis), calculation, animation controls, playback speed options, saving options, the Animation Wizard, key point controls, and simulation elements.

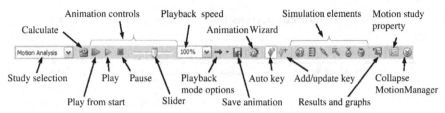

Figure S2.3: Motion toolbar.

Table S2.1: Shortcut buttons in the Motion toolbar.

Symbol	Name	Function
	Calculate	Calculates current simulation; if you alter the simulation you must recalculate before replaying
	Play from start	Reset components and play simulation; use after simulation has been calculated
	Play	Play simulation beginning at the current timebar location
	Stop	Stop animation
	Playback mode: Normal	Play from beginning to end once
	Playback mode: Loop	Continuous play from beginning to end then loop to beginning and continue playing
	Playback mode: Reciprocal	Continuous play from beginning to end then reverse—play from end to beginning
	Save animation	Save animation as AVI movie file
	Animation Wizard	Create rotate model, explode or collapse animation
	Auto key	Click to automatically place new key when you move or change components; click again to toggle

Continued

Table S2.1: Shortcut buttons in the Motion toolbar—cont'd.

Symbol	Name	Function
	Add/update key	Click to add new key or update properties of existing key
	Motor	Create motor for motion analysis
	Spring	Add spring between two components
	Damper	Add damper between two components
	Force	Create force for motion analysis
	Contact	Create 3D contact between selected components
	Gravity	Add gravity to motion study
	Results and graphs	Calculate results and create graphs
	Motion study properties	Define motion study solution parameters
	Collapse MotionManager	Collapse MotionManager window

You can use Animation to animate simple operation of assemblies, such as Rotate, Zoom In/Out, and Explode/Collapse. You may also add motors to animate simple kinematic motion of the assembly. You can use Basic Motion for approximating the effects of motors, springs, collisions, and gravity on assemblies. Basic Motion takes mass into account in calculating motion. Its computation is relatively fast, so you can use this for creating presentation-worthy animations using physics-based simulations. Both Animation and Basic Motion are available in the basic version of SolidWorks. In addition, you can use Motion Analysis (available with the SolidWorks Motion add-in from SolidWorks Premium) to accurately simulate and analyze the motion of an assembly while incorporating the effects of forces such as springs, dampers, and friction.

If you do not see the Motion Analysis option for study, you may have not activated the Motion add-in module. It must be chosen from the pull-down menu Tools > Add-Ins. In the Add-Ins window shown in Figure S2.4, click SolidWorks Motion in both boxes (Active Add-Ins and Start Up), and then click OK. You may need to restart SolidWorks to activate the Motion module.

The area to the right of the MotionManager tree is the timeline area, which is the temporal interface for animation. The timeline area displays the times and types of animation events in the motion study. It is divided by vertical gridlines corresponding to the numerical markers showing the time. The numerical markers start at 00:00:00. You may click and drag a key to define the beginning or end time of the animation or motion simulation.

After a motion analysis is completed, you will see several horizontal bars appear in the timeline area. They are Change bars for connecting key points. They indicate a change between key points, which characterize the duration of animation, view orientation, and so forth.

Switching back and forth between Motion and SolidWorks is straightforward. Click the Model tab (back to SolidWorks) and Motion Study tab (to Motion) at the bottom of the graphics area.

S2.1.3 Defining Motion Entities

As mentioned before, the basic entities of a valid Motion simulation model consist of ground part (or ground body), moving parts (or moving bodies), joints (imposed implicitly by assembly mates in SolidWorks assembly), initial conditions (usually position and velocity of a moving body), and forces or drivers for dynamic and kinematic analyses, respectively.

A ground part, or a ground body, represents a fixed reference in space. The first component brought into the assembly is usually stationary and so becomes a ground part. Parts (or

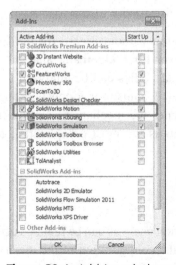

Figure S2.4: Add-Ins window.

subassemblies) assembled to the stationary components without any possibility of movement become part of the ground body. A symbol (f) is placed in front of the stationary components in the browser (or model tree).

A moving part or body represents a single rigid component moving relative to other parts (or bodies). It may consist of a single SolidWorks part or a subassembly composed of multiple parts. When a subassembly is designated as a moving part, none of its composing parts are allowed to move relative to one another within the subassembly.

An unconstrained rigid body in space has six degrees of freedom: three translational and three rotational. When you add a joint (or an assembly mate) between two rigid bodies, you remove degrees of freedom between them. Each independent movement permitted by a joint (or assembly mate) is a free degree of freedom. The free degree of freedom that a joint allows can be translational or rotational along the three perpendicular axes. It is extremely important to understand the definition and characteristics of assembly mates to generate successful motion models. In addition to standard mates such as concentric and coincident, SolidWorks provides advanced and mechanical mates, such as gears. Advanced mates provide additional ways to constrain or couple movements between bodies.

S2.1.4 Motion Simulation

The Adams/Solver employed by Motion is capable of solving typical engineering problems, such as static (equilibrium configuration), kinematic, dynamic, and the like. Static analysis is used to find the rest position (equilibrium condition), in which none of the bodies are moving, of a mechanism. A simple example of static analysis is illustrated in Figure S2.5a, in which an equilibrium position of the block is to be determined according to its own mass m, the two spring constants k_1 and k_2, and the gravity g.

Kinematics is the study of motion without regard for the forces or torque. A mechanism can be driven by a motion driver for a kinematic analysis, where the position, velocity, and acceleration of its individual bodies can be analyzed at any given time. Figure S2.5b shows a servomotor driving a mechanism at a constant angular velocity. Dynamic analysis is employed for studying the mechanism motion in response to loads, as illustrated in Figure S2.5c. This is the most complicated, common, and usually more time-consuming, analysis.

S2.1.5 Viewing Results

In Motion, the results of the motion analysis can be realized using animations, graphs, reports, and queries. Animations show the configuration of the mechanism in consecutive time frames. They provide a global view of the mechanism's behavior—for example, the motion of

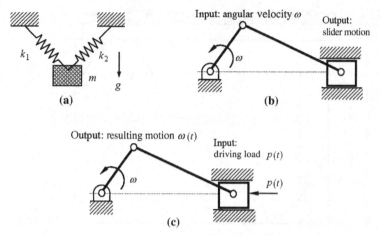

Figure S2.5: Motion analysis capabilities in SolidWorks Motion: (a) static analysis, (b) kinematic analysis, and (c) dynamic analysis.

the single-piston engine as shown in Figure S2.6 (a screen capture of the motion animation). You may also export the animation to AVI for various purposes.

In addition, you may choose an assembly mate or a part to generate result graphs—for example, position versus time of the mass center of the piston in the engine example, shown in Figure S2.7. These graphs give you a quantitative understanding of the characteristics of the mechanism.

You may also query the results by moving the mouse pointer closer to the curve in a graph and leaving it there for a short period. The results data appear next to the pointer. And you may ask

Figure S2.6: Motion animation.

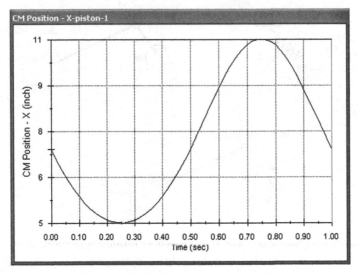

Figure S2.7: Body position versus time.

SolidWorks Motion for a report that includes a complete set of results in the form of textual data or a Microsoft Excel spreadsheet.

In addition to the capabilities just discussed, Motion allows you to check the interference between bodies during motion. Furthermore, the reaction forces calculated can be used to support structural analysis using, for example, SolidWorks Simulation.

S2.1.6 Examples

The two simple examples included in this tutorial project illustrate step-by-step details of modeling, analysis, and result visualization in Motion. We start with a very simple sliding block in Section S2.2. The second example illustrates the steps for carrying out kinematic analysis for a single-piston engine, in which the propeller is driven by a rotary motor at a constant angular speed (Sections S2.4 and S2.5). Both examples are summarized in Table S2.2.

S2.2 Sliding Block

The physical model of the sliding block is very simple. The block is made of cast alloy steel with a size of 1 in. × 1 in. × 1 in. As shown in Figure S2.8, the block travels a total of 9 in. on the slope surface due to gravity. The units system employed for this example is IPS (in.-lb$_f$-sec), in which the gravitational acceleration is 386 in./sec^2. The slider is released from a rest position at the top of the slope (that is, the initial velocity is 0). We assume no friction between the block and the ground.

Table S2.2: Project examples.

Example	FEA Model	Problem Type	Things to Learn
Sliding block		Particle dynamics	Basic operation in MotionManager Create motion model from SolidWorks assembly, define and run motion analysis, visualize simulation results Verify simulation results using analytical equations of motion
Single-piston engine		Kinematic analysis	Kinematic analysis of four-bar linkage mechanism Review assembly mates that are defined between parts and subassemblies for such a four-bar linkage Create rotary motor that drives mechanism, carry out motion simulation, visualize simulation results Verify simulation results using analytical equations of motion

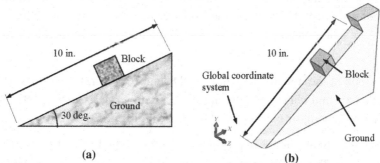

Figure S2.8: Sliding block: (a) schematic view and (b) motion model in CAD (SolidWorks).

S2.2.1 SolidWorks Parts and Assembly

For this lesson, the parts and assembly have been created in SolidWorks. There are four model files created: block.SLDPRT, ground.SLDPRT, Lesson2.SLDASM, and Lesson2withresults.SLDASM. We start with Lesson2.SLDASM, in which the block is assembled to the ground and no motion entities have been added. The assembly file Lesson2withresults.SLDASM contains the complete simulation model with simulation results.

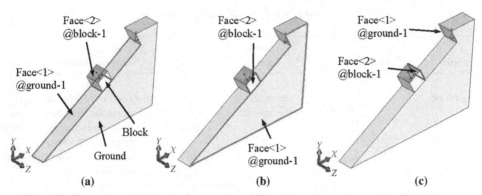

Figure S2.9: Assembly mates defined in Lesson2.SLDASM: (a) coincident1, (b) coincident3, and (c) LimitDistance1.

In the assembly models, there are three assembly mates, as shown in Figure S2.9:

Coincident1(ground<1>,block<1>)

Coincident3(ground<1>,block<1>)

LimitDistance1(ground<1>,block<1>)

The LimitDistance mate allows the block to move along the slope surface for a limited 9 in. distance. You may drag the block in the graphics area; you should be able to move it on the slope surface but not beyond it because the limitDistance mate does not allow this. Choose Edit > Undo Move Component from the pull-down menu to restore the block to its previous position.

We now take a look at the assembly mate LimitDistance1. From the SolidWorks browser, click LimitDistance1, and choose Edit Feature 📷, as shown in Figure S2.10. The mate is brought back for reviewing or editing, as shown in Figure S2.11.

Note that the distance between the two faces, Face<2>@block-1 and Face<1>@ground-1 (Figure S2.9c) is 4 in. The upper and lower limits of the distance are 9 in. and 0 in., respectively. The length of the slope surface is 10 in.; therefore, the upper limit is set to 9 in. so that the block stops when its front lower edge reaches the end of the slope surface (since the block width is 1 in.). Note that LimitDistance1 is an advanced mate in SolidWorks.

S2.2.2 Motion Model

The block and ground parts (or bodies) are assumed rigid. As mentioned earlier, a limit distance mate is defined to prevent the block from sliding out of the slope surface. The block reaches the end of the slope surface in about 0.3 sec, as shown in Figure S2.12, which is the

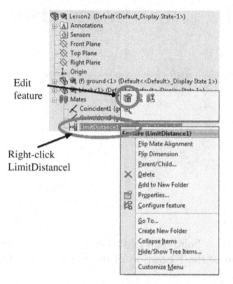

Figure S2.10: Edit LimitDistance1 mate.

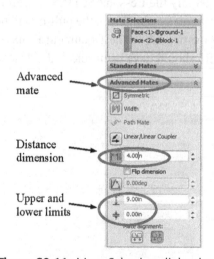

Figure S2.11: Mate Selection dialog box.

Y-position of the mass center of the block. The graph shows that it bounces back when it reaches the end; this action is due the limitDistance mate and is artificial. The *Y*-position of the mass center of the block is about 0.18 in., and travels down to about −4.31 in. at 0.306 sec. You may export the graph to an Excel file to check these numbers in SolidWorks Motion. The total vertical travel distance is 4.49 in.

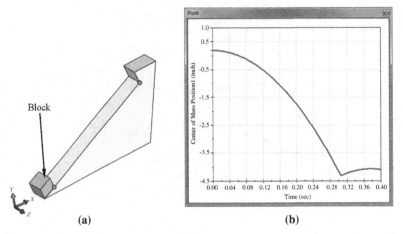

Figure S2.12: Sliding block in SolidWorks Motion: (a) reaching the end of the slope surface; (b) graph of the Y-position of the mass center of the block.

S2.3 Using SolidWorks Motion for the Sliding Block

Start SolidWorks and open assembly file Lesson2.SLDASM. Before creating any entities, it is always a good idea to check the unit system. From the pull-down menu select Tools > Options and choose the Document Properties tab in the Document Properties-Units dialog box (Figure S2.13); click Units. IPS should have been selected. Close the dialog box. Note that in

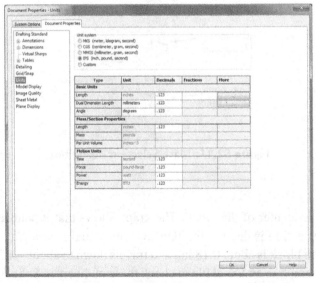

Figure S2.13: Document Properties-Units dialog box.

this units system, the gravity is 386 in./sec^2 in the negative *Y*-direction of the global coordinate system.

Click the Motion Study tab (Motion Study 1) at the bottom of the graphics area to bring up the MotionManager window.

S2.3.1 Defining Gravity

Click the Gravity button 🔲 in the Motion toolbar to bring up the Gravity dialog box. Choose Y and keep the *g*-value (386.09 in./s^2), as shown in Figure S2.14. In the graphics area, an arrow appears at the right lower corner ⬇, pointing downward to indicate the direction of the gravity.

Click the checkmark ✅ on top of the dialog box to accept the gravity. A gravity node (Gravity) should appear in the MotionManager tree.

S2.3.2 Defining Initial Position

Bring the block to the upper end of the slope surface as its initial position. The block is released from this position to simulate sliding without friction. We edit the mate LimitDistance1 and enter 0 for the distance dimension.

From the SolidWorks browser, expand the Mates branch, click LimitDistance1 (ground<1>,ball<1>), and choose Edit Feature 🔲. You should see the definition of the assembly mate in the dialog box (Figure S2.15). Enter 0.00 in. for the distance, and click the checkmark on top to accept the change. In the graphics area, the block should move to the top position on the slope face, as shown in Figure S2.15.

S2.3.3 Defining and Running the Simulation

Choose Motion Analysis from the motion study selection (directly above the MotionManager tree, as shown in Figure S2.16). Click the Motion Study Properties button 🔲 from the Motion

Figure S2.14: Gravity dialog box.

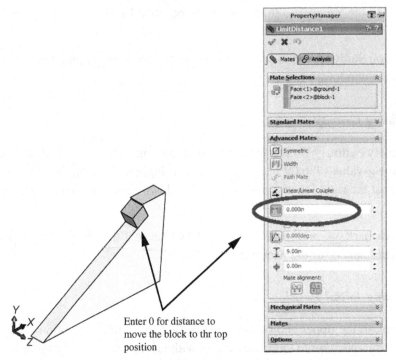

Figure S2.15: Define the initial position for the sliding block.

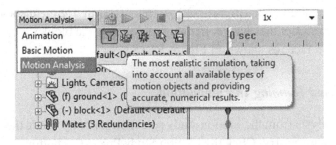

Figure S2.16: Motion analysis.

toolbar. In the Motion Study Properties dialog box (Figure S2.17), enter 500 for Frames per second, and click the checkmark at the top of the box.

Zoom in to the timeline area until you can see tenth-sec marks. Drag the end time key to the 0.4-sec mark (see Figure S2.18) in the timeline area to define the simulation duration. Click the Calculate button 🖼 on the Motion toolbar to start the motion analysis. A 0.4-sec simulation is carried out.

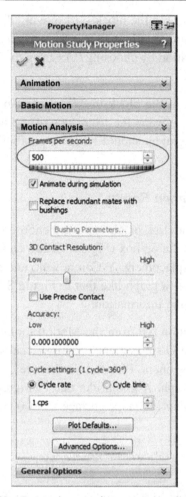

Figure S2.17: Motion study properties dialog box.

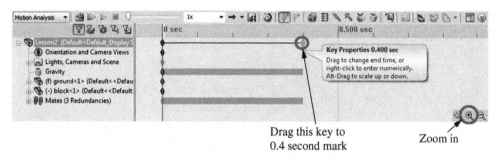

Figure S2.18: Timeline area.

After a few seconds, you should see the block start moving down along the slope surface. Since the total simulation duration is only 0.4 sec, you may want to adjust the playback speed to 10% to slow down the animation by choosing 0.1x from the scroll-down menu next to the animation slider.

Next, we graph the Y-position of the block in two ways: Y-position of the mass center of the block and then Y-distance between the front face of the block and the top end face of the slope. Both displacement graphs should reveal a parabolic curve, as is common in physics.

S2.3.4 Displaying the Simulation Results

From the MotionManager, right click block<1>, and choose Create Motion Plot, as shown in Figure S2.19. In the Results dialog box (Figure S2.20), choose Displacement/Velocity/Acceleration, select Linear Displacement, Y Component, and then click the checkmark to accept the graph. You should see a graph like that in Figure S2.12b. Next we define another graph, distance, to show the same information.

Click the Results and Plots button 📇 from the Motion toolbar. In the Results dialog box (refer to Figure S2.21), choose Displacement/Velocity/Acceleration, select Linear Displacement, and then Y Component. Pick the two vertices, as seen in Figure S2.21. Click the checkmark to accept the graph. A graph like that of Figure S2.22 should appear. From the graph, the block moves along the slope surface from 0 (the distance between two

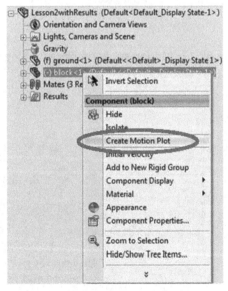

Figure S2.19: Create Motion Plot.

Figure S2.20: Results dialog box for Create Motion Plot.

coincided faces) to about -4 in. in about 0.3 sec. The block then "bounces" back because of the LimitDistance mate defined for the block. The graph does not seem to be correct since we know that the block will travel 4.5 in. in the *Y*-direction. This is because the distance the block travels is 9 in. and the slope is 30 deg. Therefore, the *Y*-distance the block travels should be $9 \times \sin 30$ deg. $= 4.5$ in.

To examine the results in more detail, you may export the graph data, for example, by right-clicking the graph and choosing Export CSV. Open the spreadsheet and examine the data. As shown in the spreadsheet exported (Figure S2.23), the time for the block to reach the lowest position is 0.306 sec, in which the *Y*-distance is about -4.5 in. This is accurate.

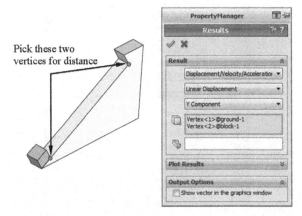

Figure S2.21: Results dialog box—choosing vertices.

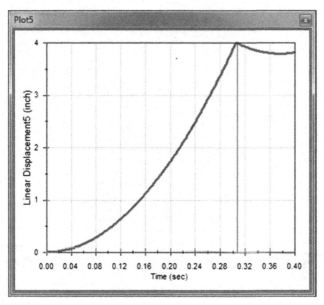

Figure S2.22: *Y*-distance graph.

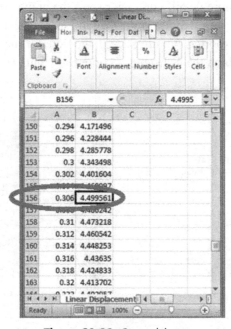

Figure S2.23: Spreadsheet.

We will carry out calculations to verify these results in the following section. Save your model by choosing File > Save from the pull-down menu.

S2.3.5 Results Verification

In this section, we verify the analysis results obtained from Motion. We assume that the block is of a concentrated mass so that particle dynamics theory applies. In addition, we assume that there is no air friction and no friction between the sliding faces.

It is well known that the equations of motion can be derived from Newton's second law for the block. By sketching a free-body diagram as shown in Figure S2.24, we have the following force equilibrium equation:

$$F = ma = mg \sin 30 = 0.5mg \qquad (S2.1a)$$

Hence, the acceleration of the block is $a = 0.5g$. The velocity and position of the block can then be obtained by integrating Eq. S2.1a over time:

$$v = at = 0.5gt \qquad (S2.1b)$$

$$s = \frac{1}{2}at^2 = 0.25gt^2 \qquad (S2.1c)$$

The Y-position of the block can be obtained as

$$P_y = -\frac{1}{2}at^2 \sin 30 = -\frac{1}{8}gt^2 \qquad (S2.2)$$

These equations can be implemented using, for example, Microsoft Excel for numerical solutions. The Y-position of the block from 0 to 3.05 sec is shown in Figure S2.25. This

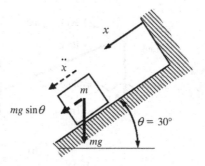

Figure S2.24: Free-body diagram.

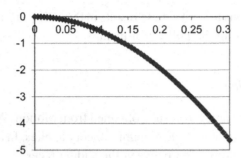

Figure S2.25: *Y*-position of the block obtained from spreadsheet calculations.

compares very well with that of SolidWorks Motion result (check spreadsheet results). At $t = 3.05$ sec, the *Y*-position of the block is -4.49 in., which again matches well with the SolidWorks Motion result.

S2.4 Single-Piston Engine

In this lesson, we learn how to create a simulation model for the single-piston engine shown in Figure S2.26a, including how to select assembly mates to connect parts for an adequate motion model. We drive the mechanism by rotating the crankshaft with a constant angular velocity, basically conducting a kinematic analysis. We start this lesson with a brief overview of the engine assembly, which was created in SolidWorks. At the end of this lesson, we will verify the kinematic simulation results using theory and computational methods commonly found in a mechanism design textbook.

Kinematically, the single-piston engine is essentially a four-bar linkage, as shown in Figure S2.26b. For an internal combustion engine, the linkage is driven by a firing load that pushes the piston, converting the reciprocal motion into rotational motion at the crank. However, in this lesson, since our goal is to conduct a kinematic analysis, we will apply a rotary motor at the crank. The rotational motion is then converted into a reciprocal motion at the piston.

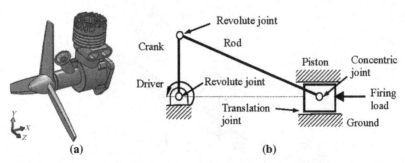

Figure S2.26: Single-piston engine: (a) CAD model and (b) schematic view of the kinematic model.

For a four-bar linkage, the length of the crank must be less than that of the rod for the mechanism to operate. This is commonly referred to as Grashof's law. In this example, the lengths of the crank and rod are 0.58 and 2.25 in., respectively, which satisfies the requirement of Grashof's law.

The unit system chosen for this example is IPS (in.-lb$_f$-sec). All parts are assumed of aluminum, 2014 alloy. No friction is present between any pair of the components (parts or subassemblies).

S2.4.1 SolidWorks Parts and Assembly

The engine example consists of four major components: case (case_asm), propeller (propeller_asm), connecting rod (connectingrod_asm), and piston, as shown in Figure S2.27a. For this lesson, the parts and assemblies have been created in SolidWorks. There are 23 model files, including 5 assemblies. Among the 5 assemblies, 3 are subassemblies mentioned previously as part of the major components of the engine. The remaining two assembly files are Lesson3.SLDASM, and Lesson3withresults.SLDASM. Note that the assembly file Lesson3withresults.SLDASM contains the complete simulation model with simulation results. You may open this file and review motion simulation results before going over this lesson.

We start with Lesson3.SLDASM, in which the engine is properly assembled with one free degree of freedom. When the propeller, which is the crank in the standard four-bar linkage, is driven by the rotary motor, it rotates and drives the connecting rod. The connecting rod pushes the piston up and down within the piston sleeve.

The engine assembly consists of three subassemblies (case, propeller, and connecting rod) and one part (piston). The case is fixed (ground body). The propeller is assembled to the case using concentric and coincident mates, as shown in Figure S2.27b. It is free to rotate along the X-direction. The connecting rod is assembled to the propeller (at the crankshaft) using concentric and coincident mates. It is free to rotate relative to the propeller (at the crankshaft) along the X-direction. Finally, the piston is assembled to the connecting rod (at the piston pin)

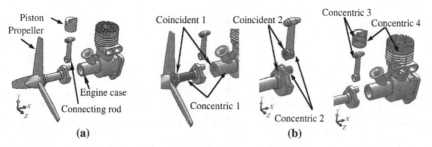

Figure S2.27: Single-piston engine: (a) exploded view and (b) constraints defined between bodies (or subassemblies).

using a concentric mate. It is also assembled to the engine case using another concentric mate. This mate restricts the piston movement along the *Y*-direction, which in turn restricts the top end of the connecting rod to moving vertically.

S2.4.2 Motion Model

We add a rotary motor to drive the propeller for a kinematic analysis. The position and velocity of the piston obtained from motion analysis are shown in Figure S2.28a and S2.28b, respectively. Note that from the position graph, the piston moves between about 1.0 and 2.1 in. vertically. The total travel distance is about 1.1 in., which can be easily verified by the radius of the crankshaft, which is 0.58 in. The piston travel distance is 2 times the radius of the crankshaft, which is 1.16 in.

S2.5 Using SolidWorks Motion for the Single-Piston Engine

Start SolidWorks and open assembly file Lesson3.SLDASM. First take a look at the exploded view by selecting the root assembly (Lesson3), and press the right mouse button. In the menu appearing (Figure S2.29) choose Explode. You should see the assembly in an exploded view similar to that of Figure S2.27. Right click the root assembly and choose Collapse to collapse it.

Click the Motion Study tab (with the default name Motion Study 1) at the bottom of the graphics area to bring up the MotionManager window.

S2.5.1 Creating a Motion Study

We create the kinematic analysis model by adding a rotary motor to the propeller; we use the Motion Analysis option to simulate the motion.

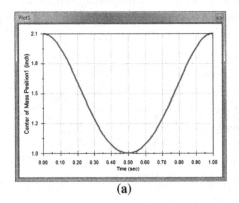
(a)

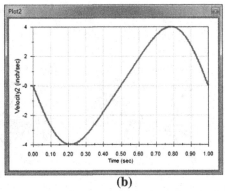
(b)

Figure S2.28: Result graphs obtained from SolidWorks Motion: (a) *Y*-position of the piston and (b) *Y*-velocity of the piston.

Figure S2.29: Display Explode view.

Choose Motion Analysis from the motion study selection (directly above the MotionManager tree, as shown Figure S2.30). Click the Motor button ⬡ from the Motion toolbar to bring up the Motor window (Figure S2.31). Choose Rotary Motor (default). Move the mouse pointer to the graphics area, and pick a circular arc that defines the rotational direction of the rotary motor—for example, the circle on the drive washer of the propeller, as shown in Figure S2.32. A circular arrow appears indicating the rotational direction of the rotary motor.

A counter-clockwise direction is desired. You may change the direction by clicking the Direction button ⬒ directly under Component/Direction. Choose Constant speed and enter 60 rpm. Click the checkmark ✅ at the top of the dialog box to accept the motor definition. You should see a RotaryMotor1 added to the MotionManager tree.

Click the Motion Study Properties button ⬚ from the Motion toolbar to see the Motion Study Properties dialog box. Change Frames per second to 100, and then click the checkmark ✅ to accept the change. Calculate and animate the motion.

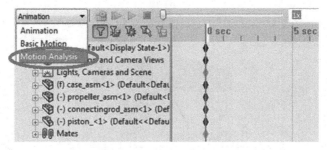

Figure S2.30: Motion Analysis.

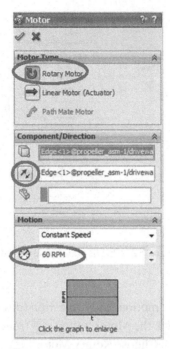

Figure S2.31: Rotary Motor.

Click the Calculate button 📷 on the Motion toolbar to carry out a motion analysis. A default 5-sec simulation is carried out. The propeller should make 5 turns (recall we entered 60 rpm for the motor earlier), and a 5-sec simulation timeline should be created in the timeline area.

You may want to hide the case_asm to see how the connecting rod and piston move. Right-click case_asm in the MotionManager tree and select Hide (Figure S2.33). The case is hidden in the graphics area. Play the animation again. You should now see the motion of the connecting rod and piston (Figure S2.34).

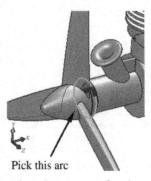

Pick this arc

Figure S2.32: Choosing an arc for the rotary motor.

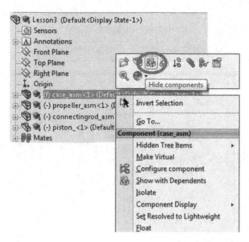

Figure S2.33: Hide engine case.

Figure S2.34: Animation without engine case.

You may want to reduce the overall simulation period to just 1 sec and increase the number of time frames to create a smoother animation. The steps are similar to those described in Section S2.2.

Calculate and play the animation. Now the propeller should rotate only one cycle. You may want to change the playback mode to loop 🔁 and play the animation continuously.

S2.5.2 Displaying Simulation Results

Click the Results and Plots button 🖳 from the Motion toolbar. In the Results dialog box (refer to Figure S2.35), choose Displacement/Velocity/Acceleration, select Center of Mass Position,

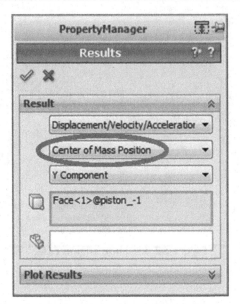

Figure S2.35: Results dialog box: Center of Mass Position.

and then Y Component. Select the piston (any surface), and click the checkmark to accept the graph.

A graph like that in Figure S2.28a should appear. It indicates that the piston moves between about 1.0 and 2.1 in. vertically. As mentioned earlier, the total travel distance of the piston is about 1.1 in., which is twice the radius of the crankshaft, 0.58 in.

Next, we create a graph for the *Y*-velocity of the mass center of the piston. Right click piston_<1> in the motion entity tree, and choose Create Motion Plot, as shown in Figure S2.36. In the Results dialog box (Figure S2.37), Face<1>@piston_−1 is listed. Choose Displacement/Velocity/Acceleration, select Linear Velocity, and then Y Component. Click the checkmark to accept the graph. A graph like that in Figure S2.28b should appear. The graph indicates that the *Y*-velocity of the piston is between about −4.0 and 4.0 in./sec vertically.

We will carry out calculations to verify these results in the next section. Save your model by choosing File > Save from the pull-down menu.

S2.5.3 Results Verification

In this section, we verify the motion analysis results using standard kinematic analysis theory. Note that in kinematic analysis, the position, velocity, and acceleration of given bodies, points, or axes of joints in the mechanism are analyzed.

In kinematic analysis, forces and torques are not involved. All bodies (or links) are assumed massless. Hence, mass properties defined for bodies do not influence the analysis results.

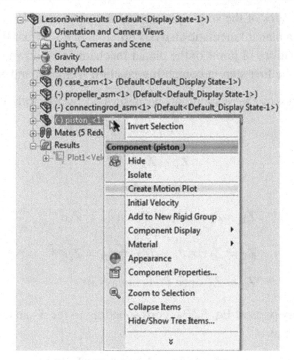

Figure S2.36: Create Motion Plot.

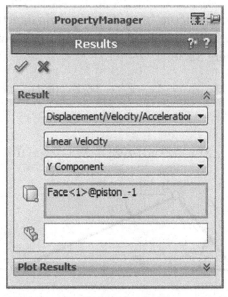

Figure S2.37: Results dialog box for Create Motion Plot.

The motion characteristics of the single-piston engine can be modeled as a slider-crank mechanism, which is a planar kinematic analysis problem of a four-bar linkage. A vector plot that represents the positions of joints of the planar mechanism is shown in Figure S2.38. The vector plot serves as the first step in computing position, velocity, and accelerations.

The position equations of the system can be described by the following vector summation:

$$\mathbf{Z}_1 + \mathbf{Z}_2 = \mathbf{Z}_3 \tag{S2.3a}$$

where

$$\mathbf{Z}_1 = Z_1 \cos \theta_A + i Z_1 \sin \theta_A = Z_1 e^{i\theta_A}$$

$$\mathbf{Z}_2 = Z_2 \cos \theta_B + i Z_2 \sin \theta_B = Z_2 e^{i\theta_B}$$

$$\mathbf{Z}_3 = Z_3 \text{ since } \theta_C \text{ is always } 0$$

The real and imaginary parts of Eq. S2.3a, corresponding to the X- and Y-components of the vectors, can be written as

$$Z_1 \cos \theta_A + Z_2 \cos \theta_B = Z_3 \tag{S2.3b}$$

$$Z_1 \sin \theta_A + Z_2 \sin \theta_B = 0 \tag{S2.3c}$$

In Eqs. S2.3b and S2.3c, Z_1, Z_2, and θ_A are given. We are solving for Z_3 and θ_B. Equations S2.3b and S2.3c are nonlinear in terms of θ_B. Solving them directly for Z_3 and θ_B is not straightforward. Instead, we will calculate Z_3 first, using trigonometric relations for the triangle ABC shown in Figure S2.38:

$$Z_2^2 = Z_1^2 + Z_3^2 - 2Z_1 Z_3 \cos \theta_A$$

Hence,

$$Z_3^2 - 2Z_1 \cos \theta_A Z_3 + Z_1^2 - Z_2^2 = 0$$

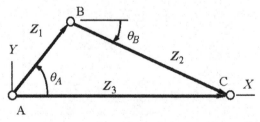

Figure S2.38: Vector plot of the slider-crank mechanism.

Solving Z_3 from the above quadratic equation, we have

$$Z_3 = \frac{2Z_1 \cos \theta_A \pm \sqrt{(2Z_1 \cos \theta_A)^2 - 4(Z_1^2 - Z_2^2)}}{2} \tag{S2.4}$$

where two solutions of Z_3 represent the two possible configurations of the mechanism shown in Figure S2.39. Note that point C can be either at C or C' for a given Z_1 and θ_A.

From Eq. S2.3c, θ_B can be solved by

$$\theta_B = \sin^{-1}\left(\frac{-Z_1 \sin \theta_A}{Z_2}\right) \tag{S2.5}$$

Similarly, θ_B has two possible solutions, corresponding to vector $\mathbf{Z}_3$.

Taking derivatives of Eqs. S2.3b and S2.3c with respect to time, we have

$$-Z_1 \sin \theta_A \dot{\theta}_A - Z_2 \sin \theta_B \dot{\theta}_B = \dot{Z}_3 \tag{S2.6a}$$

$$Z_1 \cos \theta_A \dot{\theta}_A + Z_2 \cos \theta_B \dot{\theta}_B = 0 \tag{S2.6b}$$

where $\dot{\theta}_A = \dfrac{d\theta_A}{dt} = \omega_A$ is the angular velocity of the rotation driver, which is a constant.

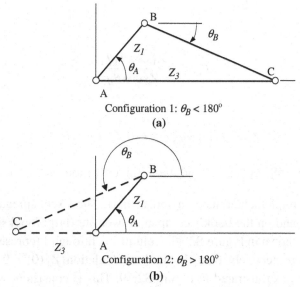

Configuration 1: $\theta_B < 180°$

(a)

Configuration 2: $\theta_B > 180°$

(b)

Figure S2.39: Two possible configurations of the mechanism.

Note that Eqs. S2.6a and S2.6b are linear functions of $\dot{Z}_3$ and $\dot{\theta}_B$. Rewrite the equations in matrix form:

$$
\begin{bmatrix} Z_2 \sin \theta_B & 1 \\ Z_2 \cos \theta_B & 0 \end{bmatrix} \begin{bmatrix} \dot{\theta}_B \\ \dot{Z}_3 \end{bmatrix} = \begin{bmatrix} -Z_1 \sin \theta_A \dot{\theta}_A \\ -Z_1 \cos \theta_A \dot{\theta}_A \end{bmatrix} \tag{S2.7}
$$

Equation S2.7 can be solved by

$$
\begin{bmatrix} \dot{\theta}_B \\ \dot{Z}_3 \end{bmatrix} = \begin{bmatrix} Z_2 \sin \theta_B & 1 \\ Z_2 \cos \theta_B & 0 \end{bmatrix}^{-1} \begin{bmatrix} -Z_1 \sin \theta_A \dot{\theta}_A \\ -Z_1 \cos \theta_A \dot{\theta}_A \end{bmatrix}
$$

$$
= \frac{1}{-Z_2 \cos \theta_B} \begin{bmatrix} 0 & -1 \\ -Z_2 \cos \theta_B & Z_2 \sin \theta_B \end{bmatrix} \begin{bmatrix} -Z_1 \sin \theta_A \dot{\theta}_A \\ -Z_1 \cos \theta_A \dot{\theta}_A \end{bmatrix}
$$

$$
= \frac{1}{-Z_2 \cos \theta_B} \begin{bmatrix} Z_1 \cos \theta_A \dot{\theta}_A \\ Z_1 Z_2 \cos \theta_B \sin \theta_A \dot{\theta}_A - Z_1 Z_2 \sin \theta_B \cos \theta_A \dot{\theta}_A \end{bmatrix}
$$

$$
= \begin{bmatrix} -\dfrac{Z_1 \cos \theta_A \dot{\theta}_A}{Z_2 \cos \theta_B} \\ \dfrac{Z_1 \left(\cos \theta_B \sin \theta_A \dot{\theta}_A - \sin \theta_B \cos \theta_A \dot{\theta}_A \right)}{\cos \theta_B} \end{bmatrix} \tag{S2.8}
$$

$$
\dot{\theta}_B = -\frac{Z_1 \cos \theta_A \dot{\theta}_A}{Z_2 \cos \theta_B} \tag{S2.9}
$$

and

$$
\dot{Z}_3 = Z_1 \left(\tan \theta_B \cos \theta_A \dot{\theta}_A - \sin \theta_A \dot{\theta}_A \right) \tag{S2.10}
$$

In this example, $Z_1 = 0.58$, $Z_2 = 2.25$, and the initial conditions are $\theta_A(0) = \theta$ and $\theta_B(0) = 0$.

The solutions can be implemented using a spreadsheet. The Excel spreadsheet file, lesson3.xls, can be found on the book's companion website (http://booksite.elsevier.com/ 9780123984609). As shown in Figure S2.40, Columns A through I represent time, Z_1, Z_2, $\dot{\theta}_A$, θ_A, Z_3, θ_B, $\dot{Z}_3$, and $\dot{\theta}_B$, respectively. Note that in this calculation, $Z_3(0) > 0$ is assumed; hence, $\theta_B(0) < 0$ (clockwise), as illustrated in Figure S2.39. This is consistent with the initial configuration we created in the Motion model.

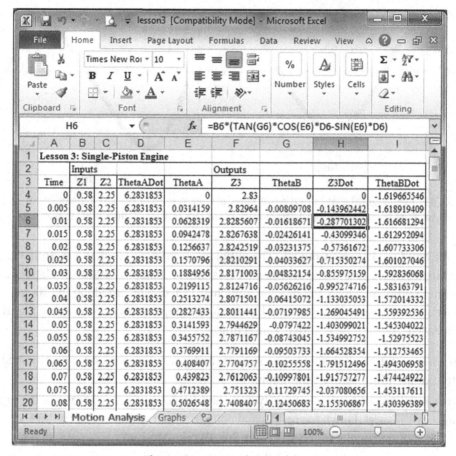

Figure S2.40: Excel spreadsheet

Figures S2.41a and S2.41b show the graphs of data in Columns F and H—that is, piston position and velocity, respectively. Comparing these figures with Figures S2.28a and S2.28b, the simulation analysis results are verified. Note that in Figure S2.41a, the piston position varies between 1.7 and 2.8 in.; that is, the distance the piston travels is about 1.1 in., which is the same as that observed in Figure S2.28a.

Note that the accelerations of a given joint in the mechanism can be formulated by taking one more derivative of Eqs. S2.6a and S2.6b with respect to time. The resulting two coupled equations can be solved using the Excel spreadsheet.

Exercises

S2.1 Use the model files provided in Section S2.2 to create the spring-mass system shown in Figure S2.42. The spring constant and the unstretched length (or free length) are $k = 20$

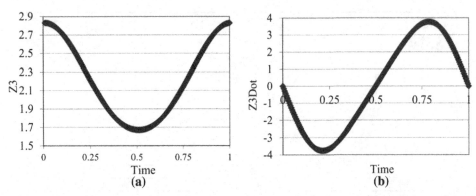

Figure S2.41: Result graphs obtained from spreadsheet calculations: (a) *Y*-position of the piston (Column F of spreadsheet) and (b) *Y*-velocity of the piston (Column H of spreadsheet).

lb$_f$/in. and $U = 3$ in., respectively. The initial position of the block is 4 in. from the top of the slope surface (use the LimitDistance1 assembly mate). You are simulating a free-vibration problem, where the block is stretched 1 in. downward along the 30 deg. slope. No friction is assumed between the block and the slope surface. We assume a gravity of $g = 386$ in./sec^2 in the negative *Y*-direction.

(a) Carry out a motion analysis for the system, and graph the position, velocity, and acceleration of the block in the *Y*-direction.

(b) Derive equations of motion for the system, and implement the solutions of the equations using a spreadsheet. Graph the results and compare them with those of (a). Are you getting accurate motion simulation results from SolidWorks Motion?

(c) Calculate the natural frequency of the system and compare your calculation with that of Motion. How do you determine the natural frequency of the system from results obtained from Motion?

S.2.2 Open the single-piston engine assembly from Section S2.3. Use the Animation Wizard to create the following animations. Note that Animation Wizard is very easy to use. To

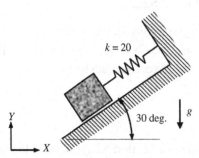

Figure S2.42: Spring-mass system.

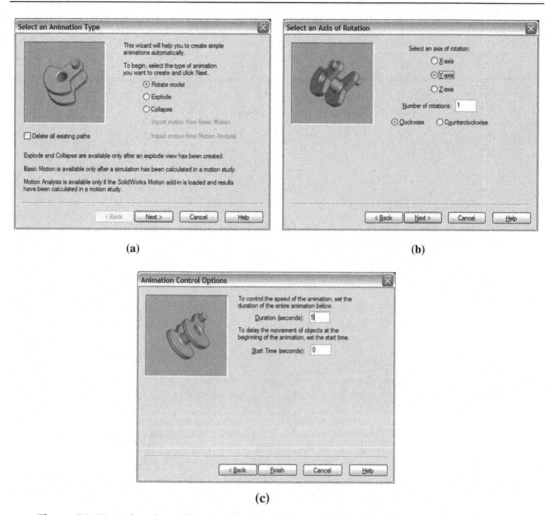

(a)

(b)

(c)

Figure S2.43: Animation Wizard: (a) select Rotate model, (b) define rotation axis, and (c) define animation time period.

access it, simply click the Animation Wizard button 🔳 on the Motion toolbar. In the Animation Wizard window (Figure S2.43), click Rotate model (default) and Next. Then follow the steps to choose rotation axis and animation time interval, as shown in Figure S2.43b and c, respectively, to create a rotation animation.

(a) Explode for 5 sec; start time: 0 sec.
(b) Rotate along the *Y*-axis for 5 sec; start time: 6 sec.
(c) Rotate along the *X*-axis for 5 sec; start time: 12 sec.
(d) Collapse for 5 sec; start time: 18 sec.

Project P3 Structural FEA and Fatigue Analysis Using Pro/MECHANICA Structure

Chapter Outline

There is no need to emphasize the importance of understanding theory and using analytical methods to solve engineering problems. In many design applications, such as structural components of complex geometry, analytical methods can only offer rough estimates on product performance at best. In many cases, engineers must rely on tools, such as finite-element analysis (FEA) software, to evaluate product performance with high accuracy in order to support design decision making. It is critical for a design engineer to learn and be able to use software tools for solving problems that are beyond hand calculations and analytical methods.

In Project P3, we will introduce Pro/MECHANICA® Structure for FEA and fatigue analysis of structural components. We include three simple examples just to get you started on learning and using the software. These three examples are a simple cantilever beam, a thin-walled tank, and a crankshaft, which should provide you with a good overview on the FEA and fatigue analysis capabilities offered by Pro/MECHANICA Structure. You may find the example files on the book's companion website (http://booksite.elsevier.com/9780123984609).

Overall the objective of this project is to enable readers to use Pro/MECHANICA Structure for basic applications. Those interested in learning more and elevating themselves to an intermediate level may want to go over more examples provided by Pro/MECHANICA Structure or review other tutorial books on the subject (e.g., *Introduction to Finite Element Analysis Using Creo Simulate 1.0*, by Randy H. Shih, ISBN: 978-1-58503-670-7; http://www.sdcpublications.com/).

Note that the lessons included in this project are developed using Pro/ENGINEER Wildfire 5.0 (Date Code M040). If you are using a different version of Pro/ENGINEER, you may see slightly different windows or dialog boxes. Since Pro/ENGINEER is fairly intuitive to use, these differences should not be too difficult to figure out.

P3.1 Introduction to Pro/MECHANICA Structure

Pro/MECHANICA Structure (or Mechanica) is a finite element analysis (FEA) tool that allows users to analyze structural performance, such as displacement, stress, natural frequency, buckling load, fatigue life, and so on. The main objective of this tutorial is to help you, as a new user, to become familiar with Mechanica and to provide you with a brief overview of using the software.

Pro/MECHANICA Structure can be started in either independent mode or integrated mode. Independent mode is started as a standalone program, separate from Pro/ENGINEER. The geometry model of the structure is either imported from CAD or created using primitive geometric entities, such as points, lines, and patches. This modeling method is less desirable than that of CAD.

The integrated mode of Mechanica is embedded into Pro/ENGINEER. It is a fully integrated module of Pro/ENGINEER. Transition from Pro/ENGINEER to Mechanica is seamless. Geometric models created in Pro/ENGINEER are automatically carried over into Mechanica. You may simply open an existing Pro/ENGINEER model (part or assembly) and choose from the pull-down menu Applications > Mechanica to enter Mechanica. In this mode, 3D solid models, part or assembly, created in Pro/ENGINEER can be directly used for creating a finite element model and carrying out an FEA. Menus and dialog boxes for the integrated mode are just like those in Pro/ENGINEER. Those who are familiar with Pro/ENGINEER, will find the integrated mode easy to learn. We will assume integrated mode in the following tutorial lessons.

P3.1.1 Overall Process

The overall process of using Pro/MECHANICA Structure for carrying out FEA consists of three main steps: preprocessing, analysis, and postprocessing, as illustrated in Figure P3.1. Key entities that constitute an FEA model include a geometric model of the structure created in Pro/ENGINEER, a finite element mesh, material properties, loads, and boundary conditions.

The analysis capabilities in Mechanica include static, modal, buckling, fatigue, prestress, and dynamic, which cover a broad range of applications encountered in engineering design. You will have to choose an analysis type that best describes the physical problem being solved.

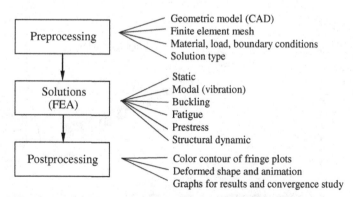

Figure P3.1: General process of using Pro/MECHANICA structure.

The analysis results can be visualized in various forms. You may use color contour or fringe plots to display stress or displacement results. You may display or animate the deformed shape or mode shape of the structure. You may also use graphs to show results, such as convergence study.

P3.1.2 User Interface

User interface of Mechanica is identical to that of Pro/ENGINEER, as shown in Figure P3.2. Pro/ENGINEER users should find it straightforward to maneuver in Mechanica. As shown in Figure P3.2, the user interface window of Mechanica consists of pull-down menus, shortcut buttons, a prompt/message window, scroll-down menu, graphics window, and a model tree window.

The graphics window displays the FEA model you are working on. The pull-down menus and the shortcut buttons at the top of the screen provide typical Pro/ENGINEER functions. The Mechanica shortcut buttons to the right provide all the functions required to create and modify FEA models. When you move the mouse pointer over these buttons, a short message describing the menu command will appear. It also shows system messages following command execution. The shortcut buttons in Mechanica and their functions are summarized in Table P3.1.

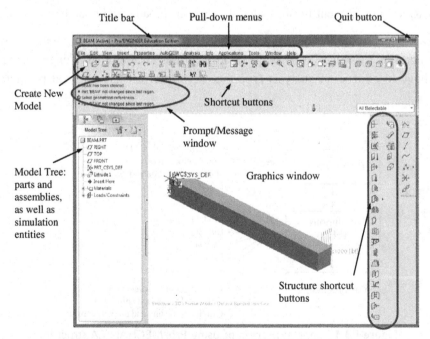

Figure P3.2: User interface of Pro/MECHANICA structure.

Table P3.1: The shortcut buttons in Pro/MECHANICA structure

Button Symbol	Name	Function
	Force/Moment Load	Create a force/moment load
	Pressure Load	Create a pressure load
	Bearing Load	Create a bearing load
	Gravity Load	Create a gravity load
	Centrifugal Load	Create a centrifugal load
	Global	Create a global temperature load
	Displacement Constraint	Create a displacement constraint
	Planar Constraint	Create a planar constraint
	Symmetry Constraint	Create a symmetry constraint
	Shell	Create a shell idealization
	Shell Pair	Create a shell pair
	Beam	Create a beam idealization

Continued

Table P3.1: The shortcut buttons in Pro/MECHANICA structure—cont'd.

Button Symbol	Name	Function
	Spring	Create a spring idealization
	Mass	Create a mass idealization
	Interface	Create an interface
	Weld	Create a weld
	Rigid Link	Create a rigid link
	Weighted Link	Create a weighted link
	Materials	Define materials
	Material Assignment	Create a material assignment to a model/volume
	Simulation Measure	Define simulation measure
	Control	Create an AutoGEM control
	Surface	Create a simulation surface region

P3.1.3 Defining FEA Entities

The basic entities of an FEA model created in Mechanica consist of geometry (in CAD), mesh, material properties, loads, and boundary conditions. These entities can be defined using Mechanica shortcut buttons, as shown in Table P3.1.

Loads can be defined using several buttons, depending on the physical conditions. These buttons allow you to create force, moments, pressure, gravity, centrifugal, thermal, and so on. Boundary conditions include displacement constraints, planar constraints (such as pin and ball constraints), and symmetry constraint. Idealization, such as converting thin-walled solids to shell and bar solids to beam, is supported. Material properties can be created and assigned to certain volume or the entire solid model. In addition, simulation measure and mesh control can be created using the shortcut buttons. Note that you may use the Surface button to create a surface region on the solid model in order to apply load or create displacement constraint as needed.

P3.1.4 Types of Analyses

Pro/MECHANICA Structure provides six analysis types for structural analysis. They are static, modal, buckling, fatigue, prestress, and structural dynamic. Static analysis, which is the most basic analysis type, analyzes structural responses to external loads, including stress, strain, and displacement. Modal analysis calculates natural frequencies and corresponding vibration mode shapes. Buckling analysis calculates buckling load factors and buckling mode shape. Both modal and buckling analyses solve eigenvalue problems mathematically. Fatigue analysis provided in Mechanica supports high-cycle fatigue using the so-called uniform material law (UML) approach, which uses a generic set of fatigue properties to model low-alloy steels, unalloyed steels, aluminum alloys, and titanium alloys. The solver technology integrated with Mechanica fatigue analysis is provided by nCode, Inc. Using prestress analysis, you can apply results from a static analysis and then calculate the natural frequencies and mode shapes of your model by carrying out prestress modal analyses. Structural dynamic analysis includes dynamic time response, dynamic frequency response, dynamic random response, and dynamic shock response.

P3.1.5 Viewing Results

Mechanica allows users to view, evaluate, and manage results for FEA analyses. Numerous approaches are provided in Mechanica for users to view analysis results, including color contour or fringe plots, animation, and graphs. Users may study the results by, for example, probing specific areas of an FEA model for displacement or stress data. Users may save the set of result windows and review or reuse them later. Also, users may prepare printed and online reports for evaluation and presentation.

P3.1.6 Examples

Three examples are included in this project, which illustrate step-by-step details of modeling, analysis, and result visualization capabilities in Mechanica. We will start with a very simple cantilever beam example. This example gives you a quick start and a quick run-through for a static FEA using Mechanica, including creating an FEA model, running an analysis, and visualizing simulation results.

The second example illustrates the steps for converting a thin-walled tank solid model to a shell model, applying load and boundary conditions to the shell model, analyzing the model with much reduced model size, and visualizing the results. In addition, we will learn to conduct a convergence study using multi pass analysis based on p-adaptation.

The third example provides details on setting up a fatigue analysis model, running fatigue analysis, and visualizing fatigue analysis results, including showing the fringe plots for both fatigue life and percentage damage. All examples and main topics to be discussed are summarized in Table P3.2.

P3.2 Simple Cantilever Beam

The purpose of this lesson is to provide you with a quick start on using Pro/MECHANICA Structure. You will learn pre- and postprocessing and analysis capabilities in Pro/MECHANICA Structure, using a simple cantilever beam example. After completing this lesson, you should be able to solve similar problems following the same procedures. In this exercise, we will use the most basic analysis option: single-pass run to carry out FEA. Discussion on the convergence study can be found in the thin-walled tank example.

P3.2.1 The Cantilever Beam Example

The cantilever beam of 1 in. $\times$ 1 in. $\times$ 10 in. is shown in Figure P3.3. The unit system chosen is the in.-lb_f-sec. (instead of the default English system in.-lb_m-sec.). The material is aluminum (AL2014), where the modulus and Poisson's ratio are $E = 1.06 \times 10^7$ psi and $v = 0.33$, respectively. The total force at the tip is $P = 1000\ lb_f$ downward.

P3.2.2 Using Pro/MECHANICA Structure

Open Pro/ENGINEER Part

Open the solid model of the cantilever beam downloaded from the book's companion website (http://booksite.elsevier.com/9780123984609). This solid model, as shown in Figure P3.4, was created using default datum planes for section sketch and extruded for a square bar. Please note the world coordinate system (WCS) shown in Figure P3.4b. All physical

Table P3.2: Examples employed in this project

Section	Example	FEA Model	Problem Type	Things to Learn
P3.2	Simple cantilever beam		Static analysis	This example helps you become familiar with the basic operation and offers a quick run-through in using Mechanica for FEA. You will learn the general process of using Mechanica to construct an FEA model, run analysis, and visualize the FEA results.
P3.3	Thin shell		Static analysis	This example illustrates the process of converting a thin-walled solid model to a shell FEA model. You will learn how to add load and boundary conditions to the shell model for a complete FEA model. You will also learn how to conduct a convergence study using p-adaptive analysis.
P3.4	Simple pendulum		Fatigue analysis	This lesson provides steps for learning fatigue analysis using a stress-based approach. You will learn how to create a fully reversed cyclic load and unified material law (UML) approach for fatigue life calculations. You will also learn how to show and interpret results, including fatigue life in load cycles and percentage damage.

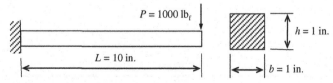

Figure P3.3: Cantilever beam example.

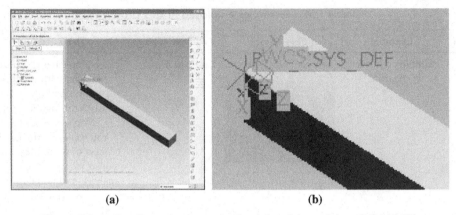

(a)   (b)

Figure P3.4: Cantilever beam example, (a) solid model and (b) WCS.

quantities to be created for the FEA model of the beam refer to the WCS. Note that the positive *X*-direction is pointing downward in the figure.

Change the unit system from default in.-lb$_m$-sec. to in.-lb$_f$-sec. in Pro/ENGINEER by choosing from pull-down menu File > Properties. In the Model Properties dialog box (Figure P3.5), change units from default in.-lb$_m$-sec. system to IPS (in.-lb$_f$-sec.).

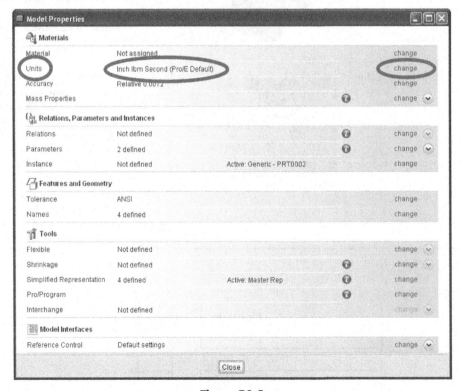

Figure P3.5

Enter Pro/MECHANICA Structure

In the Applications pull-down menu, select Mechanica. In the Mechanica Model Setup dialog box (Figure P3.6), click OK to proceed for a structural analysis (Structure is selected by default).

Note that a set of shortcut buttons (Figure P3.7) appears to the right of the Graphics area. These buttons are handy for defining finite element models, such as loads and boundary conditions (also called constraints). Those circled are the ones we will be using in this lesson. Please take a minute to get familiar with these buttons.

In addition, a default coordinate system WCS, which is identical to the default coordinate system in part mode (PRT_CSYS_DEF), is added to the analysis model automatically (Figure P3.4b).

Note that we will define all physical quantities, including constraints and loads by referring to the WCS. Pro/MECHANICA will refer to this coordinate system for FEA as well. Note that the WCS symbol is added to the model.

Preprocessing: Creating Finite Element Model

We will define load, displacement constraint (boundary condition), and material.

Choose the Force/Moment Load button (first button shown in Figure P3.7). The Force/Moment Load dialog box appears (Figure P3.8). Enter Edge_Load for Name, select Edge(s)/Curve(s) for References, choose the front edge that is parallel to the *y*-axis (while *x*-axis is pointing downward) to apply the force, and click OK on the Choose dialog box. Enter $X = 1000$ for Force Component (force component in the *X*-direction). Note that the force unit is lb$_f$. Click OK. Arrows will be shown at the tip edge of the beam to represent the load (Figure P3.9).

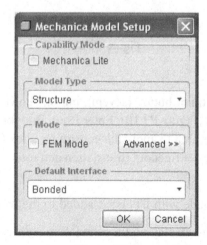

Figure P3.6

Figure P3.7

Click the Displacement Constraint button (seventh button shown in Figure P3.7). The Constraint dialog box appears (Figure P3.10). Enter Root_End for Name, choose the end face (Figure P3.11) in the Graphics area to apply the constraint, click OK in the Select dialog box (not shown), and fix all degrees of freedom (translation and rotation) at the end face (default). Click OK.

A constraint symbol (red triangle with labels) is added to the root end face of the beam.

In case you want to change the display on the simulation entities, you may choose the pull-down menu View > Simulation Display.

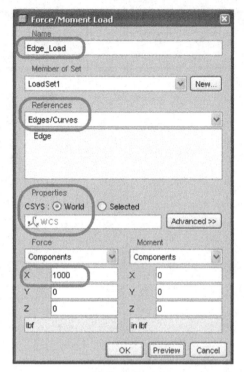

Figure P3.8

Choose various options in the Simulation Display dialog box to see changes in the Graphics area.

Next we will define material for the FEA model. Click the Material Assignment button (fifth button from bottom shown in Figure P3.7). The Material Assignment dialog box appears (Figure P3.12). Enter AL for Name, the beam part should have been selected by default (if not, pick the beam in the Graphics area), and click More for Material. The Materials dialog box appears (Figure P3.13).

In the Materials dialog box choose al2014 (the first one on the left column), move it to the right column by clicking the arrow button, and click OK. Double click AL2014 on the right column to review properties if needed. Click OK on the Material Assignment dialog box (Figure P3.12). A material symbol should appear in the part.

If you want to modify the properties of an existing material, you may choose the Edit > Properties pull-down menu from the Materials dialog box (Figure P3.13) to bring up the Material Definition dialog box (Figure P3.14) for changes. Click OK in the Material Definition dialog box to close the dialog box.

You may also create a new material type and add it back to the material library for later use. To add a new material, you may click the File > New pull-down menu to bring up the

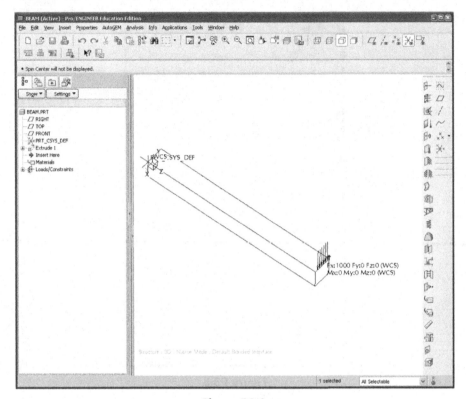

Figure P3.9

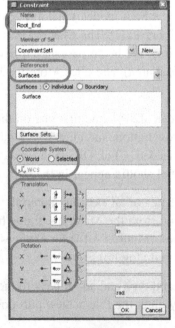

Figure P3.10

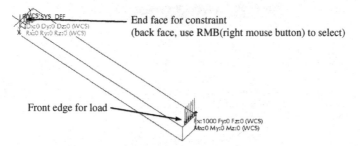

Figure P3.11

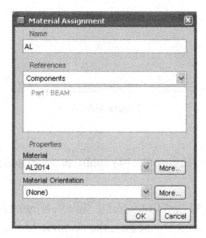

Figure P3.12

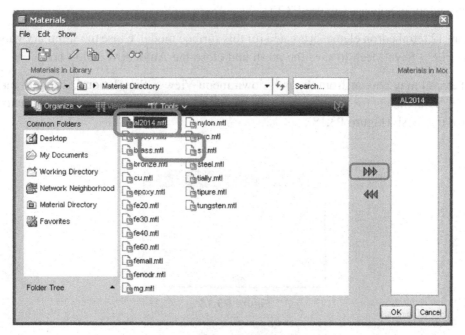

Figure P3.13

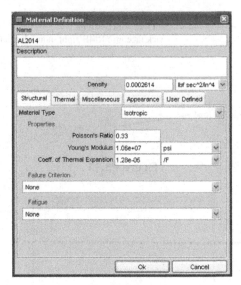

Figure P3.14

Material Definition dialog box (Figure P3.14) again, enter material name and properties, and then accept the material.

Next we will use the AutoGEM option to create finite element mesh. Click the AutoGEM button [image] (on top of the Graphics window), the AutoGEM dialog box will appear (Figure P3.15). Click Create. A mesh will be created on the solid model (Figure P3.16) and a mesh summary window appears (Figure P3.17).

There are 12 tetrahedron elements created for this surface model. Close the summary window. Choose File > Save Mesh to save the mesh and close the AutoGEM dialog box.

Note that you may choose from the pull-down menu View > Simulation Display to make necessary adjustments for mesh display; for example, shrinking the element by 10% to help visualize the mesh (Figure P3.18).

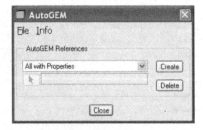

Figure P3.15

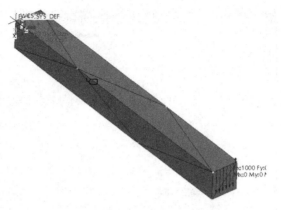

Figure P3.16

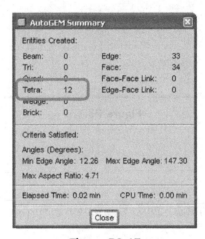

Figure P3.17

Define and Run Analysis

From the pull-down menu, choose Analysis > Mechanica Analyses/Studies. The Analyses and Design Studies dialog box appears (Figure P3.19). Choose File > New Static. In the Static Analyses Definition dialog box that appears next, enter My_Static for analysis name, select default Constraints and Loads, select default Single-Pass Adaptive as the convergence method (Figure P3.20), and click OK.

Note that there are three convergence methods in Pro/MECHANICA Structure: Quick Check, Single-Pass Adaptive, and Multi-Pass Adaptive. We will use Single-Pass for this example and use Multi-Pass for the thin-walled tanks example. Details about these options can be found on Help.

The My_Static analysis is listed in the Analyses and Design Studies dialog box. Click the Start Run button (fourth from the left) to start the analysis.

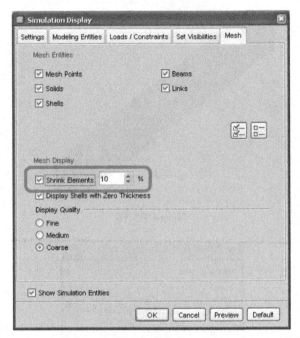

Figure P3.18

Figure P3.19

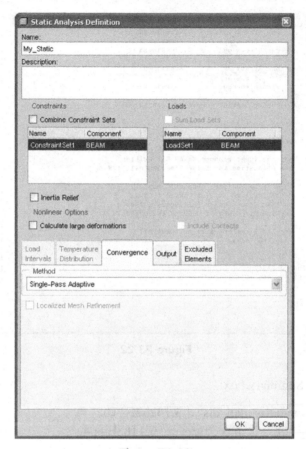

Figure P3.20

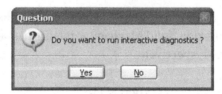

Figure P3.21

In the Question dialog box appearing next (Do you want to run interactive diagnostics?), click No (Figure P3.21) for the time being.

In the Prompt/Message window, a message "The design study has started" appears.

Click the Display Study Status button (second from the right) to check the analysis status and some key FEA results (Figure P3.22), that is, number of elements, max edge order (p-level).

Check the Summary box for information about Pass 1 and Pass 2. Note that the maximum edge order for Pass 2 is 5.

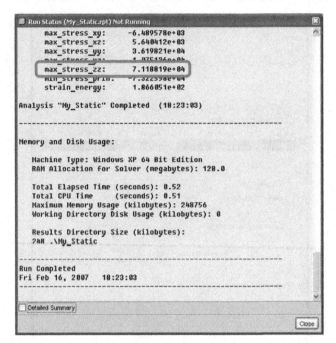

Figure P3.22

Check results in the Summary box:

The max displacement is max_dis_*x*: 3.73237E−01.
The max bending stress is max_stress_*zz*: 7.1188E+04.

Will you accept these results?

We do have analytical solutions for this simple cantilever beam example. Based on the classical beam theory we learned in Solid Mechanics, the tip displacement and the maximum bending stress at the root are, respectively,

$$y_{max} = \frac{PL^3}{3EI} = \frac{1000 \times (10)^3}{3 \times (1.06 \times 10^9) \times (1)(1)^3/12} = 0.3774 \text{ in.}$$

$$\sigma_{max} = \frac{Mc}{I} = \frac{(PL)(h/2)}{I} = \frac{(1000 \times 10)(1/2)}{(1)(1)^3/12} = 60,000 \text{ psi}$$

FEA displacement result is very close to that of classical beam theory, but not stress, why? One factor is the Poisson's ratio. The classical beam theory does not consider the effects of Poisson's ratio. In the finite element model, the beam deformation along the lateral direction due to the beam bending is adding to the bending strain and stress along the longitudinal direction due to a nonzero Poisson's ratio. If you set the Poisson's ratio to zero and rerun the analysis, the bending stress at the top edge reduces to a level that is very close to the analytical solution.

Display of Analysis Results

From the Analyses and Design Studies dialog box (Figure P3.19), click the Result button (first from the right),

The Result Window Definition dialog box appears (Figure P3.23). Enter von_Mises_Stress for Name, keep all default selections, and click OK and Show.

A new Graphics window appears, showing the beam with stress fringe (Figure P3.24).

Try to orient the beam like the one shown in Figure P3.24 (use the middle mouse button, same as in the part mode).

Does the maximum stress occur at the root? What is von Mises stress? Check your Solid Mechanics book to see how this von Mises stress is defined. Try to display the max_stress_zz also.

Next, we will display the deformed shape and show animation of the deformation.

Choose Edit > Result Window to bring the Result Window Definition dialog box back (Figure P3.23). Choose the Display Options tab (Figure P3.25), click Animate, and then OK and Show. You should see the beam deformed with varying stresses.

You may use the Play buttons on top to adjust the animation.

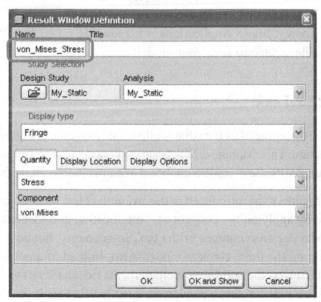

Figure P3.23

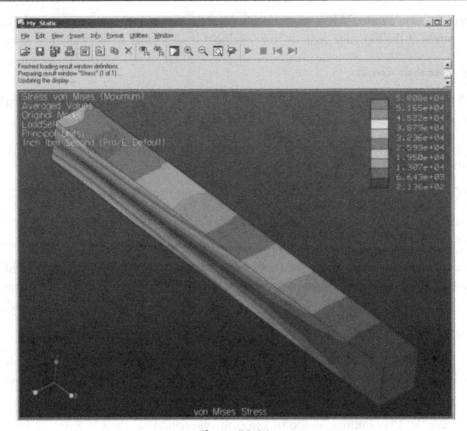

Figure P3.24

Close the Result window. Choose Applications > Standard to go back to Pro/ENGINEER part mode. Choose File > Save to save your model.

P3.3 Thin-Walled Tank

In this example we will carry out FEA for a thin-walled tank, as shown in Figure P3.26, using Pro/ MECHANICA Structure. This example is built upon the concept and procedure learned from the cantilever beam example. Please complete the beam example before going through this lesson.

There are two major points to be introduced in this example. First, we will learn how to reduce the FEA model size by applying the concept of simplification and idealization. We will use only half of the model due to symmetry of the tank in geometry, boundary condition, and load; that is, simplifying the finite element model. Also, instead of analyzing the solid thin-walled tank, we will convert the solid into shell elements (idealization) to further reduce the model size. Second, we will conduct a convergence study using multipass analysis based on the p-adaptation.

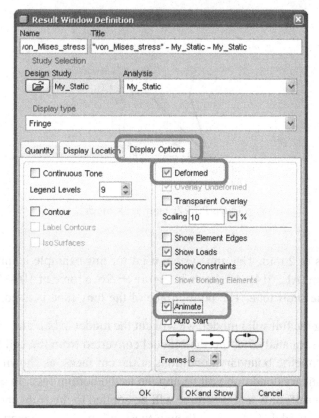

Figure P3.25

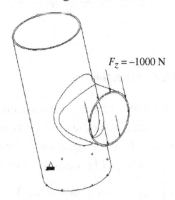

Figure P3.26: Full 3D solid model.

P3.3.1 The Thin-Walled Tank Example

The outer diameter and height of the vertical long tube are 102 mm and 240 mm, respectively. The short horizontal tube has the outer diameter and length 82 mm and 100 mm, respectively. The short tube is sitting at the halfway position of the vertical tube. The entire tank has

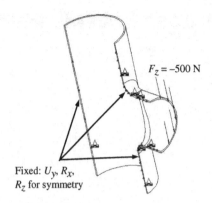

Figure P3.27: Half tank model.

a uniform thickness of 2 mm. The unit system used for this example is mmNS and the material is aluminum (AL2014). As shown in Figure P3.26, a force of 1000 N is applied at the front end face of the short tube. The bottom face of the long tube is fixed.

Instead of analyzing the full solid model, we will cut the model in half and convert the solid to shell model for FEA. In analyzing the shell model converted from the half tank solid model, we will impose symmetric boundary conditions at the cut faces, as shown in Figure P3.27. The same fixed boundary condition will be applied to the bottom face. In addition, the force will be also cut into half; only a 500 N force will be applied to the front end face of the short tube of the half tank model. Note that although the solid model is converted into shell for analysis, load and boundary conditions are defined at the faces of the solid model in Mechanica.

P3.3.2 Using Pro/MECHANICA Structure

Start Pro/ENGINEER and open the half tank model (htank.prt). You may find the model file on the book's companion website (http://booksite.elsevier.com/9780123984609). After opening the half tank model, you should see a solid model similar to that of Figure P3.28. Note the coordinate system CS0, which will be converted to WCS in Mechanica.

Verify the unit system by choosing from pull-down menu File > Properties. In the Model Properties dialog box (same as Figure P3.5), make sure the unit system chosen is mmNs (millimeter newton second).

Enter Pro/MECHANICA Structure

In the Applications pull-down menu, select Mechanica. In the Mechanica Model Setup dialog box (Figure P3.29), click OK.

Figure P3.28

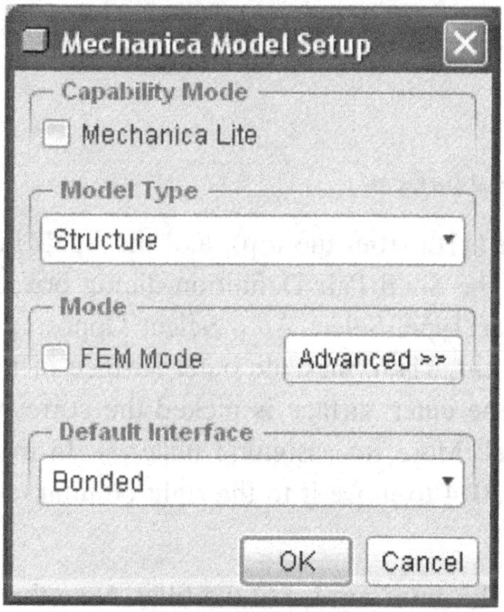

Figure P3.29

Note that a set of shortcut buttons appears to the right of the Graphics area (Figure P3.30). We will use Force/Moment Load, Displacement Constraints, New Shell Pair, and Materials buttons on the right to define the FEA model.

In addition, we will use AutoGEM, Design Study, and Results button on the top of the Graphics window to conduct FEA for the half tank model and show results.

Note that WCS, which is identical to the coordinate system in part (CS0), is added to the analysis model automatically (Figure P3.30).

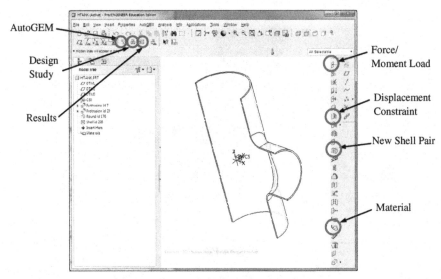

Figure P3.30

Model Idealization: Define Mid-surfaces

Click the Shell Pair button (11th from the top), and the Shell Pair Definition dialog box appears (Figure P3.31). In the Shell Pair Definition dialog box, keep the default name ShellPair1, and Constant for Type. Select Auto Select Opposing Surfaces for References and Midsurface for Placement. Then, pick the outer surface of the long tube, as shown in Figure P3.32. As soon as the outer surface is picked the corresponding inner surface is selected automatically. Click More for assigning material. In the Materials dialog box (Figure P3.33), choose AL2014 to move it to the right column, and click OK to accept the material.

Repeat the same and select the outer surface of the fillet. Note that the inner surface will not be selected automatically. You will have to rotate the model and select the inner surface of the fillet manually. This will create a pair with default name ShellPair2.

Repeat the same process to create ShellPair3 by picking the outer surface of the small tube.

Define Load

Click the Force/Moment Load button. In the Force/Moment Load dialog box (Figure P3.34) appearing next, enter Half_load for Name, pick the front end face of the small tube for load (Figure P3.35), and enter −500 for force in the Z-direction. Click the Preview button to check the load definition (Figure P3.36). Click OK to complete the load definition.

Figure P3.31

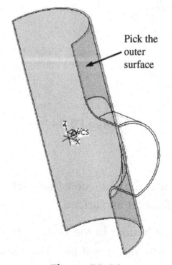

Figure P3.32

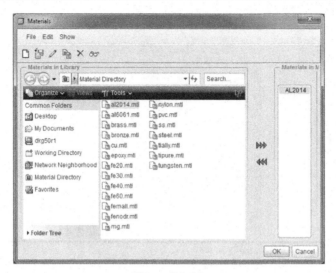

Figure P3.33

Figure P3.34

Define Displacement Constraints

We will define two constraints under the same constraint set, that is, one to ensure symmetry at the split faces and one to fix the bottom surface.

Click the Displacement Constraint button. In the Constraint dialog box (Figure P3.37) appearing next, enter Symmetry for Name, and pick the vertical face, as shown in Figure P3.38. Note that the faces on the other side will be picked automatically. Choose

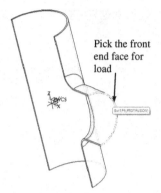

Pick the front
end face for
load

Figure P3.35

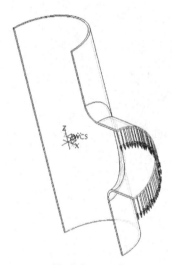

Figure P3.36

Translations *X* and *Z* to be free (Translation *Y* remains fixed), and Rotation *Y* to be free (Rotations *X* and *Z* remain fixed), and then click OK.

To add constraints at the bottom surface, click the Displacement Constraint button again. Enter Bottom for Name. Note that you need to make sure that the Member of Set is still ConstraintSet1 since we are adding additional constraints to the same constraint set. Pick the bottom surface of the long tube in the Graphics area. Click OK (all translation and rotation are fixed by default) to complete the addition of the fixed constraint. You should see constraint symbols and labels appearing in the model in the Graphics area (Figure P3.39).

Mesh the Model

Click the AutoGEM button (on top of the Graphics area; Figure P3.30), the AutoGEM dialog box will appear (Figure P3.40). Click Create. A mesh will be created on the surface model

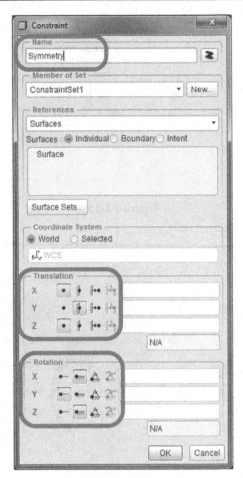

Figure P3.37

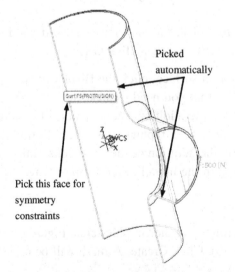

Figure P3.38

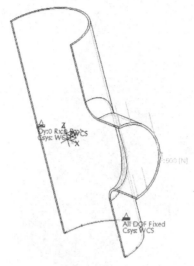

Figure P3.39

Figure P3.40

(Figure P3.41) and a mesh summary window appears (Figure P3.42). There are four triangular and nine quad-elements created for this surface model. Close the summary window. Choose File > Save Mesh to save the mesh and close the AutoGEM dialog box.

Note that you may choose from the pull-down menu View > Simulation Display to make necessary adjustment for mesh display; for example, shrink mesh 10% to make a clear view on individual elements.

Analysis

Click the Design Study button (on top of the Graphics area; Figure P3.30), and the Analyses and Design Studies dialog box will appear (Figure P3.43). Choose File > New Static to bring up the Static Analysis Definition dialog box (Figure P3.44).

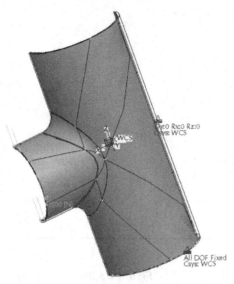

Figure P3.41

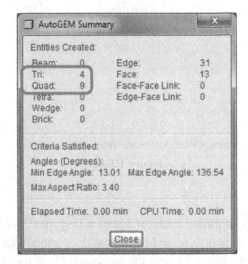

Figure P3.42

In the Static Analysis Definition dialog box, enter My_Static_Single_Pass for Name, select default Constraints and Loads (ConstraintSet1 and LoadSet1), select Single-Pass Adaptive as the Convergence Method, and click OK.

Now the My_Static_Single_Pass analysis is listed in the Analyses and Design Studies dialog box (Figure P3.45). Click the Run button (Figure P3.45) to start the analysis. Click the Status button to view the status (and summary) of the analysis.

Figure P3.43

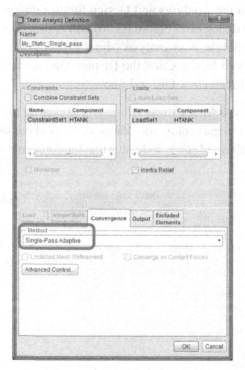

Figure P3.44

The analysis should be completed in just a few seconds. In the Analysis Summary window, you should see maximum displacement, maximum principal stress, and so on, just like that of the beam example.

Display Results

We will show a deformed shape with von Mises stress to ensure that the FEA model is adequately defined and is maintaining the symmetry characteristics.

Figure P3.45

Click the Results button in the Analyses and Design Studies dialog box (Figure P3.45), the Result Window Definition dialog box (Figure P3.46) will appear. Enter von_Mises_Stress for Name, choose My_Static_Single_pass analysis (should appear as default), choose Fringe for Display type, and choose von Mises. Click the Display Options tab to choose Deformed, Show Element Edges, and then click OK and Show (Figure P3.47).

The deformed shape with von Mises stress fringe will appear in the Graphics area (Figure P3.48). Note that the maximum von Mises stress is about 46.5 MPa located at the middle of the fillet. The deformed shape indicates that the force and displacement constraints

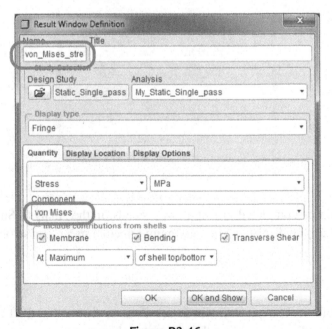

Figure P3.46

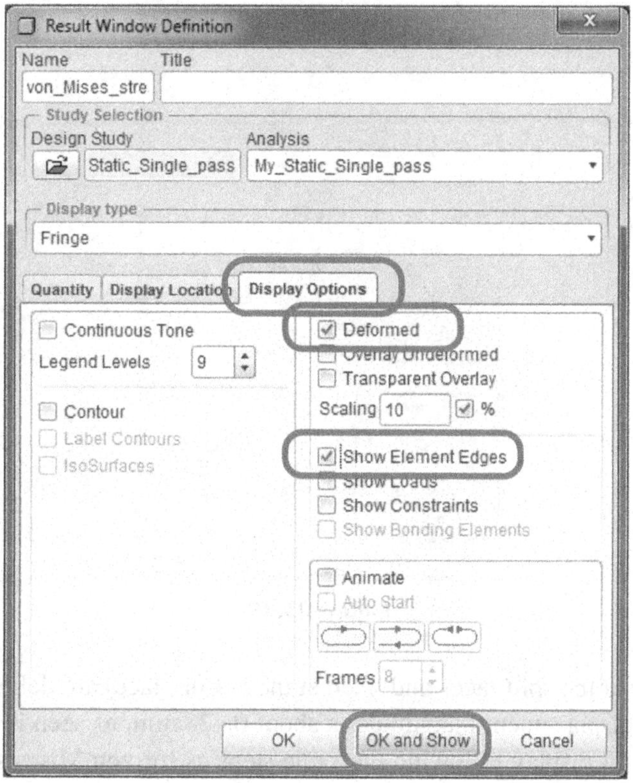

Figure P3.47

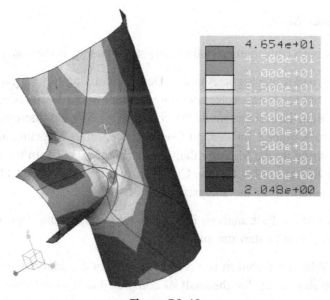

Figure P3.48

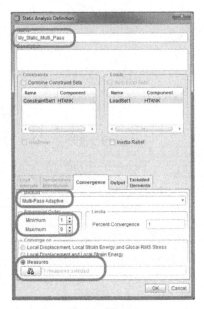

Figure P3.49

(both the symmetry at the split faces and fixed at the bottom face) are defined properly. Note that the maximum displacement magnitude is about 0.428 mm, as seen in the Summary window, or in a result display following the same steps as for von Mises stress.

Save the results window under name von_Mises, and then close the Results window.

P3.3.3 Convergence Study

Next, we will carry out a convergence study, using strain energy as the convergence criterion.

Create a new study, and in the Static Analysis Definition dialog box (Figure P3.49), enter My_Static_Multi_Pass for Name, select default Constraints and Loads (ConstraintSet1 and LoadSet1), select Multi-Pass Adaptive as the convergence Method, increase Maximum Polynomial Order to 9, enter 1 for Percent Convergence, select Measure, and click the Measure button to bring up the Measure dialog box (Figure P3.50). In the Measure dialog box, choose Strain Energy, and click OK. Click OK in the Static Analysis Definition dialog box to accept the definition for the analysis.

Now the My_Static_Multi_Pass analysis is listed in the Analyses and Design Studies dialog box. Click the Run button to start the analysis.

The analysis should be completed in just a few seconds. In the analysis summary window (Figure P3.51), you should see that the analysis completed in Pass 9. However, there are two elements not converged. This indicates that we may need to refine the mesh to receive better

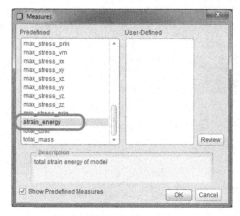

Figure P3.50

convergent results. Since the overall strain energy converges in 0.1%, which is below the required 1%, the results in general are acceptable. Note that the maximum von Mises stress is about 46.6 MPa, and the maximum displacement magnitude is 0.427 mm, which are about the same as those of single-pass adaptive analysis. In most cases, single-pass provides excellent results. We will accept these results.

Click the Results button in the Analyses and Design Studies dialog box (Figure P3.45), the Result Window Definition dialog box (Figure P3.52) will appear. Choose Static_Multi_Pass as Design Study and choose My_Static_Multi_Pass analysis. You will need to click the button under Design Study and locate the file folder. We will first display the convergence graph for the strain energy measure.

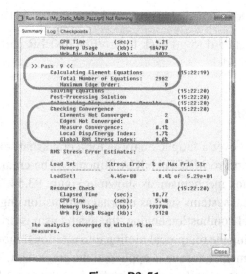

Figure P3.51

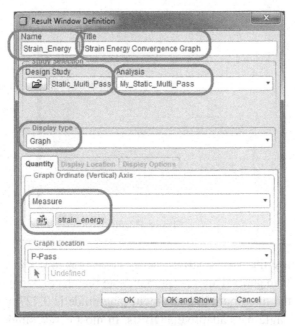

Figure P3.52

In the Result Window Definition dialog box (Figure P3.52), enter Strain_Energy for Name, and Strain Energy Convergence Graph for Title. Choose Graph for Display type. Pull down the selection of Quantity and choose Measure. Click the Measure button and choose Strain_Energy in the Measures dialog box (same as Figure P3.50). Click OK and Show. The strain energy convergence graph will appear like the one in Figure P3.53. The graph shows that the strain energy measure converges very well.

Close the Result Window Definition dialog box and the Analyses and Design Studies dialog box.

Choose Applications > Standard to go back to Pro/ENGINEER part mode. Choose File > Save to save your model.

P3.4 Fatigue Analysis

In this example we will go over a fatigue analysis for a crankshaft example using the stress-based approach provided in Pro/MECHANICA Structure. The crankshaft is one of the components in a slider-crank mechanism, as shown in Figure P3.54. This type of mechanism is often found in mechanical systems such as internal combustion engine and oil-well drilling equipment. For the internal combustion engine, the mechanism is driven by a firing load that pushes the piston, converting the reciprocal motion into rotational motion at the crankshaft. In the oil-well drilling equipment, a torque is applied at the crankshaft. The rotational motion is converted to a reciprocal motion at the piston that digs into the ground. In any case the load or

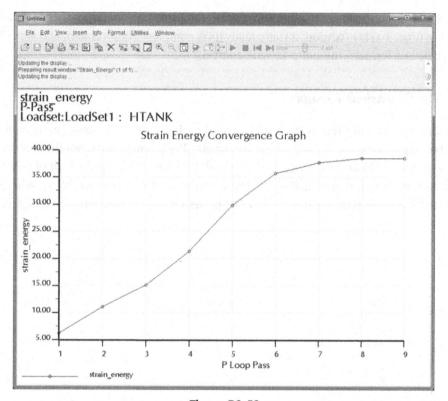

Figure P3.53

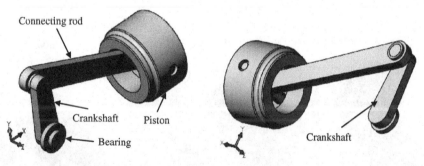

Figure P3.54: The slider-crank mechanism in two views.

reaction force acting on the crankshaft is changing direction. However, in this example we are assuming a cyclic load with constant amplitude, fully reversed along the longitudinal direction of the crankshaft. Note that this is a much simplified load case since in reality, as mentioned, the direction of the force is changing. Therefore, please keep in mind that results obtained from this simple analysis are useful only in providing a general idea of the fatigue life for the crankshaft, which is in fact, a fairly rough estimate. For more in-depth fatigue analysis, you may have to use other software tools that offer multiaxial fatigue analysis capabilities, in which load is changing, therefore, the principal direction of the stress is changing in time. To stay focused, a static analysis is completed beforehand, and we will only go over the steps for carrying out a fatigue analysis.

P3.4.1 The Crankshaft Example

The crankshaft shown in Figures P3.55a is about 4 in. long with two short shafts connecting with the bearing and the connecting rod, respectively. The distance between the centers of the two short shafts is 3.00 in. Note that there is a small fillet of 0.05 in. at the root of both shafts. The unit system chosen is the in.-lb$_f$-sec. The material is aluminum (AL2014), where the modulus and Poisson's ratio are $E = 1.06 \times 10^7$ psi and $\nu = 0.33$, respectively. A bearing load

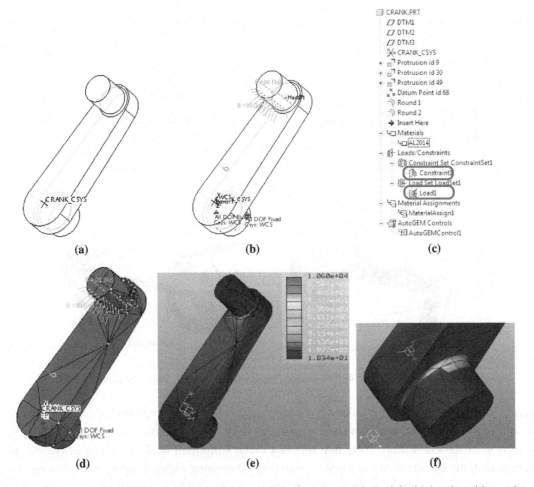

Figure P3.55: The finite element model of the crankshaft, (a) a solid model, (b) load and boundary conditions, (c) simulation entities listed in the Model Tree, (d) finite element mesh, (e) von Mises stress fringe plot, and (f) von Mises stress with a closer view.

of 100 lb$_f$ is applied at the outer cylindrical surface of the top shaft (connecting to the connecting rod) along the longitudinal direction of the crankshaft. The outer surface of the lower shaft (connecting to the bearing) is completely fixed, as shown in Figure P3.55b.

A finite element model is completely defined for this example. When you open the crankshaft model downloaded from the book's companion website (http://booksite.elsevier.com/ 9780123984609), you should see the FEA model after entering Mechanica. Expand Loads/ Constraints, then ConstraintSet1 and LoadSet1 in the Model Tree window. Right-click Constraint1 and Load1 listed in the Model Tree, as shown in Figure P3.55c, to find out more about the boundary condition and load defined for the crankshaft finite element model. Note that there are 273 tetrahedron p-elements created using the default setting, as shown in Figure P3.55d. Rerun a static analysis (Analysis1) and show von Mises stress fringe plot (Figure P3.55e). Note that the maximum von Mises stress is about 10,600 psi located in the fillet of the lower shaft, as shown in Figure P3.55f, which is largely due to the bending effect. The maximum von Mises is slightly below the material yield strength of 14,000 psi. The ultimate tensile strength is 24,000 psi.

P3.4.2 Using Pro/MECHANICA Structure

Open Pro/MECHANICA Structure Model

Open the solid model of the crankshaft from the book's companion website (http://booksite. elsevier.com/9780123984609). You may enter Mechanica to review the finite element model created. Rerun the static analysis to generate stress results for fatigue analysis. Please review the von Mises stress to make sure the maximum stress magnitude and its location are consistent with those of Figure P3.55e.

Create Fatigue Analysis Model

From the pull-down menu, choose Analysis > Mechanica Analyses/Studies. In the Analyses and Design Studies dialog box (Figure P3.56), choose from the pull-down menu File > New Fatigue.

In the Fatigue Analysis Definition dialog box (Figure P3.57), enter Fatigue for Name, enter 1e+06 for Desired Endurance, choose Constant Amplitude for Loading Type, and choose Peak-Peak for Amplitude Type. We assume that the upper limit of the high-cycle fatigue is 1,000,000 cycles, which is also where the endurance limit is measured. Peak-Peak is chosen for fully reversed cyclic load. The load magnitude is identical to that of Analysis1 (with load factors −1 and 1 for minimum and maximum, respectively, as shown in Figure P3.57).

Click the Previous Analysis tab, and select Use static analysis results from previous design study. Make sure Analysis1 is listed as Static Analysis with LoadSet1 for Load Set, as shown in Figure P3.58. Click OK to accept the selections. Fatigue is now listed in the Analyses and Design Studies dialog box.

Figure P3.56

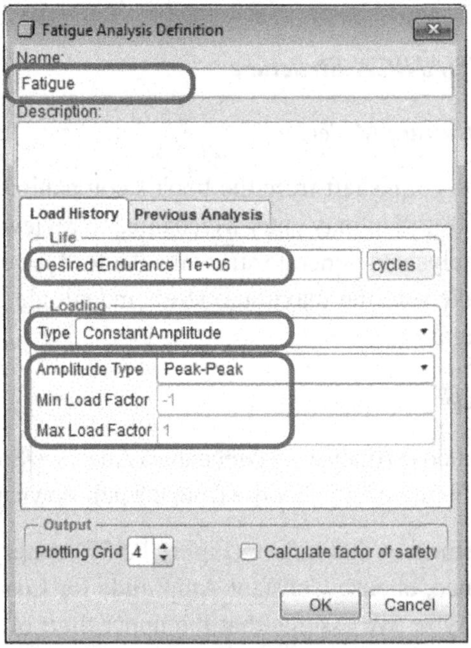

Figure P3.57

Next, we will add material properties to support the fatigue analysis. Expand the Material Assignments entity in the Model Tree, right-click MaterialAssign1, and choose Edit Definition. In the Material Assignment dialog box (Figure P3.59), AL2014 is listed. Click More to bring up the Materials dialog box (Figure P3.60). In the Materials dialog box, double click AL2014 listed in the right column to bring up the Material Definition dialog box (Figure P3.61).

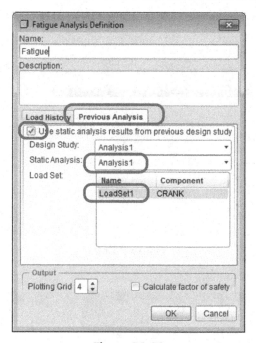

Figure P3.58

Figure P3.59

In the Material Definition dialog box (Figure P3.61), enter 14,000 and 24,000 psi for Tensile Yield Stress and Tensile Ultimate Stress, respectively. Choose Distortion Energy (von Mises) for Fatigue Criterion. In the bottom block (under Fatigue), choose Unified Material Law (UML), Aluminum Alloys, Polished, and 1, and then click OK.

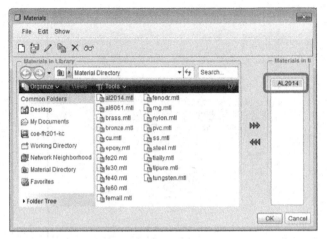

Figure P3.60

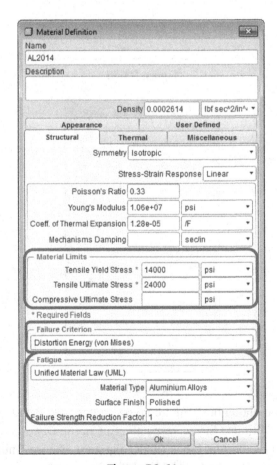

Figure P3.61

Note that Mechanica uses a generic set of fatigue properties to model low-alloy steels, unalloyed steels, aluminum alloys, and titanium alloys. These generic properties have been compiled to be known as the Unified Material Law. UML is a handy method as only tensile strength data of metals are needed for estimation of the strain–life curves.

At this point, we have completely defined a fatigue analysis model, and we are ready to conduct a fatigue life calculation.

Run Fatigue Analysis

From the pull-down menu, choose Analysis > Mechanica Analyses/Studies. The Analyses and Design Studies dialog box appears (Figure P3.62). Choose Fatigue listed in the Analyses and Design Studies dialog box, and click the Start Run button (fourth from the left) to start the analysis.

In the Question dialog box appearing next (Do you want to run interactive diagnostics?), click No.

In the Prompt/Message window, a message "The design study has started" appears. When the analysis is completed, the status of Fatigue analysis listed in the Analyses and Design Studies dialog box will become completed.

Display Results

From the Analyses and Design Studies dialog box (Figure P3.62), click the Result button (first from the right). The Result Window Definition dialog box appears (Figure P3.63). Enter Fatigue_Life for Name, choose Fringe, and Log Life, and click OK and Show.

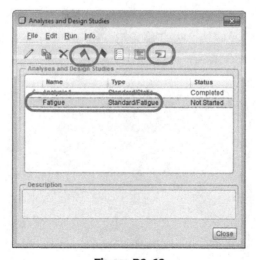

Figure P3.62

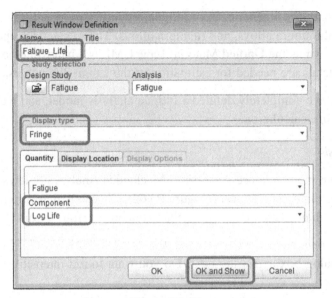

Figure P3.63

A new Graphics window appears, showing the crankshaft with a fatigue life fringe plot (Figure P3.64). The fatigue life is between $10^{5.401}$ ($\approx 252,000$) and 10^{20} cycles. The minimum life is found at the same place as that of the maximum von Mises stress, which is in the fillet of the lower shaft, as shown in Figure P3.65.

Close the result window. Choose Applications > Standard to go back to Pro/ENGINEER part mode. Choose File > Save to save your model.

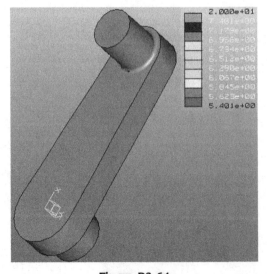

Figure P3.64

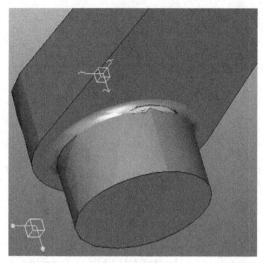

Figure P3.65

Exercises

1. Analyze the following cantilever beam using Pro/MECHANICA Structure.

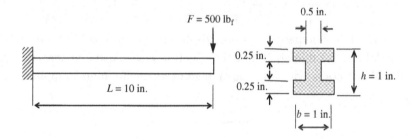

(a) Create Pro/ENGINEER part and Mechanica models, and carry out FEA for the maximum displacement and maximum stress. Submit screen captures for the maximum bending stress and maximum displacement (downward only) in the fringe plots.

(b) Solve the same problem using the classical beam theory. Compare your calculations with those of Mechanica. Are they consistent? Why or why not?

(c) Repeat (b) by letting Poisson's ratio $\nu = 0$.

2. An L-shape circular bar of diameter 1.25 in. is loaded with an evenly distributed force of 400 lbs. at its end face, as shown. Note that the elbow radius (corner of the L-shape) is 1 in. and the material is 1060 Alloy.

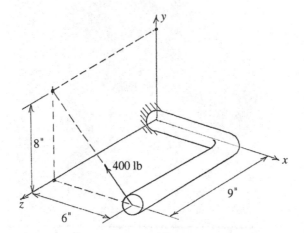

(a) Create an FEA model using Pro/MECHANICA Structure to calculate the maximum principal stress and the maximum shear stress in the bar.

(b) Calculate the maximum principal stress and maximum shear stress using analytical beam theory. Where are the maximum stresses located? Compare your calculations with those obtained from Pro/MECHANICA Structure, both values and locations. Are they close? Why or why not? Please comment on your comparison.

3. Open the full tank solid model from the book's companion website (http://booksite. elsevier.com/9780123984609) and create a finite element model that is consistent to the thin-shell model discussed in Section P3.3. Note that the load is 1,000 N downward and the bottom face is fixed, as shown in the following diagram. Create mesh using default setting, and carry out an FEA. Compare maximum displacement and von Mises stress between the full solid model and the thin-shell model of Section P3.3. Also compare the size of these two models. Please comment on the advantage of idealization and simplification in FEA modeling.

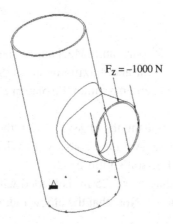

Project S3 Structural FEA and Fatigue Analysis Using SolidWorks Simulation

There is no need to emphasize the importance of understanding theory and using analytical methods to solve engineering problems. In many design applications, such as structural components of complex geometry, analytical methods can only offer rough estimates on product performance at best. In many cases, engineers must rely on tools, such as finite element analysis (FEA) software, to evaluate product performance with high accuracy in order to support design decision making. It is critical for a design engineer to learn and be able to use software tools for solving problems that are beyond hand calculations and analytical methods.

In Project S3, we will introduce SolidWorks® Simulation for FEA and fatigue analysis of structural components. We include three simple examples just to get you started on learning and using the software. These three examples are a simple cantilever beam, a thin-walled tank, and a crankshaft, which should provide you with a good overview on the FEA and fatigue analysis capabilities offered by SolidWorks Simulation. You may find the example on the book's companion website.

Overall the objective of this project is to enable readers to use SolidWorks Simulation for basic applications. Those who are interested in learning more and elevating themselves to an intermediate level, may want to go over more examples provided by Simulation or review other tutorial books on the subject (e.g., *Engineering Analysis with SolidWorks Simulation*, by Paul Kurowski, ISBN: 978-1-58503-710-0; http://www.sdcpublications.com/).

Note that the lessons included in this project are developed using SolidWorks 2011 SP2.0. If you are using a different version of SolidWorks, you may see slightly different windows or dialog boxes. Since SolidWorks is fairly intuitive to use, these differences should not be too difficult to figure out.

S3.1 Introduction to SolidWorks Simulation

SolidWorks Simulation (or Simulation) is a finite element analysis (FEA) tool that allows users to analyze structural performance, such as displacement, stress, natural frequency, buckling load, fatigue life, and so on. The main objective of this tutorial is to help you, as a new user, to become familiar with Simulation and to provide you with a brief overview of using the software.

SolidWorks Simulation is embedded in SolidWorks. It is indeed an add-in module of SolidWorks. Transition from SolidWorks to Simulation is seamless. The entire solid parts, materials, assembly mates, and so on. defined in SolidWorks are automatically carried over into Simulation. Simulation can be accessed through menus and windows inside SolidWorks. The same part or assembly created in SolidWorks is directly employed for creating FEA models in Simulation. Menus and dialog boxes for Simulation are just like those in SolidWorks. Those who are familiar with SolidWorks, will find Simulation easy to learn.

S3.1.1 Overall Process

The overall process of using SolidWorks Simulation for carrying out FEA consists of three main steps: preprocessing, analysis, and postprocessing, as illustrated in Figure S3.1. Key entities that constitute an FEA model include a geometric model of the structure created in SolidWorks, a finite element mesh, material properties, loads, and boundary conditions.

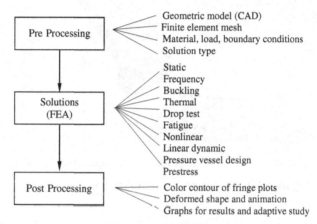

Figure S3.1: General process of using SolidWorks Simulation.

The analysis capabilities in Simulation include static, frequency, buckling, thermal, drop test, fatigue, nonlinear, linear dynamic, and pressure vessel design, which cover a broad range of applications encountered in engineering design. You will have to choose an analysis type that best describes the physical problem being solved.

The analysis results can be visualized in various forms. You may use color contour or fringe plots to display stress or displacement results. You may display or animate the deformed shape or mode shape of the structure. You may also use graphs to show results, such as h-adaptive study.

S3.1.2 User Interface

User interface of the Simulation is identical to that of SolidWorks, as shown in Figure S3.2. SolidWorks users should find it straightforward to maneuver in Simulation. As shown in

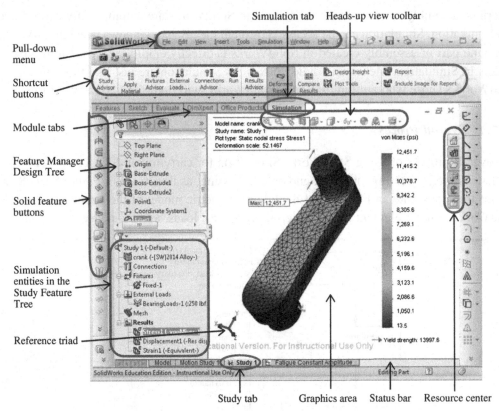

Figure S3.2: User interface of SolidWorks Simulation.

Figure S3.2, the user interface window of Simulation consists of pull-down menus, shortcut buttons, the Graphics area, the Feature Tree window, and the Study Feature Tree window.

The Graphics area displays the FEA model you are working on. The pull-down menus and the shortcut buttons at the top of the screen provide typical SolidWorks and Simulation functions. These shortcut buttons (some with drop-down selections) provide all the functions required to create and modify FEA models. When you move the mouse pointer over these buttons, a short message describing the menu command will appear. Some of the frequently used buttons in Structure and their functions are summarized in Table S3.1. You may customize these buttons by choosing from the pull-down menu Tools > Customize. The reference triad shown in Figure P3.2 is the coordinate system that all physical quantities to be created for the FEA model refer to.

Entering Simulation from SolidWorks is straightforward. If you are creating a new finite element model in Simulation, you will have to choose from the pull-down menu Simulation > Study. If you open an existing FEA model created in Simulation, you may simply click the study tab (with the study name assigned previously) at the bottom of the

Table S3.1: The shortcut buttons in Mechanism Design

Button Symbol	Name	Function
	New Study	Defines new study.
	Apply Material	Assigns material to selected items in the Simulation study tree.
	Create Mesh	Creates solid/shell mesh for the active study.
	Run	Starts the solver for the active study.
	Apply Control	Defines a mesh control for selected entities.
	Global Contact	Sets a global contact condition.
	Contact Set	Defines contact set (face, edge, vertex).
	Drop Test Setup	Defines drop test setup.
	Result Options	Edit/defines result options.
	Fixtures	Defines a fixture on the selected entities for the active structural study (static, frequency, buckling, or nonlinear study).
	Pressure	Defines a pressure on the selected entities for the active structural study (static, frequency, buckling, or nonlinear study).
	Force	Defines a force, torque, or moment on the selected entities for the active structural study (static, frequency, buckling, or nonlinear study). The specified value is applied to each selected entity.

Continued

Table S3.1: The shortcut buttons in Mechanism Design—cont'd.

Button Symbol	Name	Function
	Gravity	Defines gravity loading for the active structural study (static, frequency, or buckling).
	Centrifugal Force	Defines centrifugal forces (angular velocity/ acceleration) for the active structural study (static, frequency, buckling, or nonlinear study).
	Remote Load/ Mass	Defines remote load/mass on a set of faces with respect to a coordinate system for the active structural study.
	Rigid Connection	Defines remote load/mass on a set of faces with respect to a coordinate system for the active structural study.
	Bearing Load	Defines bearing load on set of cylindrical faces with respect to a coordinate system for the active structural study.
	Temperature	Defines a temperature on the selected entities for the active study.

Graphics area, as shown in Figure S3.2. If you do not see the Simulation pull-down menu or the study tab, you may not have activated the Simulation add-in module. To activate the Simulation module, choose from the pull-down menu Tools > Add-Ins.

In the Add-Ins window shown in Figure S3.3, click SolidWorks Simulation in both boxes (Active Add-ins and Start Up), and then click OK. You should see that Simulation is added to the pull-down menu. If not, restart SolidWorks to activate the Simulation module.

S3.1.3 Defining FEA Entities

The basic entities of an FEA model created in Simulation consist of geometry (in CAD), mesh, material properties, loads, and boundary conditions (called fixtures). These entities can be defined using Simulation shortcut buttons or right click the simulation entities listed in the Study Feature Tree.

Loads can be defined in several ways, depending on the physical conditions. Simulation allows you to create force, moments, pressure, gravity, centrifugal, thermal, and so on.

Figure S3.3: The Add-Ins window.

Boundary conditions include displacement constraints, planar constraints (such as pin and ball constraints), and symmetry constraint. Idealization, such as converting thin-walled solids to shell, or bar solids to beam, is supported. Material properties can be created and assigned to a certain volume or the entire solid model. There is an excellent material library built in Simulation. In addition, simulation measures and mesh control can be created.

S3.1.4 Types of Analyses

SolidWorks Simulation provides nine analysis types for structural analysis. They are static, frequency, buckling, thermal, drop test, fatigue, nonlinear, linear dynamic, and pressure vessel design. Static analysis, which is the most basic analysis, analyzes structural responses to external loads, including stress, strain, and displacement. Frequency analysis calculates natural frequencies and corresponding vibration mode shapes. Buckling analysis calculates buckling load factors and buckling mode shapes. Both frequency and buckling analyses solve eigenvalue problems mathematically. Thermal analysis calculates temperatures, temperature gradients, and heat flow based on heat generation, conduction, convection, and radiation conditions. Drop test studies evaluate the effect of dropping a part or an assembly on a rigid floor. Fatigue analysis provided in Simulation supports high-cycle fatigue using a stress-based approach such as an S—N diagram. Nonlinear analysis solves problems with nonlinearity caused by material behavior, large displacements, or contact conditions. Linear dynamic studies use natural frequencies and mode shapes to evaluate the response of structures to dynamic loading environments. Pressure vessel study combines the results of static studies with the desired safety factors. Each static study has a different set of loads that produce corresponding results.

S3.1.5 Viewing Results

Simulation allows users to view, evaluate, and manage results for FEA analyses. Numerous approaches are provided in Simulation for users to view analysis results, including color contour or fringe plots, animation, and graphs. In addition, users may study the results by, for example, probing specific areas of an FEA model for displacement or stress data. Users may save the set of result windows and review or reuse them later. Also, users may prepare printed and online reports for evaluation and presentation.

S3.1.6 Examples

Three examples are included in this project, which illustrate step-by-step details of modeling, analysis, and result visualization capabilities in Simulation. We will start with a very simple cantilever beam example. This example gives you a quick start and a quick run-through for a static FEA using Simulation, including creating an FEA model, running an analysis, and visualizing simulation results. You will also learn how to convert a solid model to a beam model, as well as define and run FEA for the idealized beam model.

The second example illustrates the steps for converting a thin-walled tank solid model to a shell model, applying load and boundary conditions to the shell model, analyzing the model with much reduced model size, and visualizing the results. In addition, we will learn to carry out mesh refinement for convergence study using h-adaptation.

The third example provides details on setting up a fatigue analysis model, running fatigue analysis, and visualizing fatigue analysis results, including showing the fringe plots for both fatigue life and percentage damage. All examples and main topics to be discussed are summarized in Table S3.2.

S3.2 Simple Cantilever Beam

The purpose of this lesson is to provide you with a quick start on using SolidWorks Simulation. You will learn pre- and postprocessing and analysis capabilities in SolidWorks Simulation, using a simple cantilever beam example. After completing this lesson, you should be able to solve similar problems following the same procedures. In this exercise, we will use default options for most of the selections. We will treat the beam as a solid model and use

Table S3.2: Examples employed in this project

Section	Example	FEA Model	Problem Type	Things to Learn
S3.2	Simple cantilever beam		Static analysis	This example helps you become familiar with the basic operation in Simulation. This lesson offers a quick run-through in using Simulation for FEA. You will learn the general process of using Simulation to construct an FEA model, run analysis, and visualize the FEA results. You will also learn how to convert a solid model to beam model, as well as define and run FEA for the idealized beam model.
S3.3	Thin-walled tank		Static analysis	This example illustrates the process of converting a thin-walled solid model to a shell FEA model. You will learn how to add load and boundary conditions to the shell model for a complete FEA model. You will also learn how to carry out mesh refinement for a convergence study using h-adaptation.
S3.4	Crankshaft		Fatigue analysis	This lesson provides steps for learning fatigue analysis using a stress-based approach. You will learn how to create a fully reversed cyclic load and SN diagram for fatigue life calculations. You will also learn how to show and interpret results, including fatigue life in load cycles and percentage damage.

a tetrahedron mesh for FEA. After that we will convert the solid model into beam elements, which significantly reduces the model size, and yet offers accurate solutions.

S3.2.1 The Cantilever Beam Example

The cantilever beam of 1 in. × 1 in. × 10 in. is shown in Figure S3.4. The unit system chosen is in.-;lb$_f$-sec. The material is aluminum (AL2014), where the modulus and Poisson's ratio are $E = 1.06 \times 10^7$ psi and $v = 0.33$, respectively. The total force at the tip is $P = 1000$ lb$_f$ downward.

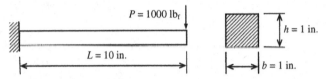

Figure S3.4: Cantilever beam example.

S3.2.2 Using SolidWorks Simulation

Open SolidWorks Part

Open the solid model of the cantilever beam downloaded from the book's companion website. This solid model, as shown in Figure S3.5, was created using default datum planes for section sketch and extruded for a square bar. Please note the reference triad shown in Figure S3.5. All physical quantities to be created for the FEA model of the beam refer to this reference triad. Note that the positive X-direction points downward in the figure.

As soon as you open the model, you may want to check the unit system chosen. From the Tools pull-down menu, choose Options. In the System Options dialog box (Figure S3.6), select the Documents Properties tab and select Units. Select IPS (inch, pound, second) then click OK. We will stay with this unit system.

Enter SolidWorks Simulation

At the bottom of the Graphics area, you should see two studies completed: Static-Solid and Static-Beam, which are FEA for solid model and beam elements, respectively. You may click these simulation tabs to browse FEA models. You may rerun FEA to check results.

We will start a new study and treat the model as solid first.

Choose from the pull-down menu Simulation > Study. Select Static from the Study dialog box, as seen in Figure S3.7 (you may change the study name to My_Study). Accept the options selected by clicking the green checkmark on top left.

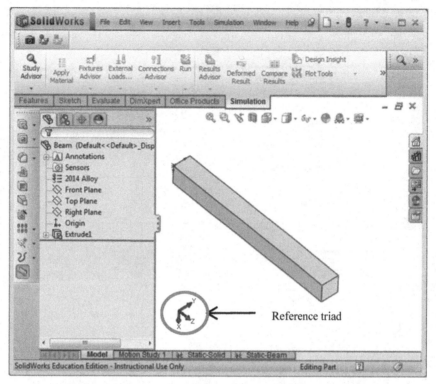

Figure S3.5: Cantilever beam example and reference triad.

A new study (My_Study) will be added to the study tab below the Graphics area, and a new set of buttons appear on top. In addition, a list of simulation entities will appear in the Study Feature Tree (right below model tree), as shown in Figure S3.8. We will use these entities (mostly right click on them) to define the finite element model, carry out FEA, and visualize FEA results. Note that material 2014 Alloy has been assigned to the beam.

Create Finite Element Model

We will define material, fixture (boundary condition), and external load, and then create finite element mesh using the right click options on the simulation entities.

Move the mouse pointer into the simulation entities. Right-click Beam, and choose Apply/ Edit Material, as shown in Figure S3.9. In the Material dialog box, 2014 Alloy is selected (one of the material type included in the material library), and material properties are listed, as shown in Figure S3.10. The unit system for the material properties may differ from the document's unit system. You can change the units displayed using Units drop-down menu (selecting Properties tab). Note that this material is already assigned. You may simply close the dialog box by clicking Close.

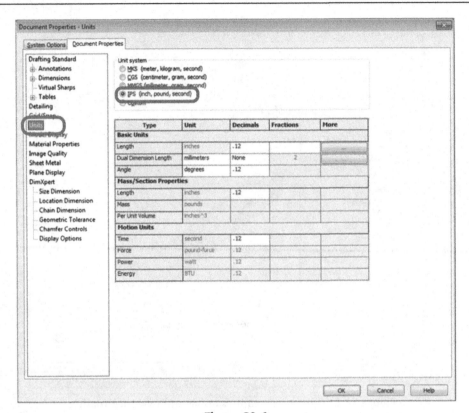

Figure S3.6

Right-click Fixtures and choose Fixed Geometry as shown in Figure S3.11. SolidWorks Simulation supports fixed geometry, roller/slider, fixed hinge, elastic support, and advanced fixtures. The Fixture dialog box will pop up in the Property Manager tab, as shown in Figure S3.12. Rotate the beam and select the end face (rear face) of the beam to apply the fixed constraint, as shown in Figure S3.13. Applying fixed geometry will constrain the selected face from all degrees of freedom (translation and rotation). The green arrow symbol of the fixture will appear at the end face. Accept the options selected by clicking the green checkmark.

Right-click External Loads and choose Force as shown in Figure S3.14. SolidWorks Simulation supports force, torque, pressure, gravity, centrifugal, bearing, and thermal loads. The Force/Torque dialog box will pop up in the Property Manager tab, as shown in Figure S3.15. Rotate the beam back and select the top edge of the front end face of the beam to apply the force. We will then define the direction of the force. Choose Selected direction (default is Normal) from the dialog box (Figure S3.15), pick the top face of the beam from the Graphics area (Figure S3.16), and enter 1,000 in the text field that shows normal to the selected face (Figure S3.15). Accept the options selected by clicking the green checkmark. Force symbol (arrows) should appear at the top edge of the front end face.

Figure S3.7

Next we will create finite element mesh.

Right-click Mesh and choose Create Mesh as shown in Figure S3.17. Accept the default mesh setup shown in Figure S3.18 by clicking the green checkmark. A finite element mesh will be created and shown in the Graphics area. You may right-click Mesh and choose Details to see a summary on the mesh created, as shown in Figure S3.19. There are total 13,366 tetrahedron elements and 20,717 nodes.

Run Analysis

We will carry out a static analysis next. Right-click My_Study and choose Run (Figure S3.20). The analysis will be completed shortly (less than a minute). Finer mesh will result in increased time to solve the problem. Coarse mesh is often recommended for the initial solve to identify potential modeling errors or problems. If Simulation shows you a message indicating that the analysis may have encountered excessive displacements, such as the one shown in Figure S3.21, simply click No to solve this problem as small displacement problem since we treat this beam as a linear elastic problem of small displacement.

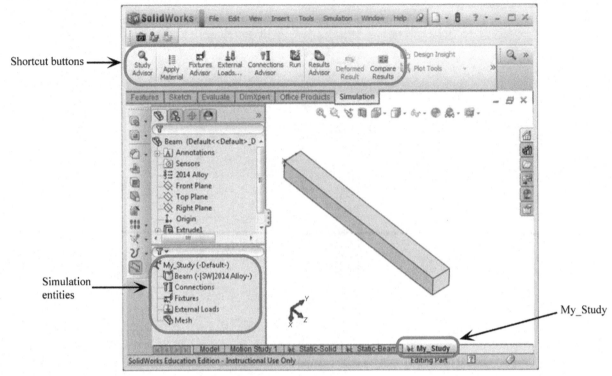

Figure S3.8

Display Results

After the analysis is completed, there are three entities added to Results, as shown in Figure S3.22. Right-click Stress1(-vonMises-), and choose Edit Definition. We will display the bending stress result, which is the normal stress along the Z-direction. In the Stress Plot dialog box shown in Figure S3.23, select SZ: Z Normal Stress, choose psi for unit, select Deformed

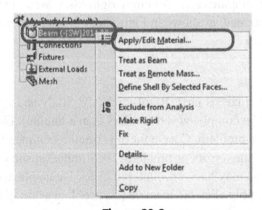

Figure S3.9

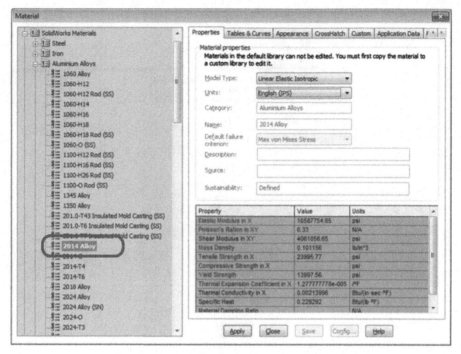

Figure S3.10

Shape, and click the green checkmark to accept the definition. Double-click the Stress1 entity to bring up the fringe stress plot in the Graphics area, like that of Figure S3.24.

Does the maximum bending stress occur at root? The maximum bending stress is about 83 ksi. Is the stress acceptable? Can we verify this result?

Similarly, right-click Displacement1, and choose displacement in the *X*-direction. The displacement fringe plot appears, like that of Figure S3.25. The maximum displacement is 0.3802 in. Again, is this result accurate? Will you accept these results?

We do have analytical solutions for this simple cantilever beam example. Based on the classical beam theory we learned in Solid Mechanics, the tip displacement and the maximum bending stress at the root are, respectively,

$$y_{\max} = \frac{PL^3}{3EI} = \frac{1000 \times (10)^3}{3 \times (1.06 \times 10^7) \times (1)(1)^3/12} = 0.3774 \text{ in.}$$

$$\sigma_{\max} = \frac{Mc}{I} = \frac{(PL)(h/2)}{I} = \frac{(1000 \times 10)(1/2)}{(1)(1)^3/12} = 60,000 \text{ psi}$$

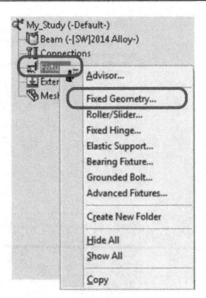

Figure S3.11

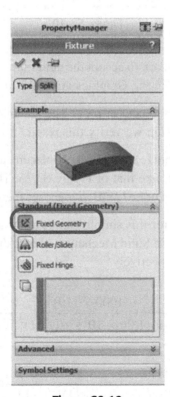

Figure S3.12

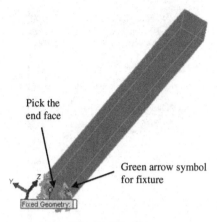

Pick the
end face

Green arrow symbol
for fixture

Figure S3.13

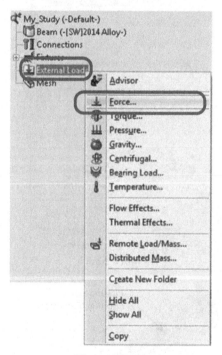

Figure S3.14

FEA displacement result is very close to that of classical beam theory, but not stress, why? One factor is the Poisson ratio. The beam theory does not consider the effects of the Poisson ratio. In the finite element model, the deformation along the lateral direction due to the beam bending is adding to the bending strain and stresses due to a nonzero Poisson's ratio.

Another factor that causes the high stress is the local stress concentration. You may probe the bending stress at nodes along the edge at the root end face. To do so, you may click Plot Tools

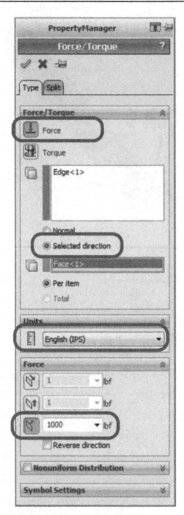

Figure S3.15

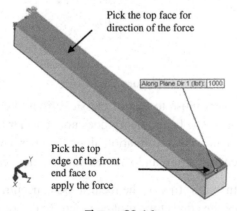

Pick the top face for direction of the force

Along Plane Dir 1 (lbf): 1000

Pick the top edge of the front end face to apply the force

Figure S3.16

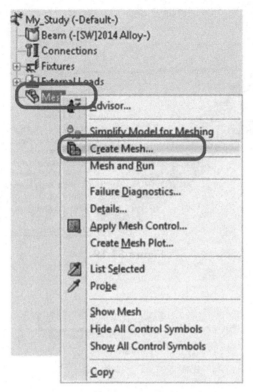

Figure S3.17

Figure S3.18

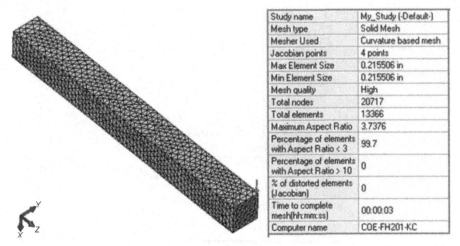

Study name	My_Study (-Default-)
Mesh type	Solid Mesh
Mesher Used	Curvature based mesh
Jacobian points	4 points
Max Element Size	0.215506 in
Min Element Size	0.215506 in
Mesh quality	High
Total nodes	20717
Total elements	13366
Maximum Aspect Ratio	3.7376
Percentage of elements with Aspect Ratio < 3	99.7
Percentage of elements with Aspect Ratio > 10	0
% of distorted elements (Jacobian)	0
Time to complete mesh(hh:mm:ss)	00:00:03
Computer name	COE-FH201-KC

Figure S3.19

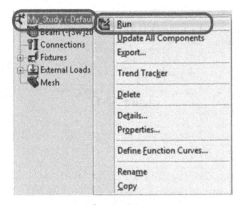

Figure S3.20

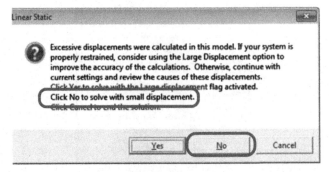

Figure S3.21

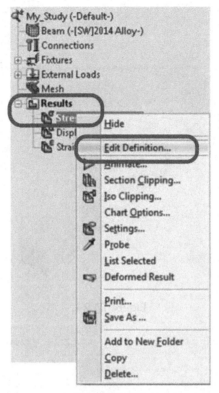

Figure S3.22

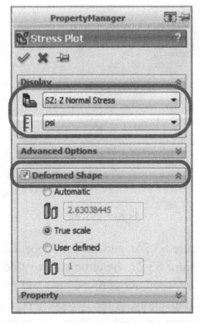

Figure S3.23

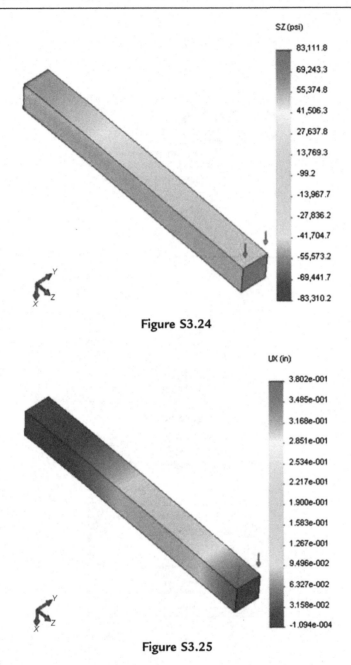

SZ (psi)

83,111.8
69,243.3
55,374.8
41,506.3
27,637.8
13,769.3
-99.2
-13,967.7
-27,836.2
-41,704.7
-55,573.2
-69,441.7
-83,310.2

Figure S3.24

UX (in)

3.802e-001
3.485e-001
3.168e-001
2.851e-001
2.534e-001
2.217e-001
1.900e-001
1.583e-001
1.267e-001
9.496e-002
6.327e-002
3.158e-002
-1.094e-004

Figure S3.25

on top of the Graphics area (Figure S3.26), and choose Probe. Click nodes along the top edge of the root end, as shown in Figure S3.27. Bending stress value at the selected nodes will appear. The stress values vary between about 65,500 psi and the maximum 83,100 psi. The difference is fairly significant. Especially, at the two corners, the stress is about 80,000 psi, which shows local stress concentration. This local stress concentration is mostly due to

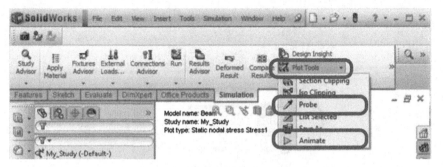

Figure S3.26

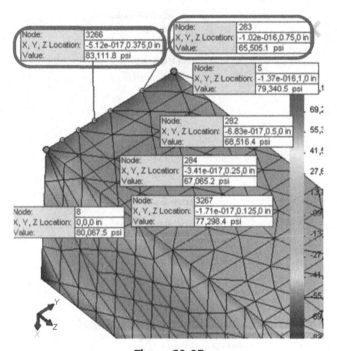

Figure S3.27

nonzero Poisson ratio. If you set the Poisson's ratio to zero and rerun the analysis, the bending stress at the top edge reduces to around 61,000 psi with minimum variation at nodes along the top edge.

Note that in Figure S3.27, stress fringe plot is shown with the finite element mesh. This was done by adjusting settings for the stress plot. You may right-click Stress1 under Results (Figure S3.28). In the Settings dialog box (Figure S3.29), choose Mesh for Boundary Options, and click the green checkmark to accept the options. Double-click Stress1 to show stress result with finite element mesh, as shown in Figure S3.30.

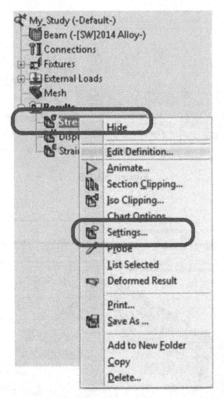

Figure S3.28

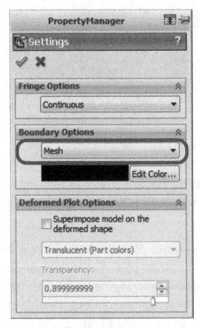

Figure S3.29

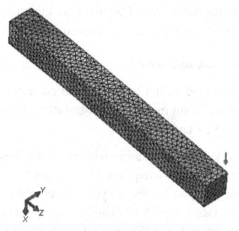

Figure S3.30

S3.2.3 Idealization: Converting Solid Model to 1D Beam

SolidWorks Simulation is capable of converting a long slender solid object in to a beam component, which significantly reduces the model size for FEA. For many applications, the beam model provides FEA results similar to those of 3D solid models, but with a much reduced analysis time. For some applications, it is almost impossible to carry out FEA due to the complexity of the geometric model without converting the solid objects into beam components.

Convert Solid Model to Beam

Create a new Static Study for the beam with the name Study1. Right-click on the part in Study Feature Tree and choose Treat as Beam, as shown in Figure S3.31. Then right-click on Joint Group appearing under the part in the Study Feature Tree and choose Edit, as shown in

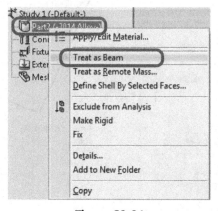

Figure S3.31

Figure S3.32. In the Edit Joints menu click Calculate, as shown in Figure S3.33. This will create joints at critical locations; in this case, at the ends.

Define Boundary Conditions, Load, and Solve

Follow the same steps as discussed earlier to apply fixed constraint and a 1000 lb$_f$ vertical load on the opposite end. For a beam model all constraints and loads must be defined on joints, as shown in Figures S3.34 and S3.35.

Right-click Study1 and choose Run. Simulation will create the mesh and carry out FEA for this beam model. The beam model only requires 46 elements (created by using the default mesh setup) to solve the study (Figure S3.36). As you can see, the results match exactly beam theory in both bending stress (see Figure S3.37) and displacement (Figure S3.38). Note that for this example, one beam element is sufficient to provide exact solutions since cubic functions are employed for finite element shape functions, and for a cantilever beam with a point load the displacement is a cubic function of its location.

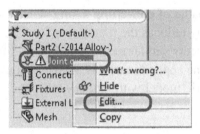

Figure S3.32

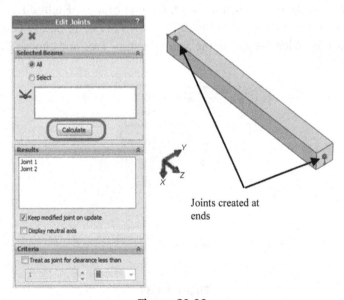

Joints created at ends

Figure S3.33

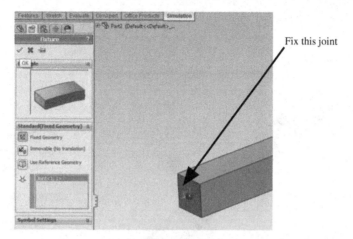

Fix this joint

Figure S3.34

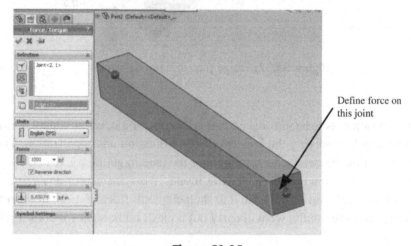

Define force on
this joint

Figure S3.35

Study name	Study 1 (-Default-)
Mesh type	Beam Mesh
Total nodes	48
Total elements	46
Time to complete mesh(hh:mm:ss)	00:00:00
Computer name	LAB-FH146-003

Figure S3.36: Mesh details.

S3.3 Thin-Walled Tank

In this example we will carry out FEA for a thin-walled tank, as shown in Figure S3.39, using SolidWorks Simulation. This tutorial example is built on the concept and procedure learned from the cantilever beam example described previously. Please complete the beam example before going through this example.

Figure S3.37: Fringe plot of bending stress.

There are two major points to be introduced in this example. First, we will learn how to reduce the FEA model size by applying the concept of simplification and idealization. That is, we will use only half of the model due to symmetry of the tank in geometry, boundary condition, and load; that is, simplifying the finite element model. Also, instead of analyzing the solid thin-walled tank, we will convert the solid into shell elements (idealization) in order to further reduce the model size. Secondly, we will carry out a mesh refinement for convergence study using h-adaptation.

S3.3.1 The Thin-Walled Tank Example

The outer diameter and height of the vertical long tube are 102 mm and 240 mm, respectively. The short horizontal tube has outer diameter and length 82 mm and 100 mm, respectively. The short tube is positioned halfway up the vertical tube. The entire tank has a uniform thickness of 2 mm. The unit system used for this example is MMGS (millimeter, gram, second) and the material is aluminum (AL2014). As shown in Figure S3.39, a force of 1000 N is applied at the front end face of the short tube. The bottom face of the long tube is fixed.

Instead of analyzing the full solid model, we will cut the model into half and convert the solid to shell model for FEA. In analyzing the shell model converted from the half tank solid model, we will impose symmetric boundary conditions at the outer edges of the cut faces, as

Figure S3.38: Fringe plot of vertical displacement.

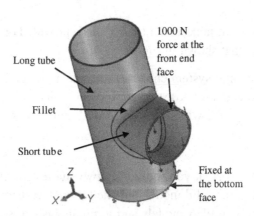

Figure S3.39: Full 3D solid model.

shown in Figure S3.40. The fixed boundary condition will be applied to the outer edge of the bottom face. In addition, the force will be cut into half; only a 500 N force will be applied to the half tank model.

There are at least two approaches to convert a solid thin-walled solid model to a shell model. One approach is to insert an offset surface from either the outer or inner surface of the solid model as the mid-surface, and mesh only the inserted mid-surface for FEA. This is essentially employing a part-modeling technique in SolidWorks. The second approach is to treat the solid

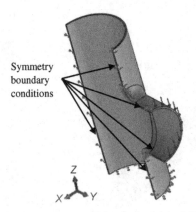

Figure S3.40: Half tank shell model.

as shell, which is a capability provided in SolidWorks Simulation. We will use the latter approach in this lesson.

S3.3.2 Using SolidWorks Simulation

Open SolidWorks Part

Start SolidWorks. Open the half tank model (htank shell) provided on the book's companion website. The model is like that shown in Figure S3.40.

You may want to check the unit system chosen. From the Tools pull-down menu, choose Options. Select the Documents Properties tab and select Units. Select MMGS (millimeter, gram, second) then click OK. We will stay with MMGS unit system.

Enter SolidWorks Simulation

At the bottom of the Graphics area, you should see two studies completed: Static-Solid and Static-Shell, which are FEA for solid model and shell model, respectively. You may click these simulation tabs to browse FEA models and rerun analysis to see the results.

We will start a new study and convert the solid model to shell.

Choose from the pull-down menu Simulation > Study. Select Static from the Study dialog box, and change the study name to My_Study (similar to what you learned in the beam example). Accept the options selected by clicking the green checkmark at the top left.

A new study (My_Study) will be added to the simulation tab below the Graphics area, and a new set of buttons appear on top. In addition, a list of simulation entities will appear in the Study Feature Tree (directly below the model tree). We will use these entities (mostly right-click on them) to define the finite element model, carry out FEA, and visualize FEA results.

Model Idealization: Define Shell Surfaces

In the Study Feature Tree window, right-click htank shell and select Define Shell by Selected Faces, as shown in Figure S3.41.

In the Shell Definition dialog box (Figure S3.42), select Thin for the type of surface, select the outer surfaces of the part, change the Shell Thickness to 2 mm, and select Top for Offset (expand the window for selection), as shown in Figure S3.42. Click the green checkmark to accept the definition. Recall that we picked the outer surfaces as the geometry for the shell

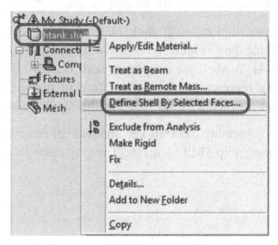

Figure S3.41

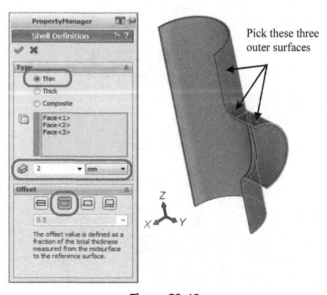

Figure S3.42

model. That is why we choose Top as the offset, which implies that the surfaces we picked are the top (or outer) surfaces of the shell model. Note that in Simulation, the finite element mesh will be created on the surfaces we picked (instead of the mid-plane of the thickness). The formulation of the stress, displacement, and strain is adjusted to compensate for the mesh being at the outer surfaces of the thin wall rather than acting as the mid-plane of the thickness.

Another thing we have to ensure is when we create a mesh we may need to adjust the top/bottom face of the shell elements to be consistent with the offset we defined. Note that a gray color of the mesh represents the top face of the shell. The bottom face is an orange color. In this case, the shell on the outer surface must be gray in color.

In the Study Feature Tree window, right-click Mesh and choose Create Mesh (Figure S3.43). In the Mesh dialog box (Figure S3.44), use default setup and click the checkmark (Figure S3.44). A shell mesh appears with orange color on the outside (shown in Figure S3.45), representing the bottom face of the shell elements, which is inconsistent to the offset we defined. We will flip the shell elements.

Select all three surfaces by pressing the Shift key and pick all three in the Graphics area. Right-click Mesh and choose Flip Shell Elements. The gray color should now appear on the

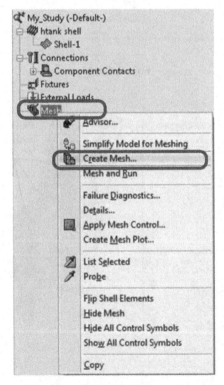

Figure S3.43

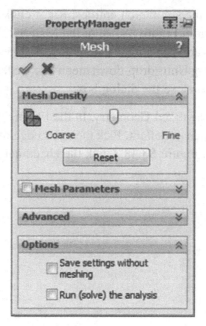

Figure S3.44

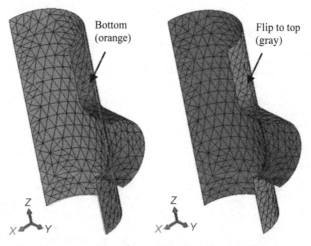

Figure S3.45

outside of the mesh, as shown in Figure S3.45. We have properly idealized the solid model to shell elements. There are about 780 triangular shell elements created. We are ready to define material, load, and boundary conditions for a complete FEA model.

Create Finite Element Model

We will use the right-click options on the simulation entities to create the FEA model.

Move the mouse pointer into the simulation entities. Right-click htank shell, and choose Apply/Edit Material, as shown in Figure S3.46. In the Material dialog box, choose 2014 Alloy. The material properties are listed in the dialog box, as shown in Figure S3.47. You may change the units displayed using Units drop-down menu like we learned in the beam example. Click Apply and then Close to close the dialog box.

Right-click Fixtures and choose Fixed Geometry. In the Fixture dialog box shown in Figure S3.48, Fixed Geometry is chosen as default. Pick the outer edge of the bottom face for a fixed geometry fixture, as shown in Figure S3.48. Click the checkmark to accept the fixture.

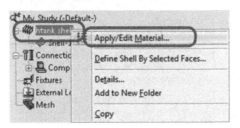

Figure S3.46

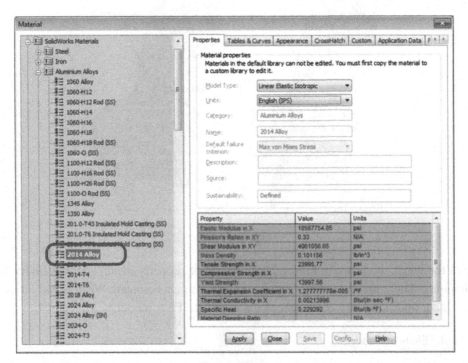

Figure S3.47

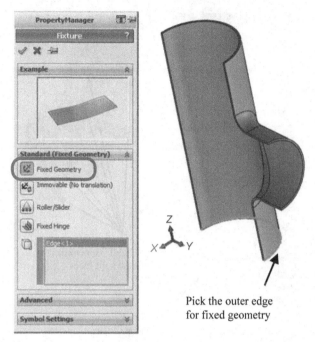

Pick the outer edge
for fixed geometry

Figure S3.48

Next we will define the symmetry constraint. Right-click Fixtures again and choose Advanced Fixtures. For nonshell parts you can directly use the Symmetry option for a symmetry condition, but for shell surfaces you must manually define the symmetry constraint. We will use the Right Plane as the reference to define the symmetry constraint.

In the Fixture dialog box shown in Figure S3.49, choose Use Reference Geometry. Pick the outer edges of the split faces, as shown in the figure. You should select all seven edges. Click the box for Direction and select Right Plane for direction. Now we need to choose Translation and Rotation to create the symmetry condition correctly. For Translation select the Normal To. For Rotation select Along Plane Direction 1 and Along Plane Direction 2, as shown in the figure. Click the green checkmark to accept.

Next, we will add a downward force of 500 N at the outer edge of the front end face.

Right-click External Loads and choose Force. In the Force/Torque dialog box shown in Figure S3.50, Force is chosen as default. Pick the outer edge of the front end face of the short tube to apply the force, as shown in the figure. We will then define the direction of the force. Choose Selected direction (default is Normal) from the Force/Torque dialog box, pick Front Plane from the model tree for direction, enter 500 in the text field that shows normal to the selected plane, and click Reverse direction to change the direction of the force to downward. Accept the options by clicking the green checkmark. Force symbol (arrows) should appear at the outer edge of the front end face.

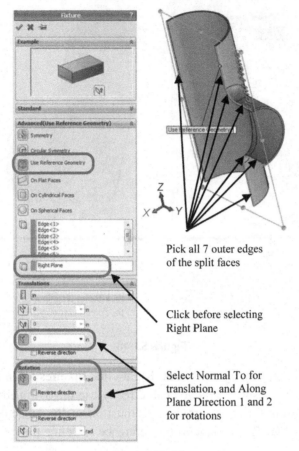

Pick all 7 outer edges of the split faces

Click before selecting Right Plane

Select Normal To for translation, and Along Plane Direction 1 and 2 for rotations

Figure S3.49

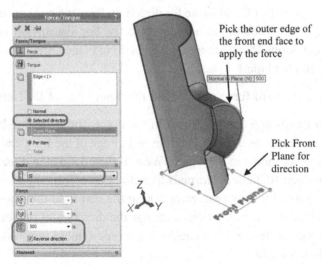

Pick the outer edge of the front end face to apply the force

Pick Front Plane for direction

Figure S3.50

Run Analysis and Display Results

We will carry out a static analysis next. Right-click My_Study and choose Run. The analysis will be completed quickly (less than a minute).

After the analysis is complete, expand the Results entity and double-click Stress1(-vonMises-) to bring up the fringe stress plot in the Graphics area, like that of Figure S3.51. Note that the stress shown is von Mises stress at the bottom face, as noted in the caption on top of the plot. The maximum stress calculated for the mesh of the default setting is about 40 MPa. You may want to show the maximum stress by right-clicking Stress1(-vonMises-), choosing Chart Options, and picking proper settings. Are we going to see a different stress on the top face?

Right-click Stress1, and choose Edit Definition. In the Stress Plot dialog box shown in Figure S3.52, choose Top, and click the checkmark to accept the change. Double-click Stress1(-vonMises-) to bring up the stress fringe plot. The stress plot will appear like that of Figure S3.53. Use the Probe capability to find the stress on the top face at the same place where the maximum von Mises stress of the bottom face is located. As shown in the figure, the von Mises stress probed is about 31 MPa, which is smaller than that of the bottom face.

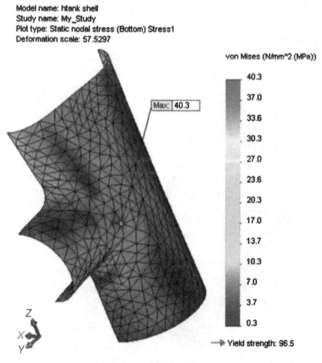

Figure S3.51

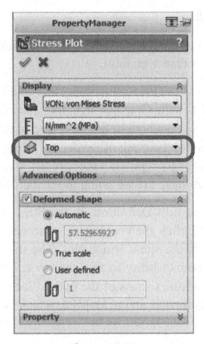

Figure S3.52

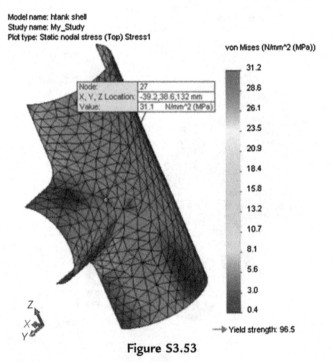

Figure S3.53

Is the stress result accurate? We will refine the mesh around the high stress area and conduct more analyses to see if the stress results converge.

Before that, choose File and then select Save to save the model.

S3.3.3 Mesh Refinement

We will apply mesh control at the fillet surface and carry out FEA to see if the maximum von Mises stress is significantly affected by mesh.

In the Study Feature Tree window, right-click Mesh, and choose Mesh Control (Figure S3.54). In the Mesh Control dialog box shown in Figure S3.55, a default setting is shown (with element size about 2.9 mm). Enter 2.5 mm for element size, and accept the change. Right-click Mesh and choose Create Mesh. Note that you may need to flip the shell elements to make the shell top face (in gray color) on the outside surface. There are 2,399 elements created (you may see a slightly different number). Rerun the analysis and show von Mises stresses on the top and bottom faces, as shown in Figure S3.56 (bottom face) and Figure S3.57 (top face). They are 47.9 and 35.4 MPa at the bottom and top faces, respectively.

Repeat the same steps but enter element sizes 2.0, 1.5, and 1.0. Rerun analyses and show von Mises stress plots. The maximum stresses and number of elements in each refinement are summarized in Table S3.3.

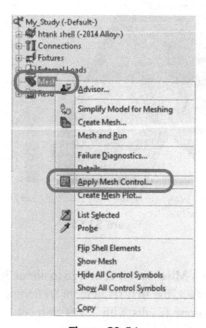

Figure S3.54

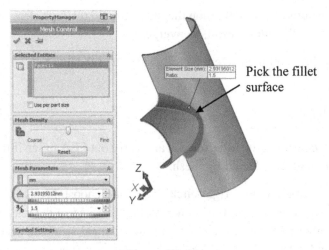

Pick the fillet surface

Figure S3.55

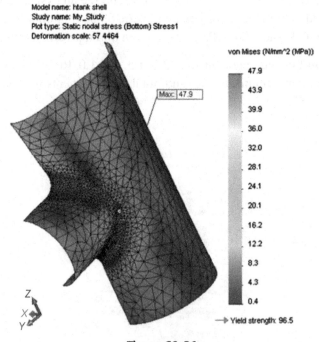

Figure S3.56

As shown in Table S3.3, the von Mises stress converge to about 36 and 48 MPa at the top and bottom faces, respectively. In fact, the first refinement (refinement 1 with element size 2.5 mm) provides already accurate results with 2,399 elements. There is no need to further refine the mesh since there is no significant improvement on stress results. The stress results

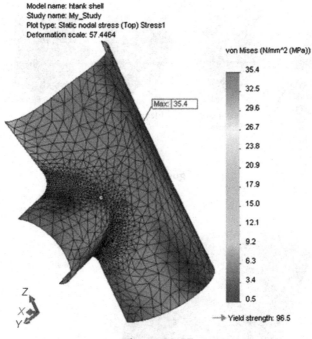

Model name: htank shell
Study name: My_Study
Plot type: Static nodal stress (Top) Stress1
Deformation scale: 57.4464

von Mises (N/mm^2 (MPa))

Max: 35.4

35.4
32.5
29.6
26.7
23.8
20.9
17.9
15.0
12.1
9.2
6.3
3.4
0.5

→ Yield strength: 96.5

Figure S3.57

Table S3.3: Mesh refinement for (von Mises) stress convergence study

Refinement	Mesh Setup	Number of Elements	Top Face (MPa)	Bottom Face (MPa)
Current	Default	783	31.1	40.3
1	Element size: 2.5 mm	2399	35.4	47.9
2	Element size: 2.0 mm	3170	35.8	48.0
3	Element size: 1.5 mm	4956	35.8	48.7
4	Element size: 1.0 mm	9271	36.4	48.4

converge, hence they are considered accurate. The refined meshes on the fillet surface are shown in Figure S3.58.

Note that if we use the solid model to carry out a static study, the maximum von Mises stress is found to be about 47.6 MPa located at the inner face (Figure S3.59) just like that of the shell model. The shell model provides as good results as those of the solid model. However, with the element size specified as 2.5 mm, there are 9,443 tetrahedron finite elements created for this solid mode, with a total of 57,147 degrees of freedom. For the shell model (Refinement 1), there are 2,399 triangular elements and a total of 29,250 degrees of freedom, which is about half of that the solid model.

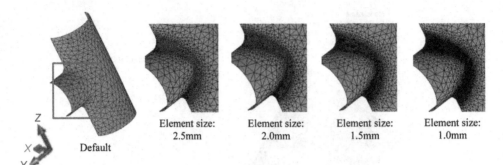

Figure S3.58

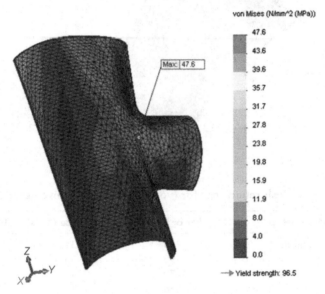

Figure S3.59

S3.4 Fatigue Analysis

In this example we will go over a fatigue analysis for a crankshaft example using the stress-based approach provided in SolidWorks Simulation. The crankshaft is one of the components in a slider-crank mechanism, as shown in Figure S3.60. This type of mechanism is commonly found in mechanical systems, for example, an internal combustion engine and oil-well drilling equipment. For the internal combustion engine, the mechanism is driven by a firing load that then pushes the piston, converting the reciprocal motion into a rotational motion at the crankshaft. In the oil-well drilling equipment, a torque is applied at the crankshaft. The rotational motion is converted to a reciprocal motion at the piston that digs into the ground. In any case, the load or reaction force acting on the crankshaft is changing direction. However, in this lesson, we are assuming a cyclic load with constant amplitude, fully reversed along the

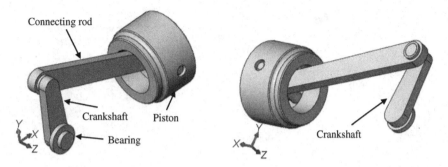

Figure S3.60: The slider-crank mechanism in two views.

longitudinal direction of the crankshaft. Note that this is a much simplified load case since in reality, as mentioned, the direction of the force is changing. Therefore, please keep in mind that results obtained from this simple analysis are useful only in providing a general idea of the fatigue life for the crankshaft, which is in fact, a fairly rough estimate. For more in-depth fatigue analysis, you may have to use other software tools that offer multiaxial fatigue analysis capabilities, in which load is changing and, therefore, the principal direction of the stress is changing in time. To stay focused, a static analysis is completed beforehand, and we will only go over the steps for carrying out a fatigue analysis.

S3.4.1 The Crankshaft Example

The crankshaft shown in Figure S3.61a is about 4 in. long with two short shafts connecting with the bearing and the connecting rod, respectively. The distance between the centers of the two short shafts is 3.00 in. Note that there is a small fillet of 0.05 in. at the root of both shafts. The unit system chosen is in.-lb$_f$-sec. The material is aluminum (AL2014), where the modulus and Poisson's ratio are $E = 1.06 \times 10^7$ psi and $\nu = 0.33$, respectively. A bearing load of 250 lb$_f$ is applied at the outer cylindrical surface of the top shaft (connecting to the connecting rod) along the longitudinal direction of the crankshaft. The outer surface of the lower shaft (connecting to the bearing) is completely fixed, as shown in Figure S3.61b.

A finite element model is completely defined for this example. When you open the crankshaft model provided (available at publisher's web site), you can review the FEA model by clicking Static Analysis tab at the bottom of the Graphics area. Right-click Fixed-1 and BearingLoads-1(:250 lb$_f$:) listed in the Study Feature Tree, as shown in Figure S3.61c, to find out more about the boundary condition and load defined for the crankshaft finite element model. Right-click Mesh to recreate a finite element mesh by choosing the Fine mesh option. There are about 11,500 tetrahedron elements created for this model, as shown in Figure S3.61d. Rerun a static analysis and show von Mises stress fringe plot (Figure S3.61e). Note that the maximum von Mises stress is about 12,800 psi located in the fillet of the lower shaft,

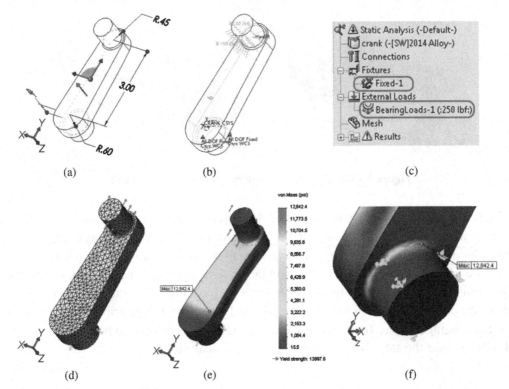

Figure S3.61: The finite element model of the crankshaft, (a) Physical sizes, (b) Load and boundary conditions, (c) Simulation entities listed in the Study Feature Tree, (d) Finite element mesh, (e) von Mises stress fringe plot, and (f) von Mises stress with a closer view.

as shown in Figure S3.61f, which is largely due to the bending effect. Note that the maximum von Mises is slightly below the material yield strength of 14,000 psi.

S3.4.2 Using SolidWorks Simulation

Open SolidWorks Simulation Model

Open the solid model of the crankshaft from the book's companion website. Click the Static Analysis tab to review the finite element model created. Rerun the static analysis to generate stress results for fatigue analysis. Please review the von Mises stress to make sure the maximum stress magnitude and its location are consistent with those of Figure S3.61e.

Create Fatigue Analysis Model

From the pull-down menu, choose Simulation > Study.

In the Study dialog box (Figure S3.62), choose Fatigue for Type, click the Constant Amplitude button (left at the bottom), enter Fatigue Constant Amplitude for Name, and accept the options selected by clicking the green checkmark.

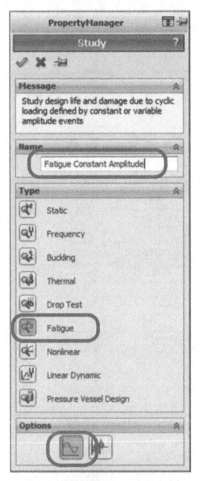

Figure S3.62

Right-click Fatigue Constant Amplitude (the root entity listed in the Study Simulation Tree), and choose Properties, as shown in Figure S3.63, to bring up the Fatigue dialog box (Figure S3.64). In the Fatigue dialog box, choose Equivalent stress (von Mises), click Infinite life and enter 1,000,000 cycles, and then click OK. We assume that the upper limit of the high-cycle fatigue is 1,000,000 cycles, which is also where the endurance limit is measured. Also, we do not have to select any criterion for mean stress correction since we assume that the cyclic load is fully reversed.

Next, we will define the load for fatigue analysis. We will assume a constant amplitude cyclic load that is fully reversed. Right-click Loading listed in the Study Feature Tree and choose Add Event (Figure S3.65). In the Add Event dialog box, simply click the green checkmark to accept all default options. Basically, we chose a load that repeats 1,000 times, fully reversed ($LR = -1$), and the load magnitude is identical to that of the Static Analysis (with scale 1).

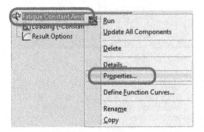

Figure S3.63

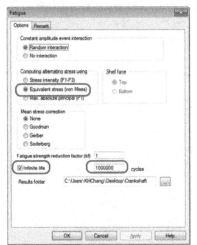

Figure S3.64

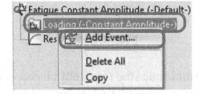

Figure S3.65

As soon as you click the checkmark in the Add Event dialog box, a new entity crankshaft is added to the Study Feature Tree (Figure S3.66). Right-click crankshaft and choose Apply/ Edit Fatigue Data, as shown in Figure S3.67. We will define an S—N diagram for the chosen material AL2014. In the Material dialog box, 2014 Alloy is chosen, as shown in Figure S3.68. Choose Log—Log for Interpolate, click Derive from material Elastic Modulus, and then choose Based on ASME Austenitic Steel curves. A set of data should appear. Choose psi for Units (if it is not chosen already), and click View to see the S—N diagram.

An S—N diagram should appear like that of Figure S3.69. Note that if you draw a horizontal line at a stress level of about 12,800 psi (maximum von Mises stress), intersect the

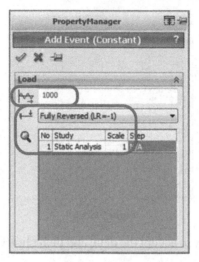

Figure S3.66

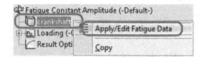

Figure S3.67

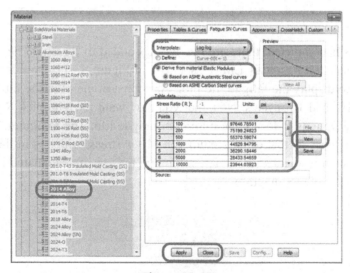

Figure S3.68

horizontal line with the S−N curve, and project the intersecting point down to the abscissa for the number of cycles, which is roughly 2.5×10^5, as shown in Figure S3.69. We will check later to see if Simulation provides a fatigue life similar to this number for the crankshaft.

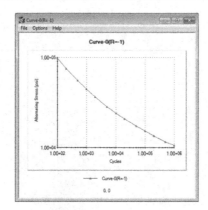

Figure S3.69

For the time being, click Apply to accept the data, and then click Close to close the dialog box.

At this point, we have completely defined a fatigue analysis model, and we are ready to conduct a fatigue life calculation.

Run Fatigue Analysis

Right-click Fatigue Constant Amplitude and choose Run (Figure S3.70). The analysis will be completed in just a few seconds.

Display Results

After the analysis is completed, there are two entities added to Results. They are Results1(-Damage-) and Results2(-Life-), as shown in Figure S3.71. Double-click Results2(-Life-) to display the fatigue life plot. As shown in the plot, the fatigue life is between 2.65×10^5 and 10^6 cycles. Note that you may want to right-click Results2(-Life-) and choose Chart Options to locate the minimum fatigue life. The minimum life should be found at the same place as that of maximum von Mises stress, which is in the fillet of the lower shaft, as shown in Figure S3.72. The minimum fatigue life calculated is very close to our estimate using the S—N diagram.

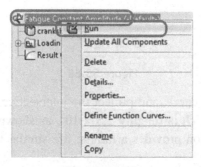

Figure S3.70

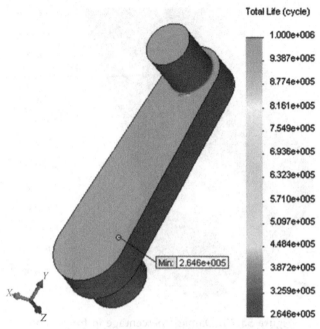

Figure S3.71: Fatigue life results in fringe plot

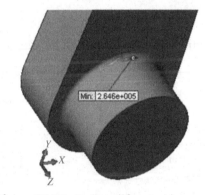

Figure S3.72: Fatigue life at a closer view

Double-click Results1(-Damage-), a damage fringe plot appears like that of Figure S3.73. The damage percentage ranges from 0.1% to 0.378%. The maximum damage 0.378% is found at the same location in the fillet as that of maximum stress and shortest fatigue life. The damage plot shows the percentage damage after a block of cyclic loads is applied. In our case, one block is 1,000 cycles (Figure S3.66). Note that the damage percentage can be converted to fatigue life. For example, at the fillet of the lower shaft where the damage is 0.378%, the fatigue life is 1/0.378% times 1,000 (one block); that is, 26,460, which is the same result shown in Figure S3.72.

We have completed the lesson. Save your model.

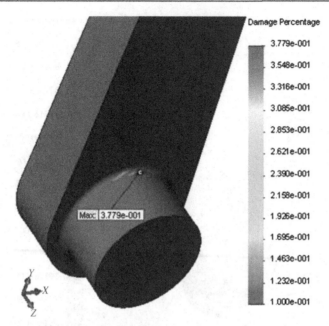

Figure S3.73: Damage percentage in fringe plot

Exercises

1.   Analyze the following cantilever beam using SolidWorks Simulation.

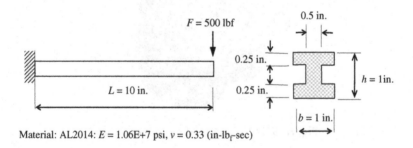

Material: AL2014: $E = 1.06E+7$ psi, $v = 0.33$ (in-lb$_f$-sec)

(a)   Create SolidWorks part and Simulation models, and then carry out FEA for the maximum displacement and maximum stress. Submit screen captures for the maximum bending stress and for the maximum displacement (downward only) in fringe plots.

(b)   Solve the same problem using the classical beam theory. Compare your calculations with those of Simulation. Are they consistent, why or why not?

    (c)   Repeat (b) by letting Poisson's ratio $\nu = 0$. Comment on the impact of the nonzero Poisson's ratio to the stress results.

2.   Continue with Problem 1 by converting the solid model into beam component.

    (a)   Use the default mesh setup to carry out FEA. Do the maximum bending stress and maximum displacement match with those of classical beam theory? How many beam elements were created?

    (b)   Adjust mesh setup to create only one element for FEA. Do the maximum bending stress and maximum displacement match with those of classical beam theory? Do you need more than one beam element for this example?

3.   An L-shape circular bar of diameter 1.25 in. is loaded with an evenly distributed force of 400 lbs at its front end face, as shown. Note that the elbow radius (corner of the L-shape) is 1 in. and the material is 1060 Alloy.

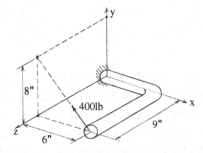

    (a)   Create an FEA model using SolidWorks Simulation to calculate the maximum principal stress and the maximum shear stress in the bar.

    (b)   Calculate the maximum principal stress and maximum shear stress using analytical beam theory. As expected, the maximum stress occurs at the root. Where are the maximum stresses located in the beam cross-section at the root? Compare your calculations with those obtained from SolidWorks Simulation, both values and locations. Are they close? Why or why not? Please comment on your comparison.

4.   Open the full tank solid model from the book's companion website and create a finite element model that is consistent with the thin-shell model discussed in Section S3.3. Note that the load is 1,000 N downward and the bottom face is fixed, as shown in the following diagram. Create mesh using the default setting, and carry out an FEA. Compare maximum displacement and von Mises stress between the full solid model and the thin-shell model of Section S3.3. Also compare the size of these two models in terms of number of finite elements and number of degrees of freedom. Please comment on the advantage of idealization and simplification in FEA modeling.

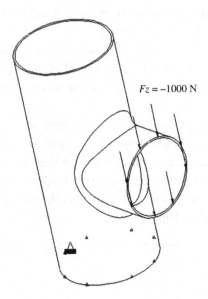

$F_Z = -1000$ N

5. Carry out a fatigue analysis for the same crankshaft example for a different load scenario. In this case, we assume the load event to be Zero Based (that is, a repeated cyclic load) with a scale of 0.65; choose Soderberg for mean stress correction. Rerun fatigue analysis. Does the fatigue life result make sense? How do you verify it? Note that the equivalent alternating stress of Soderberg criterion can be obtained by using

$$\sigma_A = \frac{S_y \sigma_a}{S_y - \sigma_m}$$

where σ_a and σ_m are alternating and mean stresses, respectively. S_y is the material yield strength.

Index

Note: Page numbers with "*f*" denote figures; "*t*" tables.